Project Finance for Construction

The world of construction is intrinsically linked with that of finance, from the procurement and tendering stage of projects right through to valuation of buildings. In addition to this, things like administrations, liquidations, mergers, take-overs, buy-outs and floatations affect construction firms as they do all other companies.

This book is a rare explanation of common construction management activities from a financial point of view. Whilst the practical side of the industry is illustrated here with case studies, the authors also take the time to build up an understanding of balance sheets and P&L accounts before explaining how common tasks like estimating or valuation work from this perspective.

Readers of this book will not only learn how to carry out the tasks of a construction cost manager, quantity surveyor or estimator, they will also understand the financial logic behind them, and the motivations that drive senior management. This is an essential book for students of quantity surveying or construction management, and all ambitious practitioners.

Anthony Higham is a Senior Lecturer and Chartered Quantity Surveyor at the University of Salford, UK. He has delivered both undergraduate and postgraduate modules in quantity surveying and commercial management for the last 10 years.

Carl Bridge is a Chartered Management Accountant at the University of Bolton, UK, and Head of Accounting within the university's Business School. He joined the university in 1989 as a Senior Management Accountant within the central finance office and moved into an academic management role in 2008.

Peter Farrell is a Reader in Construction Management at the University of Bolton, UK, and Programme Leader for the university's MSc degree in construction project management. He has delivered undergraduate and postgraduate modules in construction management, commercial management and quantity surveying for 20 years.

Project Finance for Construction

Anthony Higham, Carl Bridge
and Peter Farrell

Routledge
Taylor & Francis Group

LONDON AND NEW YORK

First published 2017
by Routledge
2 Park Square, Milton Park, Abingdon, Oxon OX14 4RN

and by Routledge
605 Third Avenue, New York, NY 10017

Routledge is an imprint of the Taylor & Francis Group, an informa business

British Library Cataloguing-in-Publication Data
A catalogue record for this book is available from the British Library

Library of Congress Cataloging in Publication Data
Names: Higham, Anthony, author. | Bridge, Carl, 1963– author. |
  Farrell, Peter, 1955– author.
Title: Project finance for construction / Anthony Higham, Carl Bridge and
  Peter Farrell.
Description: Abingdon, Oxon; New York, NY : Routledge is an imprint
  of the Taylor & Francis Group, an Informa Business, [2017] | Includes
  bibliographical references and index.
Identifiers: LCCN 2016028249 | ISBN 9781138941298 (hardback : alk.
  paper) | ISBN 9781138941304 (pbk. : alk. paper) | ISBN 9781315673769
  (ebook)
Subjects: LCSH: Construction industry—Great Britain—Finance. |
  Construction industry—Great Britain—Management. | Construction
  industry—Finance. | Construction industry—Management.
Classification: LCC HD9715.G72 H525 2017 | DDC 690.068/1—dc23
LC record available at https://lccn.loc.gov/2016028249

ISBN: 978-1-138-94129-8 (hbk)
ISBN: 978-1-138-94130-4 (pbk)
ISBN: 978-1-315-67376-9 (ebk)

Typeset in Times New Roman
by Apex CoVantage, LLC

Contents

Figures

Tables

Preface

This book is aimed at practitioners, and under- and post-graduate students in construction. Whilst of particular interest to quantity surveyors and others who deal with project finance, it is relevant to sister disciplines such as construction managers, architects, architectural technologists, building surveyors and civil engineers. It is argued those professionals in construction who do not gain at least a good appreciation of finance exclude themselves from decision-making tables.

The text takes an overview of project finance from inception through to final account and life cycle. Early chapters deal with project appraisals from the perspective of employers and design teams. Methods of establishing budgets are detailed, and options for procurement methods explored; it may be in modern methods of procurement that contractors and lower-tier suppliers support design teams in forming budgets. Value management is explored in the context of considering the cost of alternative design solutions.

The internal financial control systems of contractors are examined, from initial bid stages through to the construction phase and beyond. The two separate issues of how to control costs and cashflow are described. The cost of employing labour directly and subcontracting are compared. There are sections on payment procedures, bonds, insurances, retentions and discounts.

Insight is given into the role of the Chartered Accountant at the head office of designers and contractors. Whilst having no construction expertise, accountants often have seats on boards of directors and drive many business decisions that impact upon the work of professionals at project level. Investment appraisal techniques, raising finance and corporate accounts are all examined. Examples are provided of accounting ratios that are used to make judgements about the success of companies. Construction professionals and part-time students may often find themselves in the midst of liquidations, mergers, acquisitions or take-overs; this may be at their own companies or companies higher or lower down supply chains. Insight is given into these issues, from the viewpoint of partners or company directors.

References are made to both the JCT (2011) Standard Form of Building Contract and the New Engineering Contract (NEC3). Throughout the text there are discussion points, exercises and tasks. Model answers are given at the end of chapters.

Acknowledgement: the authors are grateful to Dr Jason Challender who has provided valuable feedback during production of the book.

1 Pre-contract financial management

1.1 Project appraisal and developing the business case

It can be argued that the main role of the construction industry is to produce buildings or structures, mindful of cost, time, quality, health and safety and environmental constraints. Indeed many have argued that construction is fundamentally a production focused industry, whereby the outputs of the construction process are new or rehabilitated buildings, roads, bridges, railways or something similar. It could be argued that production is at the core of the service the construction industry provides to its customers and to society.

However, production is a very resource intensive process, not only in terms of the materials used to the construct buildings, or indeed the labour and plant involved in undertaking the construction process, but also in the amount of money customers need to commit to develop bespoke products. As a result, detailed project appraisals are often regarded as a critical stage in the evolution of construction projects. At this point the customer or employers will be faced with several fundamental decisions that will ultimately determine the success or otherwise of projects.

Immediately prior to this stage, employers should have identified and considered the need for the project at a very strategic level within their organisation, and aligned this with the overall strategic direction of their business. Figure 1.1 shows how this process will have required employer organisations to evaluate the potential benefits of projects and strategic fit with their overall organisation.

Project appraisals give employers and their senior management teams the opportunity to assess and question the potential options identified by project teams before resources are committed and production processes are commenced. It is project appraisals that will ultimately decide not only whether proposed projects are financially viable, but it will also influence the future direction of projects in terms of budget, project objectives and alignment with the overall business case of employers developed as part of project strategic definition.

Optional appraisal

For cost consultants or private quantity surveyors, option appraisals often represent the first major input they will have into projects. Forming a key part of overall project appraisals, option appraisals are undertaken at RIBA stage 0 'strategic

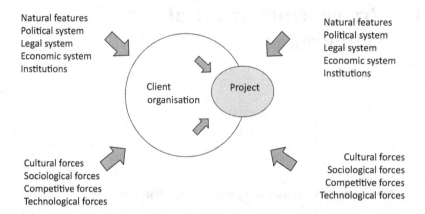

Natural features
Political system
Legal system
Economic system
Institutions

Natural features
Political system
Legal system
Economic system
Institutions

Client organisation   Project

Cultural forces
Sociological forces
Competitive forces
Technological forces

Cultural forces
Sociological forces
Competitive forces
Technological forces

Figure 1.1 Project/business interface considerations at strategic definition.

definition'. During this stage in the evolution of projects, the RIBA plan of work suggests "*the project will be strategically appraised and defined before a detailed brief is created*" (RIBA, 2013). Introduced in 1963 the RIBA plan of work has been the definitive UK model for organising the process of briefing, designing, constructing, maintaining, operating and using building projects into a number of key stages (RIBA, 2013). The latest iteration of the RIBA plan of work, the RIBA plan of work 2013, is designed to align with current industry best practice in terms of digitalisation and sustainable development.

At this point, it is highly likely option appraisals will be undertaken by project teams; this is likely to require "*a review of a number of sites or alternative options such as extensions, refurbishment or new build*" (RIBA, 2013). In terms of the role of cost consultants, these reviews will involve detailed financial evaluation of options available to employers, through the production of 'order of cost estimates', which in essence are assessments of the affordability of all options and the establishment of realistic cost limits.

Option appraisals will typically be applied at project level, and will either consider the viability of developing a building type on various sites, or will consider the viability of varying levels of quality or levels of refurbishment for proposed buildings to establish how these align with employer overall budgets, something that will have been identified as part of business case development.

For the rest of this chapter, the example of a budget hotel development will be used to demonstrate the key processes in pre-contract financial management.

Budget Hotel Development Scenario: AZX Property Ltd has recently completed the purchase of an old 1960s office building with a view to developing a budget hotel. Structural analysis of the existing building discovered it was unsuitable for rehabilitation. Consequently the existing office building will be demolished, the site remediated and a 5-storey, 64-bedroom hotel constructed on the site.

The building will have a GFA of 2,225m², with a standard double bedroom having a GFA of 14.70m².

Discussion point 1.1: How could the findings from a detailed option appraisal influence the employer's non-financial decisions about the project?

1.2 Introduction

The chapter considers the initial exposure of quantity surveyors to financial management, through their involvement in pre-construction financial advice. It is focused towards the role of employer facing quantity surveyors, traditionally known as private practice quantity surveyors or professional quantity surveyors (PQS) or more recently cost consultants. Quantity surveyors that are contractor or specialist contractor facing again have a variety of titles including Contractors' Quantity Surveyors, Commercial Managers or, if they are members of the Royal Institution of Chartered Surveyors, they can adopt the title *Construction Surveyor*. Some authoritative sources use the word 'client' to describe those responsible for commissioning projects, but the term 'employer' is most often used in construction contracts; this textbook therefore uses the word 'employer'. The chapter evaluates quantity surveyors' responsibilities for developing initial project appraisals, providing advice on project budgets, including identifying the absolute maximum budget (the cost limit) for projects. As designs evolve, quantity surveyors will continue to provide detailed reports for employers. These reports provide details of anticipated expenditure, and corrective action required to ensure budgets are not exceeded. Quantity surveyors will further advise about the possible effect of construction phase risks. Towards the end of the chapter, the role of quantity surveyors in preparing bid documents is examined, including the development of unit rates by estimators tasked with forecasting the constructor's costs. Finally the chapter explores how cost consultants advise employers as projects move to the next fundamental stage, the *commitment to construct*. The commitment to construct marks the critical decision of employers to proceed with the agreement of contract terms and the appointment of the successful bidder. Importantly for quantity surveyors, the agreement to proceed to construction marks the transition from pre-contract to post-contract financial management. In summary, chapter 1 aims to develop readers' understanding of the different elements of pre-contract financial management, how these elements of the role impact on projects and finally how quantity surveyors manage this critical phase of project development. Principally, the chapter will explore:

- How initial project appraisals are prepared, and the impetus for cost consultants will be reviewed. This section will introduce the idea of overall project budgets and look at the techniques used to prepare the 'order of costs estimate'.
- The process of preparing cost forecasts, through the use of sequential formal elemental cost plans. The link with design information will also be explored.

- How cost consultants use various sources of information to produce pre-contract financial advice.
- The role of indices and historic data in pre-contract financial management.
- How approximate quantities and composite rates are generated and used in elemental cost plans.
- The development of tender documentation and bid documents.
- The importance of post-tender financial management and reporting from cost consultants to employers to ensure they make informed decisions before committing to proceed with projects.

1.3 Order of cost estimate

The order of cost estimate represents the first attempt by cost consultants to estimate the cost of proposed buildings. Often given as a budget range rather than a clearly defined cost, the order of cost estimate is produced, either as part of option appraisals or shortly afterwards in stage 1 'preparation and brief' of the RIBA plan of work. At this time the project team is tasked with defining *"project objectives, including Quality Objectives and Project Outcomes, Sustainability Aspirations, Project Budget, other parameters or constraints and develop initial project brief"* (RIBA, 2013).

The *New Rules of Measurement (NRM)* Volume 1 *Order of Cost Estimating and Cost Planning for Capital Building Works*, produced by the Royal Institution of Chartered Surveyors (RICS, 2012a), asserts that the order of cost estimate is produced by cost consultants or private quantity surveyors to both define the cost limit for projects whilst also evaluating whether or not the proposed project is feasible. The constituent elements of the order of cost estimate are shown in Table 1.1.

Drafted in response to industry concerns about a lack of consistency and robustness in the pre-contract financial management services provided by construction professionals, the 'RICS New Rules of Measurement: Order of Cost Estimating and Cost Planning of Capital Building Works' (referred to as NRM1) has come to be the cornerstone of good cost management of capital building works projects – enabling more effective and accurate cost advice to be given to clients and other project team members, whilst facilitating better cost control (Binge, 2014).

Discussion point 1.2: Why do you think it is critical for quantity surveyors to determine a cost limit for the project?

1.3.1 Developing the order of cost estimate

The accuracy of estimates that cost consultants or quantity surveyors produce at this stage is highly variable and depends on the level of detail of information

Table 1.1 Order of cost estimate framework

Constituent
Facilitating works
Building works estimate
Contractor preliminaries
External works (additional allowance)
Sub-total
Main contractors overheads and profit
Works cost estimate
Project/design team fees
Other development/Project costs
Base cost estimate
Risk allowance estimate
Design development risk estimate
Construction risk estimate
Employer change risk estimate
Employer other risk estimate
Cost limit (excluding Inflation)
Tender inflation estimate
Cost limit (inc. tender inflation)
Construction Inflation Estimate
Cost limit (inc. construction inflation)
VAT assessment

Source: RICS (2012a).

Note: The rows in bold correspond to group elements as identified in NRM1. As a result, the costs against these items are sub-totals from the constituent elements.

provided by project teams. In an ideal situation, the full range of information identified in NRM1 (RICS, 2012a, p. 21) would have been provided to include:

- Drawings of floor and roof plans;
- Elevation details for all the main facades;
- Section drawings;
- Information relating to the storey heights;
- Information relating to MEP (Mechanical, Electrical and Plumbing) installations;
- Structural design for building frame and foundations.

Should this level of information be available at the very early stage of projects, then quantity surveyors would be able to take-off detailed elemental quantities. Taken alongside specifications emerging from project teams, quantity surveyors may be able to develop reasonably accurate initial forecasts of employers' likely expenditure on projects, whilst also establishing maximum costs, termed the 'cost limits', for schemes. However, in reality this 'ideal situation' hardly, indeed if ever, occurs, leaving quantity surveyors with the difficult task of predicting employers' likely expenditure and establishing cost limits for projects with minimal information. Where such information is not available and therefore elemental level analysis cannot be used, NRM makes provision for replacing elemental unit quantities with the gross floor area of buildings. In this situation appropriate rates, usually based on comparative projects, can be used to form the basis of the order of cost estimate. For this reason, NRM1 (RICS, 2012a, pp. 20–22) directs quantity surveyors to use one of three techniques for producing the order of cost estimate:

- Functional unit method;
- Floor area method;
- Elemental method.

The elemental method is preferred as this provides the most accurate estimate of project cost, but it may not be possible since data from other members of project teams is not always available. So on the majority of projects, quantity surveyors will revert to either floor area or functional unit methods to establish project estimates. Due to the highly unreliable nature of these techniques, the final outcome will often be reported as a budget range, with the maximum cost in that range defined as the cost limit for projects. The next sections show how these different techniques can be used to prepare the order of cost estimate for the hotel project outlined earlier.

Discussion point 1.3: Why do you think the quantity surveyor reports initial budgets as a cost range rather than a single figure?

Functional unit method

The functional unit method, in terms of estimating accuracy, is the most unreliable technique available to quantity surveyors. As a result this would only be applied to projects if drawings were not available and quantity surveyors could not establish the gross floor area of buildings. A functional unit is a unit of measurement representing the prime use of buildings, or parts of buildings if developments proposed have a mixture of intended uses, for example if an employer is looking to develop a mixed use scheme which included leisure facilities, retail floor space and maybe office space at the higher levels.

Using the example of the hotel development, the functional unit method can be used to establish the construction costs for the project. However, to achieve this, we would first need to identify:

- The prime unit of analysis – in the case of a hotel it would be *cost per bedroom.*
- The number of the prime unit – so how many bedrooms will the hotel provide, in this case 64.
- Historical cost data relating to prime unit – this could include:

    ○   Commercially available price books and rates databases;

○   The quantity surveyor's own cost data library;
○   The Building Cost Information Service (BCIS) database.

To develop the order of cost estimate for the hotel project using the functional unit method, a range of functional unit rates have been sourced from the Building Cost Information Service (BCIS); these are included in Table 1.2. The costs have been adjusted so they reflect tenders in Northern England, and a tender date of 1st Quarter (1Q) 2017 is assumed.

The range detailed in Table 1.2 is extremely wide, since the functional unit costs range from £36,215 to £133,062. It must be remembered that this data potentially includes the full range of hotel specifications (1* to 5*). For this reason, the quantity surveyor must analyse the data to ensure the proposed project is correctly positioned in this cost range. Invariably this will require more information from the employer relating to the anticipated quality and complexity of the proposed hotel project. In this example it is assumed the employer has narrowed the project scope only slightly to specify that the hotel will be a budget level facility with a 2* rating. As a result, the quantity surveyor will need to develop a range of cost, identifying the minimum and maximum. These are calculated as follows:

*   Number of rooms x minimum cost for the hotel (in this case the lowest rate was used). So: £36,215 x 64 = £2,317,760
*   Number of rooms x maximum cost for the hotel (in this case the lower quartile was used). So: £53,361 x 64 = £3,415,104

To fully determine the cost limit for the project, the quantity surveyor will use the data to develop a full order of cost estimate, illustrated in Table 1.3, for the scheme. The order of cost estimate reveals the upper cost limit for the hotel project has been estimated at £5,142,766. Obviously as the quantity surveyor receives more information from both the design team and the employer, this estimate would be refined and the cost limit would be amended to reflect both the additional information and associated increases in estimate accuracy.

Discussion point 1.4: Identify the potential limitations of using a functional unit cost to determine the cost target for the proposed hotel.

Table 1.2 BCIS functional unit data

Building function	£/functional unit						
New build Hotels	Mean	Lowest	Lower quartile	Median	Upper quartile	Highest	Sample size
No of bedrooms	70,725	36,215	53,361	58,344	90,342	133,062	19

Source: BCIS (2016).

Table 1.3 Order of cost estimate for hotel based on functional unit data

Constituent	%	Minimum (£)	Maximum (£)
Facilitating works		0	0
Building works estimate		2,317,760	3,415,104
Contractor preliminaries	Included	0	0
External works (additional allowance)	Say	75,000	125,000
Sub-total		**2,392,760**	**3,540,104**
Main contractors overheads and profit	Included	0	0
Works cost estimate		**2,392,760**	**3,540,104**
Project/design team fees	10%	239,276	354,010
Other development/Project costs		0	0
Base cost estimate		**2,632,036**	**3,894,114**
Risk allowance estimate			
Design development risk estimate	7%	184,243	272,588
Construction risk estimate	7%	184,243	272,588
Employer change risk estimate	3%	78,961	116,823
Employer other risk estimate	3%	78,961	116,823
Cost limit (excluding Inflation)		**3,158,443**	**4,672,937**
Tender inflation estimate	1.19%	37,586	55,608
Cost limit (inc. tender inflation)		**3,196,029**	**4,728,545**
Construction Inflation Estimate	8.76%	279,972	414,221
Cost limit (inc. construction inflation)		**3,476,001**	**5,142,766**
VAT assessment		Excluded	Excluded

Note: The rows in bold correspond to group elements as identified in NRM1. As a result, the costs against these items are sub-totals from the constituent elements.

Floor area method

The floor area method, also known as the GIFA rate method, is the most popular tool used by quantity surveyors to develop feasibility budgets and determine project cost limits. NRM stresses the importance of using elemental unit quantities as the most accurate and therefore preferred approach to estimating the cost of proposed projects. This is often not possible when details are unavailable, as design has not fully commenced. In spite of this, it is reasonably feasible to expect that designers will have ascertained a rough layout for buildings, which allows quantity surveyors to establish the gross internal floor areas. The gross internal floor area (GIFA) of the building is defined in the *RICS Code of Measuring Practice* (RICS, 2007) as "the area of a building measured to the internal face of the perimeter walls at each floor". At the time of writing the *Code of Measuring Practice* is being phased out and replaced with the *International Property Measurement Standard*. It is not expected that the definition or use of the GIFA unit will change as part of this process.

Returning to the hotel development, this second example uses the floor area method to establish an initial estimate for the proposed project. Once again to utilise the technique, the quantity surveyor would need to identify a suitable information source providing *costs per m² GIFA* for hotel buildings. As before sources of potential information include:

- Commercially available price books;
- The quantity surveyor's own cost data library;
- The Building Cost Information Service (BCIS) database; these can be located in the 'average prices' section.

As before to determine the cost limit for the project, the quantity surveyor will use the data to develop a full order of cost estimate as shown in Table 1.4 for the scheme. The costs have been adjusted so they reflect tender prices in Northern England, again assuming the project will be tendered in the first quarter of 2017.

The cost data extracted from BCIS suggests the cost of constructing the hotel could range from £979 up to £2,587 per square metre based on the gross floor area of the building. As with functional cost, the data encapsulates the full range of hotel specifications (1* to 5*). Once more the quantity surveyor must analyse the data to ensure the proposed project is correctly positioned within this range. Invariably this will require more information from the employer relating to the quality and complexity of the proposed hotel. Once again, it is assumed the hotel will be a budget facility with a 2* rating. Accordingly the quantity surveyor will develop a range of costs, identifying the minimum and maximum levels of expenditure. These are calculated as follows:

- Gross floor area of the hotel x minimum cost for the hotel (in this case the lowest rate was used). So: £979 x 2,225m² = £2,178,275
- Gross floor area of the hotel x maximum cost for the hotel (in this case the lower quartile was used). So: £1,416 x 2,225m² = £3,150,600

To determine the cost limit for the hotel the quantity surveyor has used the data to develop a full order of cost estimate for the scheme as shown in Table 1.5.

From the analysis, it can be seen that the improved information has allowed the quantity surveyor to establish the gross internal floor area of the building, rather than simply relying on the number of bedrooms. As a result of improved information it has been possible to refine the upper cost limit for the project, leading to a reduction in the anticipated maximum cost from £5.20m down to £4.76m. As the quantity surveyor receives information from the design team and employer

Table 1.4 BCIS £/m² GIFA data

Building function	£/m² gross internal floor area						
New build hotels	Mean	Lowest	Lower quartile	Median	Upper quartile	Highest	Sample size
Hotels	1,633	979	1,416	1,536	1,883	2,587	22

Source: BCIS (2016).

Table 1.5 Order of cost estimate for hotel based on floor area (£/m²) data

Constituent	%	Minimum (£)	Maximum (£)
Facilitating works		0	0
Building works estimate		2,178,275	3,150,600
Contractor preliminaries	Included	0	0
External works (additional allowance)	Say	75,000	125,000
Sub-total		**2,253,275**	**3,275,600**
Main contractors overheads and profit	Included	0	0
Works cost estimate		**2,253,275**	**3,275,600**
Project/design team fees	10%	225,328	327,560
Other development/Project costs		0	0
Base cost estimate		**2,478,603**	**3,603,160**
Risk allowance estimate			
Design development risk estimate	7%	173,502	252,221
Construction risk estimate	7%	173,502	252,221
Employer change risk estimate	3%	74,358	108,095
Employer other risk estimate	3%	74,358	108,095
Cost limit (excluding Inflation)		**2,974,323**	**4,323,792**
Tender inflation estimate	1.19%	35,394	51,453
Cost limit (inc. tender inflation)		**3,009,717**	**4,375,245**
Construction Inflation Estimate	8.76%	263,651	383,272
Cost limit (inc. construction inflation)		**3,273,369**	**4,758,517**
VAT assessment		Excluded	Excluded

Note: The rows in bold correspond to group elements as identified in NRM1. As a result, the costs against these items are sub-totals from the constituent elements.

the estimate will be refined and the cost limit amended to reflect the increased accuracy of the forecast.

Discussion point 1.5: What are the advantages of using a GIFA rate (£/m² gross floor area) when compared to the unit rate approach?

Elemental method

The elemental method represents the most accurate approach to producing the initial feasibility estimate for projects. This fact is fully recognised and documented in NRM1, which stresses the importance of using elemental analysis when developing cost forecasts for proposed projects. The method is the most demanding in terms of design and specification development. So whilst it is highly probable that designers will be able to provide rough layouts for buildings with accompanying elevation sketches, it is unlikely they will have developed their conceptual ideas sufficiently to allow quantity surveyors to achieve the level of measurement accuracy needed to produce comprehensive cost plans at this stage. As a result, any attempt at the elemental analysis will usually be a simple comparative cost plan based on the gross floor area of buildings. For this reason, the elemental method has not been considered in depth at this point in the chapter. Instead a comprehensive explanation of the elemental cost planning process is provided in section 1.4.

1.3.2 Reporting the order of cost estimate

The final stage in the production of initial feasibility estimates (order of cost estimate) is to communicate this professionally and clearly to employers. It is essential quantity surveyors ensure this part of the process is handled with care. The order of cost estimate report represents the first estimate employers see relating to the cost of buildings, as a result, this will be the figure they are most likely to remember.

It is also important for quantity surveyors to be wary of employer expectations and ensure they are managed effectively. Employers may have a vision of their buildings that far exceeds their budget, given that employers are likely to have *'the aspirations of a prince and the budget of a pauper'*. It is prudent to report cost limits, stating explicitly what has been included and what has been excluded from estimated project costs and therefore what falls outside the overall cost limit. Detailed guidance on reporting the order of cost estimate is provided at the end of Part 2 of the NRM1 (RICS, 2012a, p. 32).

Once initial budgets and cost limits have been reported to employers, in consultation with professional teams there will be an evaluation of overall viability and affordability. This decision will result in employers either committing to proceed with designs, or alternatively taking the decision to abandon projects before expensive design commences. Assuming employers are happy to proceed with designs, this is signified by them signing a *commit to invest*. The *commit to invest* is the first of three critical project milestones, illustrated in Figure 1.2 and explained thus:

A   *Commit to invest* – the point at which employers decide in principle to invest in projects and authorise design teams to proceed with conceptual designs.
B   *Commit to construct* – the point at which employers authorise the construction of projects. This allows teams to negotiate with successful bidders and move towards signing contracts and commencement of the construction phase.
C   *Available to use* – the point at which project teams authorise buildings to be occupied and used by employers.

Once *commit to invest* is signed the design phase of projects will commence, and during this time quantity surveyors will continue to provide pre-contract financial management. As the design develops, so too will the precision and level of detail of estimates, and the cost limit defined at this stage in the project will,

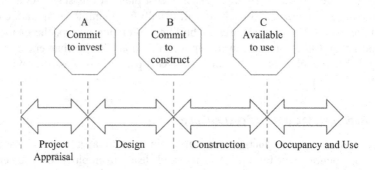

Figure 1.2 Major project milestones.

it is hoped, not be exceeded. It is therefore important quantity surveyors remain reasonably restrained in advice provided.

Discussion point 1.6: Identify the various types of information you would need, and therefore expect your employer to provide at the 'business case' stage of projects.

1.4 Cost planning

Once projects move beyond the *commit to invest* stage employers are committed to developing schemes to the end of the pre-contract phase. The full design of projects will now proceed. As design teams start to develop conceptual schemes approved by employers at project appraisal stage, the role of quantity surveyors is now one of cost prediction and cost control. Quantity surveyors will be producing increasingly detailed estimates for overall project costs. These estimates will be aligned to the key employer reporting stages which are typically aligned with phases of design development. There is a need to ensure the budget of employers is adhered to, and pre-determined cost limits approved by employers prior to signing the *commit to invest* are not exceeded.

As a result quantity surveyors will adopt a financial management technique known as *design to cost*. Associated with production, design to cost is a financial management technique that tracks innovative (bespoke or new) products through their design phase with the aim of ensuring they do not exceed budgets allocated for their development. *Design to cost* therefore acts as a benchmarking technique to see how far the development of projects or buildings have progressed, whilst ensuring designs remain financially balanced. Moreover the technique allows design teams to pre-empt potential problems by identifying areas where projects are likely to exceed the available budget of employers. Design teams may then identify creative solutions to overcome any difficulties. When implemented in construction, the design to cost management approach is widely known as *cost planning*. Although some textbooks will argue cost planning is in reality only one aspect of much more complex pre-contract cost control systems, the accepted industry best practice guidance, NRM1, defines this process simply as *cost planning*. This text will adopt the same terminology.

The cost plan is defined in NRM1 as:

> *The critical breakdown of the cost limit for the building into cost targets for each element of the building. It provides a statement of how the design team proposes to distribute the available budget amongst the elements of the building, and a frame of reference from which to develop the design and maintain cost control. It also provides both a work breakdown structure and a cost breakdown structure which by codifying can be used to redistribute work in elements to construction works packages for the purpose of procurement.*
>
> (RICS, 2012a, p. 10)

From the definition provided in NRM1 it is clear that cost plans form the principal tool for the pre-contract financial management of construction projects. Not only do cost plans provide breakdowns of overall project budgets into core cost centres, referred to as *elements*, they also provide a point of reference against which design teams can develop schemes. There is the opportunity to ensure costs are fully managed and controlled, against overall budgets. Table 1.6 illustrates a breakdown of budgets into individual cost centres that form part of the *building works estimates*.

Cost planning, however, must not be seen as a single activity that only happens at the start of design phases. Cost plans may appear to be like a photograph; a view of the financial position of a project captured in a single point of time. In reality, cost planning is a continuous process that spans the full design phase of

Table 1.6 Constituent elements of a cost plan

Constituent
Building works estimate
Main contractor preliminaries estimate
Sub-total
Main contractors overheads and profit estimate
Works cost estimate
Project/design team fee estimate
Consultants fees
Main contractor's pre-construction fee estimate
Main contractor's design fee estimate
Sub-total
Other development/Project cost estimate
Base cost estimate
Risk allowances estimate

(*Continued*)

Table 1.6 (Continued)

Design development risk estimate
Construction risk estimate
Employer change risk estimate
Employer other risk estimate
Cost limit (excluding Inflation)
Tender inflation estimate
Cost limit (inc. tender inflation)
Construction Inflation Estimate
Cost limit (inc. construction inflation)
VAT assessment

Source: RICS (2012a, p. 37).

Note: The rows in bold correspond to group elements as identified in NRM1. As a result, the costs against these items are sub-totals from the constituent elements.

projects. Alternatively, cost planning can be likened to a digital recording, since it is an evolving process consisting of numerous snapshots of projects that are linked together to provide a continuous image; in this case, a real-time view of cost mapped against the available budget of employers. The fundamental objectives of cost plans are defined by NRM1 (RICS, 2012a, p. 36) as:

• To ensure employers received value for money;
• To make both employers and designers aware of the cost consequences of their design or proposal;
• To provide advice to designers that enables them to arrive at a practical and balanced design within the scope of the budget;
• To keep project expenditure within the cost limit approved by the employer;
• Provide robust cost information upon which the employer can make informed decisions.

To make this process useful for employers, cost plans are captured and reported at key project milestones. These are aligned to the RIBA plan of work, as shown in Figure 1.3, and constitute the formal cost plan requirements of the NRM1, although the exact reporting requirements may be amended by employers.

NRM provides for three 'formal cost plans' or reports to employers, whilst each formal cost plan is really a progression of the order of cost estimate. As design information evolves and a clearer picture of projects emerge, each cost plan will provide a more comprehensive and therefore accurate view of the expenditure profile of projects than the previous. The three formal cost plans are:

• *Formal cost plan 1* – Prepared when the scope of work is fully defined and key criteria specified but no detailed designs have commenced.
• *Formal cost plan 2* – Prepared when design development is complete.
• *Formal cost plan 3* – The final cost plan produced is based on completed technical designs, specifications and detailed information now available to quantity surveyors.

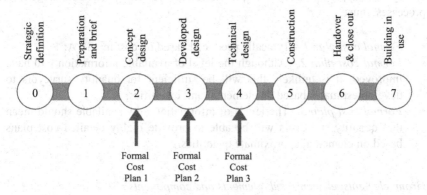

Figure 1.3 Formal cost plan reporting stages mapped to RIBA plan of work.

At each point quantity surveyors are required to prepare formal cost plans and provide employers with detailed reports, updating them on the financial position of projects and of any anticipated financial risks identified, as the forward horizons of projects are scanned.

1.4.1 Preparing cost plans: An overview

Each of the three formal cost plans represents a progression and refinement of the order of cost estimates provided to employers at the outset of projects. As a result, each cost plan is a more detailed, logical progression from the first; so with each cost plan the level of accuracy achieved, when compared to tender and final account figures, should be enhanced, thereby reducing the levels of variability in estimates. As a result of this evolutionary process, neither formal cost plan 2 nor formal cost plan 3 will require quantity surveyors to produce new documents from a zero start. Improvements in accuracy will result from the depth of analysis each iteration provides. With this in mind, the NRM sets out the methods that can be used when preparing formal cost plans. These include:

* The elemental method based on elemental unit quantities;
* The elemental method based on approximate quantities;
* If neither of the above are possible due to lack of design information, then the elemental method based on cost/m² GIFA should be used.

The usual progression from the initial order of cost estimate will ideally be through the elemental method based on elemental unit quantities, and onwards towards very detailed cost plans based on elemental approximate quantities. However, in practice this is not always possible. It is more realistic to expect quantity surveyors to apply all three techniques during pre-contract financial management processes, thus:

- *Formal cost plan 1* – Typically produced based on cost/m² GIFA;
- *Formal cost plan 2* – Although the level of available information will have improved, it is unlikely this will be sufficient for quantity surveyors to develop estimates based on elemental unit quantities;
- *Formal cost plan 3* –The levels of information now available should mean that quantity surveyors will be able to provide highly detailed cost plans based on elemental approximate quantities.

Group elements, elements, sub-elements and components

In addition to detailing the main approaches to the development of cost plans, section 4 of NRM1 provides a breakdown of the elements that make up a project. This breakdown of the building works estimates into elements allows design teams and employers to ensure costs are comprehensively controlled. However, given the varying approaches to estimating, NRM1 also provides four principal levels of detail that can be incorporated into cost plans. These levels of detail are outlined comprehensively in the fourth section of NRM1 called the 'tabulated rules of measurement' (RICS, 2012a, pp. 75–327). The tabulated rules of measurement set out the different levels of description and measurement that quantity surveyors should use. These levels of measurement are defined below:

- *Group Element* – The main headings used to describe the facets of cost plans; examples of group elements include substructure, superstructure, internal finishes etc.
- *Element* – A major part of the building that is applicable to a variety of building types; for example all buildings regardless of their use will have external walls.
- *Sub-Element* – Part of an element; cost targets are set at sub-element level in a similar way to those set for elements.
- *Component* – This is a measurable item that forms part of both an element and a sub-element.

To further help you understand the differences between these levels of measurement, Table 1.7 provides an example of the levels of measurement in relation to the roof of the budget hotel project.

The decision of applicability rests with the quantity surveyor, but this will be largely dependent on both the levels of design information available and the method of production selected. For example it would be impossible to measure sub-elements without comprehensive technical designs and without using elemental approximate quantities. As a result the levels of measurement expected at

Table 1.7 Tabulated measurement rules

Level 1: group element	Level 2: element	Level 3: sub-element	Level 4: component
2. Superstructure	2.3 Roof	1. Roof structure	1. Roof structure – pitched roofs 2. Extra over roof structure (dormer) 3. Roof structure – flat roofs
		2. Roof coverings	1. Roof coverings, non-structural, screeds, insulation and the like 2. Extra over roof coverings (dormers) including cladding 3. Eaves, verge treatment 4. Edge treatment to flat roof 5. Flashings
		3. Glazed roofs	1. Glazed roof
		4. Roof drainage	1. Gutters 2. Rain water pipes

Source: RICS (2012a, pp. 85–87).

Note: The rows in bold correspond to group elements as identified in NRM1. As a result, the costs against these items are sub-totals from the constituent elements.

each cost planning stage would be significantly more detailed and comprehensive. For example:

- *Formal cost plan 1* – Would usually be prepared on the basis of group elements or more realistic elements, measured using elemental unit quantities, m²/GIFA or a combination of both, with elemental quantities developed wherever possible.
- *Formal cost plan 2* – Expanding from formal cost plan 1, it would be expected elemental unit quantities will be applied and the analysis will be broken down into sub-elements.
- *Formal cost plan 3* – This would be almost a bill of quantities in terms of measurement precision, so it would be expected that quantity surveyors will break costs down to component level.

Applying this to practice, quantity surveyors should work to the most detailed level possible given available information. So for the budget hotel, at the end of the concept design phase, the quantity surveyor is likely to have an idea of the number of doors to be provided in the building. They should also have an idea of the likely level of specification (e.g. fire doors, the fire rating and maybe some finishing details). This would allow the quantity surveyor to apply elemental units to the scheme, but the lack of detailed specification would prevent the application of approximate quantities until further information was provided by the design team. The next sections of text build on this simple example and provide illustrations of how each of these techniques could be applied to the budget hotel project.

Discussion point 1.7: When would you use: group elemental, elemental, sub-element and component level data in an order of cost estimate or formal cost plan?

Preparing a cost plan using floor areas (£/m² GIFA)

This is often referred to as a *comparative cost plan* because the analysis is largely based on the selection and use of a similar project. This is the simplest, and in some respects, the least labour-intensive technique available to quantity surveyors for the production of elemental cost plans. Whilst the technique is presented here in the context of formal cost plan 1, some quantity surveyors will also apply the technique when developing the order of cost estimate, in an attempt to produce a more refined budget cost at the commencement of projects.

At this stage in the development of projects, it is likely quantity surveyors will have only received initial, conceptual designs for schemes from architects. Although project appraisals and order of cost estimates will have identified potential directions for projects, for example new construction rather than refurbishment, or construction on site A rather than site B, the appearance and layout of buildings will not have been evaluated. As a result, the conceptual design stage represents the point at which project layout and appearance are resolved. It is likely quantity surveyors will receive multiple designs and will be asked to produce several different cost plans for these designs. Each alternative design is likely to include 1:100 or 1:200 scale plans for each floor, elevation designs and brief specifications for major elements. As a result, it is unlikely measurement (quantification) of buildings will be possible. The task of quantity surveyors at this point is essentially to evaluate the affordability of designs, compare them to cost limits determined earlier and identify cost targets for each element of projects.

At this stage in the evolution of projects the scarcity of information and the lack of detailed measurement, over and above simple floor areas, means the cost data available to quantity surveyors is limited. As the name 'comparative' suggests, quantity surveyors should aim to identify previous projects that, with adjustment, will provide a useful basis for the development of formal cost plan 1. This information is accessible in many commercially available price books and magazines aimed at construction professionals and quantity surveyors should have a library of similar past projects from which they can formulate their own forecasts of cost. The Building Cost Information Service (BCIS), a subsidiary of the RICS, is a subscription-based service that provides a database of cost information based on real life completed projects. As part of the subscription, quantity surveyors provide cost information about projects that they have been involved in, and in return they are able to get cost information about projects provided by others.

The BCIS database provides a variety of levels of useful information, including 'average prices' that can be used to develop the order of cost estimate and 'analyses' that provide a more detailed breakdown of individual projects. To ensure uniformity and to maximise the usability of the cost analysis reports within the BCIS database, each project is analysed and submitted by subscribers using a standard format and set of measurement rules. These measurement rules are termed the *Standard Form of Cost Analysis 4th Edition* (BCIS, 2014). Care should be taken not to mix these up with NRM as they are designed to serve a different function. The *Standard Form of Cost Analysis* has been designed to transform project data into a standard format for upload onto the database, not to provide a base for preparing new cost plans. Yet, by having these rules it is possible for quantity surveyors to analyse and use cost analysis reports as a basis for their comparative cost plans.

Discussion point 1.8: How would you distinguish between an elemental cost analysis and an elemental cost plan?

Example cost plan using floor areas (£/m² GIFA)

To develop formal cost plan 1 for the hotel project using floor areas, the BCIS database has been searched to find a cost analysis that matches as closely as possible the attributes of the proposed project, including:

- The overall size of the proposed building (2,225m²);
- The number of storeys (5);
- The footprint of the building (area of each floor – 445m²);
- The wall to floor ratio of the building (perimeter/area);
- The number of bedrooms (64);
- The individual storey height;
- The main construction specification;
- The anticipated level of quality (2* hotel).

Using these parameters, a search of the BCIS database reveals nine suitable projects. Further consideration of these allows selection of the most appropriate project, in this case a hotel constructed in London from which the cost plan presented in Table 1.8 has been developed. From the analysis you can see that additional design information has allowed the quantity surveyor to establish a more detailed breakdown for the main construction works, thereby ensuring the *building works estimate* for the project remains within the budget range of £2.25 – £3.28m reported in the order of cost estimate. Using £/m² based on the gross floor area, the building works estimate has been calculated at £3.15m.

Table 1.8 Formal cost plan 1 for the hotel based on floor area (£/m^2)

	Element	Total cost	Cost per m^2	EUQ	EUR	%
0	**Facilitating works**	**£0**	**£0**			**0%**
1	**Substructure**	**£175,775**	**£79**			**6%**
2	**Superstructure**	**£1,248,225**	**£561**			**40%**
2.1	Frame	£315,950	£142			10%
2.2	Upper floors	£155,750	£70			5%
2.3	Roof	£71,200	£32			2%
2.4	Stairs and ramp	£31,150	£14			1%
2.5	External walls	£280,350	£126			9%
2.6	Windows and external doors	£271,450	£122			9%
2.7	Internal walls and partitions	£26,700	£12			1%
2.8	Internal doors	£95,675	£43			3%
3	**Internal Finishes**	**£175,775**	**£79**			**6%**
3.1	Wall finishes	£102,350	£46			3%
3.2	Floor finishes	£22,250	£10			1%
3.3	Ceiling finishes	£51,175	£23			2%
4	**Fittings, Fixtures and Equipment**	**£60,075**	**£27**			**2%**
4.1	Fittings, fixtures and equipment	£60,075	£27			2%
5	**Services**	**£818,800**	**£368**			**26%**
5.1	Sanitary installations		See 5.6			0%
5.2	Services Equipment		See 5.6			0%
5.3	Disposal Installations		See 5.6			0%
5.4	Water Installations		See 5.6			0%
5.5	Heat Source		See 5.6			0%
5.6	Space heating and air conditioning	£760,950	£342			24%
5.7	Ventilation		See 5.6			0%
5.8	Electrical installations		See 5.6			0%
5.9	Fuel installations	£0	£0			0%
5.10	Lift and conveyor installations	£51,175	£23			2%
5.11	Fire and lightening protection	£0	£0			0%
5.12	Communications and security installations		See 5.6			0%
5.13	Specialist installations	£0	£0			0%
5.14	Builders work in connection with services	£6,675	£3			0%
6	**Complete buildings and building units**	**£0**	**£0**			**0%**
7	**Work to existing buildings**	**£0**	**£0**			**0%**
	Sub-total	**£2,478,650**				**79%**
8	**External Works**	**£311,500**	**£140**			**10%**
8.1	Site preparation works	£140,175	£63			4%
8.2	Roads, paths and pavings		See 8.1			0%
8.3	Soft landscaping, planting and irrigation systems		See 8.1			0%
8.4	Fencing, railings and walls		See 8.1			0%
8.5	External fixtures		See 8.1			0%
8.6	External drainage	£100,125	£45			3%
8.7	External services	£71,200	£32			2%
8.8	Minor building works and ancillary buildings	£0	£0			0%
	BUILDING WORKS ESTIMATE	**£2,790,150**				**89%**

	Element	Total cost	Cost per m²	EUQ	EUR	%
9	**Main contractors preliminaries**	**£342,650**	**£154**			**11%**
10	**Main Contractors Overheads and Profit**	**£0**	**£0**			**0%**
10.1	Main contractors overheads	£0	£0			0%
10.2	Main contractors profit	£0	£0			0%
	TOTAL CONTRACT SUM (excluding risks and design fees)	**£3,132,800**	**£1,408**			**100%**

Note: The rows in bold correspond to group elements as identified in NRM1. As a result, the costs against these items are sub-totals from the constituent elements.

Selection of the London hotel was based on the many similarities in core areas such as specification, building function and layout and design quality. Looking at the London project, it related to a popular brand of 2* hotel (similar to the proposed), it also had a comparable layout and bedroom size to the employer's proposed project. Additional areas of similarity include number of storeys and the comparable although slightly larger gross internal floor area. Therefore although the London project is slightly larger and it is located in London, this project provided the most appropriate comparator for the employer's scheme.

Location variance and tender price movements

The cost plan presented in Table 1.8 does not fully reflect the realities of the project in Lancashire. The data included has simply been taken from BCIS and applied to the proposed scheme. This is what economists call *nominal*. Nominal prices are monetary values presented in their original year of capture. Therefore the data provided in Table 1.8 represents 2011 prices not the current costs of constructing a comparable building today. To remove this ambiguity, we need to use *real* monetary values, or values consistent with each other both in terms of time and location. To help the quantity surveyor in this regard, BCIS publishes a series of indices; the key indice for predicting price movements is the Tender Price Index (TPI), which records macro level change in the economy. A separate indice called location factors records micro level change between geographic locations. Both indices provide these figures on a quarterly basis. Other commercial organisations provide similar indices for use by quantity surveyors; however these are often not easily interchangeable as projects evolve, so it is perhaps better to use these standardised BCIS indices.

The BCIS Tender Price Index (TPI) provides records of historic price movements and predictions of future tender price increases or decreases. Forecasts of future price movements are made by bringing together a range of key economic data and market intelligence, sourced from government departments such as the Treasury, the Department for Business, Energy and Industrial Strategy, the Bank of England, and other eminent economic forecasting organisations. Forecasts of future indices are indicated with the letter 'f' (forecast). Those which are recently published and subject to change as more data becomes available are indicated with the letter 'p' (provisional). The BCIS TPI was set at 100 in 1985; this is referred to as the base date.

The index allows surveyors to then calculate price movement on a quarterly basis, using a simple formula given in NRM1 (RICS, 2012a, p. 42):

$$P = ((\text{index } 1 - \text{index } 2) \div \text{index } 1) \times 100$$

Where:
Index 1 = the index at the base date of the cost data
Index 2 = index at the current estimate base date
P = percentage addition or subtraction

So for example the London project had a tender base date of June 2011, giving it a TPI index of 223 (2Q:2011) it is anticipated the proposed project will have a tender date of February 2017 giving it a TPI index of 286(f). That represents a price movement of 28.25% calculated using the formula thus: $P = ((223-286) \div 223) \times 100$.

It is important at this stage to explain an important distinction in economics and finance between *price* and *cost*. Although this chapter has used the word *cost* extensively, when we are *cost planning* we are actually attempting to forecast the likely tender price and ideally the likely final account price for projects under consideration. However, in other areas of practice, such as dealing with fluctuations under contracts, it would be important for surveyors to consider changes in construction costs, those are the costs incurred by main contractors at the first tier in construction supply chains, related to movements in the cost of labour, materials and plant.

For this reason the BCIS service also produce a separate index known as the *General Building Cost Index (GBCI)*. This index provides evidence of cost movements, so as costs rise, the index number will increase and, as costs fall, the index number will decrease. Some would expect the GBCI and TPI to be highly positively correlated, whereby an increase in one would cause an increase of near identical proportion in the other, as contractors pass on increases in costs to employers through higher tender prices. However, the highly competitive nature of construction prevents such a correlation.

When work is scarce, such as immediately after the market crash in 2008, contractors may seek to absorb increases in cost in an attempt to win work. In some situations contractors desperate for turnover would often submit a below-cost bid (known as a suicide bid) simply to keep trading, in the hope that lucrative design changes would ease the project to profitability. Equally when work is more abundant and contractors are managing swelling forward order books such as in the early part of the 21st century (2000–2007), contractors will often seek to increase their prices and still win work. This can be seen graphically in Figure 1.4 which shows a plot of quarterly price changes over the period from 2000 to 2020.

It should be noted that as the financial crisis impacted on the construction industry in mid-2008 tender prices dipped considerably. As a result, the period between 2007 and 2010 shows that construction costs are rising whilst tender prices are falling. Suggesting scarcity of work during this period pressured contractors to

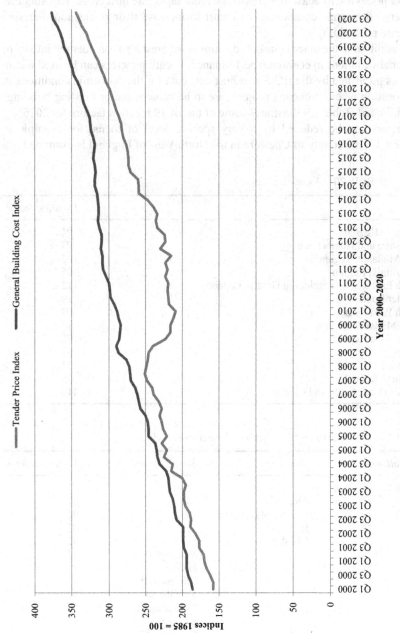

Figure 1.4 Line diagram indicating Tender Price and General Building Cost Indices.

absorb the inflationary pressure on input costs. As a result margins of contractors reduced, and some contractors will have bid below cost simply to remain trading. Yet as the economy is forecast to recover in the period from 2018 to 2020, tender prices are forecast to increase at a more rapid rate than costs. This suggests as demand returns, contractors will start to increase their prices and therefore increase profitability.

In addition to the macro market adjustment we must also consider the impact of regional variations in economic performance. Location factors can be used within indices published by the BCIS to adjust cost data for the economic conditions in the location where proposed projects are to be constructed or existing buildings rehabilitated. Table 1.9 illustrates some of the BCIS location factors for 2016.

Factors can be reduced by a very specific level of focus, for example in Table 1.10, the county of Cheshire in the North West of England is examined.

Table 1.9 BCIS 2016 location factors

Area	Location factor
Northern Region	94
Yorkshire and Humberside	93
East Midlands Region	102
East Anglia Region	96
South East Region (excluding Greater London)	112
Greater London	129
South West Region	99
West Midlands Region	96
North West Region	92
Wales	91
Scotland	91
Northern Ireland	55
Islands (Man, Scilly and Channel)	110

Table 1.10 BCIS 2016 location factors, hierarchy of levels

Overall area	County	Town or city	Location factor
North West Region			92
	Merseyside		92
	Greater Manchester		92
	Lancashire		92
	Cheshire		92
		Warrington	91
		Halton	91
		Ellesmere Port	93
		Vale Royal	92
		Macclesfield	98
		Chester	93
		Crewe and Nantwich	91
		Congleton	90

The final point to make about the adjustment for price movements relates to when these should be included in cost plans. The order of cost estimate and the three separate formal cost plans should all use cost data adjusted for inflation and location. However, NRM1 states that these adjustments should only be made to the base date of reports. Therefore as projects progress, cost data used will require continuous updating or *rebasing* as it is termed in BCIS. Quantity surveyors must, however, also add their prediction of both tender and construction inflation. NRM stipulates that tender inflation is to be taken from the date of the cost plan to the anticipated base date of the tender period (usually the mid-point of the tender period). Construction inflation is to be taken from the tender period to the mid-point (time not expenditure) of construction projects.

Using this information, the cost plan presented in Table 1.8 can now be reproduced as detailed in Table 1.11. The cost plan has been rebased to the time it was produced, which for this example was 2Q:2016.

Table 1.11 Formal cost plan 1 for the hotel based on floor area ($£/m^2$) rebased on 2Q:2016

	Element	Total cost	Cost per m^2	EUQ	EUR	%
0	**Facilitating works**	**£0**	**£0**			**0%**
1	**Substructure**	**£153,525**	**£69**			**6%**
2	**Superstructure**	**£1,092,475**	**£491**			**40%**
2.1	Frame	£275,900	£124			10%
2.2	Upper floors	£135,725	£61			5%
2.3	Roof	£62,300	£28			2%
2.4	Stairs and ramp	£26,700	£12			1%
2.5	External walls	£244,750	£110			9%
2.6	Windows and external doors	£238,075	£107			9%
2.7	Internal walls and partitions	£24,475	£11			1%
2.8	Internal doors	£84,550	£38			3%
3	**Internal Finishes**	**£153,525**	**£69**			**6%**
3.1	Wall finishes	£89,000	£40			3%
3.2	Floor finishes	£20,025	£9			1%
3.3	Ceiling finishes	£44,500	£20			2%
4	**Fittings, Fixtures and Equipment**	**£53,400**	**£24**			**2%**
4.1	Fittings, fixtures and equipment	£53,400	£24			2%
5	**Services**	**£716,450**	**£322**			**26%**
5.1	Sanitary installations		See 5.6			0%
5.2	Services Equipment		See 5.6			0%
5.3	Disposal Installations		See 5.6			0%
5.4	Water Installations		See 5.6			0%
5.5	Heat Source		See 5.6			0%
5.6	Space heating and air conditioning	£665,275	£299			24%
5.7	Ventilation		See 5.6			0%
5.8	Electrical installations		See 5.6			0%
5.9	Fuel installations	£0	£0			0%
5.10	Lift and conveyor installations	£44,500	£20			2%
5.11	Fire and lightening protection	£0	£0			0%

(*Continued*)

Table 1.11 (Continued)

	Element	Total cost	Cost per m²	EUQ	EUR	%
5.12	Communications and security installations		See 5.6			0%
5.13	Specialist installations	£0	£0			0%
5.14	Builders work in connection with services	£6,675	£3			0%
6	**Complete buildings and building units**	**£0**	**£0**			**0%**
7	**Work to existing buildings**	**£0**	**£0**			**0%**
	Sub-total	**£2,169,375**	**£975**			**79%**
8	**External Works**	**£271,450**	**£122**			**10%**
8.1	Site preparation works	£122,375	£55			4%
8.2	Roads, paths and pavings		See 8.1			0%
8.3	Soft landscaping, planting and irrigation systems		See 8.1			0%
8.4	Fencing, railings and walls		See 8.1			0%
8.5	External fixtures		See 8.1			0%
8.6	External drainage	£86,775	£39			3%
8.7	External services	£62,300	£28			2%
8.8	Minor building works and ancillary buildings	£0	£0			0%
	BUILDING WORKS ESTIMATE	**£2,440,825**	**£1,097**			**89%**
9	**Main contractors preliminaries**	**£300,375**	**£135**			**11%**
10	**Main Contractors Overheads and Profit**	**£0**	**£0**			**0%**
10.1	Main contractor's overheads	£0	£0			0%
10.2	Main contractor's profit	£0	£0			0%
	TOTAL CONTRACT SUM (excluding risks and design fees)	**£2,741,200**	**£1,232**			**100%**

Note: The rows in bold correspond to group elements as identified in NRM1. As a result, the costs against these items are sub-totals from the constituent elements.

Discussion point 1.9: Why would quantity surveyors not adopt a national measure of inflation to make these adjustments such as the retail price index (RPI) or consumer price index (CPI)? Why do you think it is important for quantity surveyors to also consider regional variance?

Elemental unit quantities and elemental unit rates

The preparation of formal cost plans is progressive and highly dependent on the amount of design information available at the time of their production. As a result, it is difficult to identify which cost planning techniques will used and when; this will be largely governed by the speed of design and the method of procurement employers adopt.

NRM1 stresses the importance of using elemental unit quantities as the most accurate and therefore the preferred approach to estimating the cost of proposed projects. It is therefore important for quantity surveyors to move away from cost plans derived from gross floor area measurement as soon as information release permits.

As more design information is developed it is possible to undertake some measurement of individual elements. Although the technical design information may not yet be available, designs may be sufficient to enable take-off of elemental quantities, known as *elemental unit quantities* or EUQ for short. These are typically measured at the sub-element or possibly element level of buildings. Examples of elemental unit quantities include:

- The area of external walls;
- The number of internal doors;
- The area of roof coverings, etc.

Even with partial design information, it would often be possible for quantity surveyors to measure these elements of buildings. The first stage in this process would be to develop EUQs at element or sub-element level. These should be measured using the *tabulated rules of measurement* provided in part 4 of NRM1 (RICS, 2012a, pp. 37–329). EUQ can range from areas serviced by heating systems based on net floor areas, or counts for the number of sanitary fittings. Quantity surveyors should seek to measure each major type of the element under consideration. In terms of floor finishes, for example, quarry tiles, carpets, non-slip sheet flooring and laminate or real wood flooring would all be measured separately.

The next stage in the process is to then develop costs for these elements. Rather than using £/m² rates as in earlier examples, quantity surveyors should this time develop *elemental unit rates* (EUR). Elemental unit rates are also known as *composite rates* as they will not be broken down into the level of detail expected in unit rates for bills of quantities. Composite rates are a collection of unit rates that have been brought together to provide quantity surveyors with a collection of price data that is more accurate than the £/m² rates, but also reflects the on-going lack of complete information that makes accurate measurement and unit rate costing possible.

Discussion point 1.10: In what circumstances would you say it is appropriate for quantity surveyors to use element unit quantities and element unit rates to build up element costs?

Example of a cost plan developed using elemental units

Referring back to the hotel project, it has been assumed the architect prepared a series of dimensioned plan drawings, along with drawings of the main elevations.

Unfortunately these drawings cannot be provided here, so the quantities provided in Table 1.12 are fictitious. Nevertheless the process the quantity surveyor would adopt when preparing a cost plan using elemental quantities is explained and demonstrated.

The EUQ and EUR provided in the cost plan illustrated in Table 1.12 are based on a hotel cost model obtained from a commercially available price book. The cost model, developed for an 8,400m^2 hotel in Manchester, provides a very reliable basis from which to prepare the cost plan for the budget hotel scheme this chapter has reviewed. To achieve this, appropriate pro-rata adjustments have been made for quantity changes between the 8,400m^2 cost model and the 2,225m^2 hypothetical project. As a quantity surveyor would do in practice, the outcome of this process has subsequently been checked and benchmarked against data from BCIS to ensure the validity of the cost plan presented. Furthermore the rates have been rebased for tender price movements and location.

The cost plan in Table 1.12 illustrates how additional design information allows the quantity surveyor to establish a more detailed breakdown of the main construction work costs. In some areas, such as substructure and roof covering, the example has been taken beyond element level to consider specific component

Table 1.12 Formal cost plan 1 for the hotel based on elemental units

	Element	Total cost	Cost per m^2	EUQ	Unit	EUR
0	**Facilitating works**	**£0**	**£0**			
1	**Substructure**	**£182,450**	**£82**	**445**	**m^2**	**£410**
	Excavation, ground beams, filling to levels, lift pits, ground slab	£120,150	£54	445	m^2	£270
	Rotary bored piles	£48,950	£22	445	m^2	£110
	Under slab drainage	£13,350	£6	445	m^2	£30
2	**Superstructure**	**£920,490**	**£414**			
2.1	Frame	£71,200	£32	445	m^2	£160
2.2	Upper floors	£391,600	£176	1780	m^2	£220
2.3	Roof	£57,850	£26	445	m^2	£130
2.3.1	Roof structure					
	Flat roof, pre-cast concrete roof slab	£71,200	£32	445	m^2	£160
	Extra over for forming upstands and copings	£4,450	£2	1	item	£4,450
2.3.2	Roof Covering					
	Single-ply roof membrane, insulation, rainwater outlets	44,500	£20	445	m^2	£100
	Mansafe system	£2,225	£1		Item	£2,225
2.4	Stairs and ramp	£18,000	£8	2	nr	£9,000
2.5	External walls	£213,030	£96	1578	m^2	£135
2.6	Windows and external doors	£65,000	£29	250	m^2	£260
2.7	Internal walls and partitions	£47,560	£21	1160	m^2	£41

	Element	Total cost	Cost per m²	EUQ	Unit	EUR
2.8	Internal doors	£56,250	£25	90	nr	£625
3	**Internal finishes**	**£218,020**	**£98**			
3.1	Wall finishes	£121,000	£54.38	5500	m²	£22
3.2	Floor finishes	£53,460	£24	1980	m²	£27
3.3	Ceiling finishes	£43,560	£20	1980	m²	£22
4	**Fittings, fixtures and equipment**	**£262,400**	**£118**			
4.1	Fittings, fixtures and equipment	£262,400	£117.93	64	nr	£4,100
5	**Services**	**£976,160**	**£439**			
5.1	Sanitary installations	£77,400	£34.79	1800	m²	£43
5.2	Services equipment	£300,000	£135		Item	£300,000
5.3	Disposal installations	£12,160	£5	64	nr	£190
5.4	Water installations	£39,600	£18	1800	m²	£22
5.5	Heat source	£180,000	£81	1800	m²	£100
5.6	Space heating and air conditioning	£0	£0	0	m²	£0
5.7	Ventilation	£0	£0	0	m²	£0
5.8	Electrical installations	£234,000	£105	1800	m²	£130
5.9	Fuel installations	£0	£0	1800	m²	£0
5.10	Lift and conveyor installations	£70,000	£31	2	nr	£35,000
5.11	Fire and lightening protection	£3,600	£2	1800	m²	£2
5.12	Communications and security installations	£54,000	£24	1800	m²	£30
5.13	Specialist installations	£0	£0		m²	
5.14	Builders work in connection with services	£5,400	£2	1800	m²	£3
6	**Complete buildings and building units**	**£0**	**£0**			
7	**Work to existing buildings**	**£0**	**£0**			
	Sub-total	**£2,559,520**	**£1,150.35**			
8	**External works**	**£271,450**	**£122**			
8.1	Site preparation works	£122,375	£55			
8.2	Roads, paths and pavings		See 8.1			
8.3	Soft landscaping, planting and irrigation systems		See 8.1			
8.4	Fencing, railings and walls		See 8.1			
8.5	External fixtures		See 8.1			
8.6	External drainage	£86,775	£39			
8.7	External services	£62,300	£28			
8.8	Minor building works and ancillary buildings	£0	£0			
	BUILDING WORKS ESTIMATE	**£2,830,970**	**£1,272.35**			

(*Continued*)

Table 1.12 (Continued)

	Element	Total cost	Cost per m²	EUQ	Unit	EUR
9	**Main contractors preliminaries**	**£368,026**	**£165.40**	**13%**	**%**	**£368,026**
10	**Main contractors overheads and profit**	**£0**	**£0**			
10.1	Main contractor's overheads	£0	£0			
10.2	Main contractor's profit	£0	£0			
	TOTAL CONTRACT SUM (excluding risks and design fees)	**£3,198,996**	**£1,438**			

Note: The rows in bold correspond to group elements as identified in NRM1. As a result, the costs against these items are sub-totals from the constituent elements.

level considerations, where the NRM1 suggests these are significant. For example NRM1 measurement rules dictate the structure of the roof is to be measured using the plan area. Conversely the roof covering is measured based on the area to be covered (in other words the quantity surveyor (QS) must take the roof's pitch into account).

The increased accuracy achieved through the use of elemental units has increased the accuracy of the forecast cost from the arbitrary cost plan illustrated in Table 1.11. With a revised forecast cost of £3.14m the project continues to remain within the budget range determined at the option evaluation phase (£2.25 – £3.28m), and is therefore still below the building cost limit for the project.

Adding this data, along with the information to the summary elemental cost plan, using the framework outlined in Table 1.6, we can now provide the employer with an overall budget summary for the project as illustrated in Table 1.13. Again, the project remains under the overall cost limit of £4.76m (see Table 1.5) agreed by the employer at the commencement of the design phase of the project.

Preparing a cost plan using approximate quantities

The final approach quantity surveyors can adopt when preparing cost plans is to use approximate quantities, mindful that elemental cost plans are not produced in isolation, but are an extension of previous plans. Elemental cost plans based on complete approximate quantities will usually only be produced towards the end of design phases, when the information available to quantity surveyors will be reaching levels required to send projects out to the market place for constructors to bid for work. The term *approximate quantity* seems to be the cause of confusion for many. It is often used to simply differentiate it from a firm quantity which has been measured and included in the tender documents (bills of quantities), as approximate quantities are often forecasts based on incomplete information.

For some projects such as major civil engineering schemes, approximate quantities will also form the basis of tenders. In these situations re-measurement will always form part of post-contract financial management. Although less common when constructing buildings, the JCT Standard Form of Building Contract with

Table 1.13 Summary of formal cost plan 1 for the hotel

Constituent	%	Cost
Building works estimate		£2,830,970
Main contractor preliminaries estimate	13%	£368,026
Sub-total		£3,198,996
Main contractors overheads and profit estimate	Included	£0
Works cost estimate		**£3,198,996**
Project/design team fee estimate		**£319,900**
Consultants fees	10%	£319,900
Main contractor's pre-construction fee estimate	0%	£0
Main contractor's design fee estimate	0%	£0
Sub-total		£3,518,896
Other development/Project cost estimate	0	**£0**
Base cost estimate		**£3,518,896**
Risk allowances estimate		**£703,779**
Design development risk estimate	7%	£246,323
Construction risk estimate	7%	£246,323
Employer change risk estimate	3%	£105,567
Employer other risk estimate	3.00%	£105,567
Cost limit (excluding Inflation)		**£4,222,675**
Tender inflation estimate	**1.19%**	**£50,250**
Cost limit (including tender inflation)		£4,272,925
Construction Inflation Estimate	**8.76%**	**£374,308**
Cost limit (including construction inflation)		£4,647,233
VAT assessment		£0

Note: The rows in bold correspond to group elements as identified in NRM1. As a result, the costs against these items are sub-totals from the constituent elements.

Quantities (JCT, 2011) still makes express provision for the use of approximate bills of quantities. This is usually the case when "*the quantity of an item or group of items of work cannot be accurately ascertained at the time of preparing the tender documents*" (Ramus *et al.*, 2006, p. 238). On some occasions, perhaps due to accelerated employer timeframes, it is often difficult for consultants to fully complete designs before tender. In these situations, quantity surveyors will often include provisional sums for work, or more realistically develop approximate quantities for elements, which will be re-measured post-contract when works have been completed.

The final elemental cost plan produced for projects, usually formal cost plan 3, will often adopt the use of approximate quantities to provide employers with the most accurate pre-tender cost forecast for projects. Unlike earlier cost plans, which may have been derived from the overall budget, or known costs for similar completed projects, that are refined by quantity surveyors using their own experience and professional judgements, elemental cost plans produced using approximate quantities represent the first complete attempt to measure defined quantities from drawings (or to extract them from Building Information Models) using the 'tabulated rules of measurement' contained in part 4 of NRM1.

Given the levels of quantification achievable at this stage in projects, this elemental cost plan will provide design teams and employers with more accurate pictures of both the overall cost of projects and the distribution of costs through the various elements, sub-elements and components that make up buildings. This document may almost constitute a full bill of quantities, with detailed take-offs fully supporting the dimensions reported. Novice quantity surveyors arriving in the cost consultant's office fresh from finishing their studies (or on a placement year) are often astounded by both the size and complexity of this type of cost plan. It is normal for this document to contain hundreds of pages of measurement and cost data at this stage in the project's development.

Discussion point 1.11: How would you differentiate between elemental units of measurement and approximate quantities?

Example of a cost plan developed using approximate quantities

Referring once again to the budget hotel project, the architect is in the final stages of design development, and final technical drawings are available. Once again these drawings cannot be provided here, so the quantities are fictitious; however, the process the quantity surveyor would adopt when preparing a cost plan based on elemental quantities is explained and demonstrated.

Given the complexity and scale of a cost plan prepared using approximate quantities it is impossible to provide a fully broken down example. Therefore excerpts from the larger cost plan have been provided in Tables 1.14 and 1.15 to illustrate the process and the levels of detail required. The quantities provided

Table 1.14 Approximate quantities for internal walls and partitions

	Element	Total cost	Cost per m²	Approx. quantity	Unit	Composite rate
2	Superstructure	£920,490	£414			
2.7	Internal walls and partitions	£33,674	£15.13			
2.7.1	Walls and partitions					
2.7.1.1	Internal walls; blockwork, 100 thick	£3,096		129	m²	£24
2.7.1.1	Internal walls; blockwork, 140 thick	£800		25	m²	£32
2.7.1.2	Extra over internal walls for forming openings in walls for doors and the like door opening 1000mm	£180		12	nr	£15
2.7.1.3	Fixed partitions, acoustic metal stud, 100 thick	£27,846		714	m²	£39
2.7.1.4	Extra over fixed partitions for forming openings in walls for doors and the like door opening 1000mm	£1,752		146	nr	£12

Table 1.15 Approximate quantities for sanitary installations

	Element	Total cost	Cost per m²	EUQ	Unit	EUR
5	Services	£976,160	£439			
5.1	Sanitary installations	£279,230	£126			
5.1.1	Sanitary appliances					
5.1.1.1	WC	£1,200		8	nr	£150
5.1.1.1	Urinals	£988		4	nr	£247
5.1.1.1	Wash basin	£1,020		6	nr	£170
5.1.1.1	Belfast sink	£822		3	nr	£274
5.1.2	Pods					
5.1.2.1	Shower room pod – Details to follow from hotel franchise	£275,200		64	nr	£4300

are derived from a hotel cost model published in a commercially available price book with appropriate pro-rata adjustment for building GFA. The rates have been rebased to take account of tender price movements and location factors.

Summary

This section of the textbook has shown how pre-contract financial management develops through the option selection and design phase of projects. The phases of pre-contract financial management implemented by cost consultants/quantity surveyors from the very outset of projects are illustrated. The 'order of cost estimate' and option appraisals both form a key part of employers' overall appraisal of projects. The chapter highlights that the order of cost estimate represents the *"determination of possible costs of a building early in the design stage in relation to the employer's functional requirements"* (RICS, 2012a, p. 14). The order of cost estimate is prepared. Whilst allowing employers to evaluate potential options, locations or levels of refurbishment for given buildings, the order of cost estimate also establishes project cost limits at Stage 1: Preparation and brief of the RIBA plan of work (2013).

As the design develops the cost consultant/quantity surveyor will move forwards with pre-contract financial management, producing various formal cost plans for the project. As evidenced earlier a cost plan provides a statement of how the available budget will be allocated to the various elements of the building. It provides a frame of reference to be used when developing the design to ensure that costs are fully controlled as the project moves forwards through the RIBA work stages. It has also been shown that cost planning is performed in stages of increasing detail as more design information becomes available (RICS, 2012a, p. 50). Formal cost plans are prepared in the following stages of the RIBA plan of work (2013):

- Stage 2: Concept Design   formal cost plan 1
- Stage 3: Developed Design   formal cost plan 2
- Stage 4: Technical Design   formal cost plan 3

Throughout this process cost consultants/quantity surveyors should forecast costs, but they are reliant on design information provided by design teams. The main purposes or benefits of cost planning are revealed in NRM1 (RICS, 2012a, p. 50) to include:

- Ensuring that employers are provided with value for money;
- Making employers and designers aware of the cost consequences of their desires and/or proposals;
- Providing advice to designers that enables them to arrive at practical and balanced designs within budget;
- Keeping expenditure within cost limits approved by employers;
- Providing robust cost information upon which employers can make informed decisions.

As soon as the cost planning phase of projects has been concluded, cost consultants/quantity surveyors will be tasked with developing bid documents for projects. The format these take is highly dependent on the procurement route selected by employers. Section 1.5 of this chapter moves on to consider the types and content of bid documents as well as evaluating cost planning and cost forecasting from the perspective of contractors as they commence their work to prepare detailed bids for projects.

Exercise 1.1

An employer is considering the development of a new 20,000m² retail warehouse on a retail park close to the M1 in Nottingham. It is expected the project will be procured using a design and build contract with work commencing on-site in third quarter (3Q):2019. Regardless of the year to attempt this exercise, your cost analysis should be based on 4Q:2018. Prepare an order of cost estimate for the project using both the data provided in Table 1.16 and listed below.

- Professional fees are assumed at 10%;
- Risk is to be allowed for at a rate of 5%;
- Facilitating works – allow £100,000.

Table 1.16 BCIS cost data for retail warehouses

Building Function	£/m²							
Retail warehouses generally	Mean	Lowest	Lower quartile	Median	Upper quartile		Highest	Sample size
£/m² GIFA	751	381	578	669		773	2,254	51

Source: BCIS (2016).

Exercise 1.2

Using the data in Table 1.17, adjust the price data provided in Table 1.16 for movement in price and regional differences.

Exercise 1.3

Discuss the statement: 'The preparation of orders of cost estimate (OCE) is fraught with difficulties for cost consultants, making the OCE worthless; can the impact of such difficulties be minimised?'

Exercise 1.4

Using the project data in Tables 1.17 and 1.18, complete the excerpt from an elemental cost plan provided below (fill in the gaps). You need to decide whether elemental unit quantities (EUQ) and elemental unit rates (EUR) data should be used, and compute the total cost and cost per square metre for the project.

Table 1.17 BCIS tender and regional indices data

	BCIS Data	New Project
Tender Price Index	276	319 (F)
Locational Index	100	96

Table 1.18 Cost plan excerpt

Element	Total cost (£)	Cost per m² GIFA (£)	Elemental unit quantity	Elemental unit rate (£)
2.1 Frame	446,556	372.13		
2.2 Upper floors	64,143	71.27	900m²	
2.3 Roof	37,763		318m²	118.75
2.4 Stairs		86.02	12nr	6,773.75
2.5 External walls				
2.6 Windows and external doors				
2.7 Internal walls and partitions				
2.8 Internal doors				
2 Superstructure				

DATA

- Project information: GIFA 1,200m²
- 4-storey office building
- Footprint area 300m²
- Element data:

  ○   2.5 External Walls 2,876m² £468,898.32
  ○   2.6 Windows 142nr £63,900
  ○   2.7 Internal Walls 587m² £80,544
  ○   2.8 Internal Doors 35nr £180 each

Exercise 1.5 – To complete this exercise it is assumed you have BCIS access

Using your answer from Exercise 1.1, for the development of a new 8,000m² retail development, transform the order of cost estimates into an elemental cost plan (formal cost plan 1) to provide a budget breakdown for the design of the scheme. You should also download cost analysis *28177 Trade and Retail Warehouses, Cardinal Point.*

1.5 Preparing and pricing bid documents

After design work has been completed, and projects are reaching the end of RIBA plan of work stage 4 (assuming traditional lump sum procurement is used), the role of quantity surveyors moves from cost planning and control, or tender and final account forecasting, to preparation of bid documents. For traditionally procured projects (see chapter 2) this change will happen between technical design and construction. For other forms of procurement such as design and build, the cost control phase could be significantly shortened and bid document production could happen as early as conceptual design (RIBA plan of work stage 2). This will have a significant impact on the way bid documents are produced.

1.5.1 Types of bid documents and approaches to measurement

The types of bid documents issued to main contractors will largely depend on the procurement approach adopted by employers. The most common procurement routes used in the UK construction sector, identified in the recent NBS Contract and Law Survey 2015 (NBS, 2015) include:

- Traditional lump sum bill of quantities;
- Specification and drawing;
- Design and build.

As a result, this section will discuss the tender documents commonly issued with these three main forms of procurement. A full discussion of how the bids for projects are developed will be provided in chapter 4. This section is limited to focus on the development of unit rates for the pricing of measured works elements of projects.

Traditional lump sum bill of quantities

Bills of quantities evolved as a way of eliminating the need for all contractors bidding for projects to compute their own quantities. Employers commission quantity surveyors to prepare bills that are made available to all bidders. This approach led to the formalisation of the rules for the preparation of bills of quantities. The first Standard Method of Measurement (SMM) was introduced in 1922, and the last using this name was SMM7 in 1988 and revised in 1998. NRM2 – Detailed Measurement for Building Works (RICS, 2012b) replaced SMM7, although the latter is still used by some practitioners. NRM2 is designed to be a simplification of measurement rules in SMM7. The popularity of bills of quantities as part of bid documentation has reduced significantly over the last 30 years, with employers becoming increasingly reluctant to bear the risk of inaccurate measurement. Alternatively therefore, the risk of measurement is passed to supply chains.

Assuming employers are willing to adopt bill of quantities as the procurement approach, JCT Practice Note 6 (JCT, 2012) identifies that the tender documents issued to contractors should include:

- Invitation to tender (ITT);
- Instructions to tenders – date and time for return, site visit details, *et al.*;
- Tender evaluation criteria;
- Confirmation of receipt form;
- Conditions of proposed contract;
- Bill of quantities (including prelims, measured works, provisional sums and daywork bills);
- Specification document (also known as the preambles);
- Full set of tender drawings;
- Pre-construction health and safety information;
- Form of tender and return envelope;
- Other appendix documents e.g. asbestos survey *et al.*

In terms of measured work sections, contractors develop their unit rates and commence pricing the works as discussed in section 1.5.2.

Specification and drawings

Often seen as a very similar approach to procurement to traditional lump sum with a bill of quantities, specification and drawings is an established method of tendering for construction projects. Firms bidding for projects are provided with full specifications alongside a series of drawings on which to base their estimate. Unlike with the bill of quantities discussed above, responsibility for measurement now falls on bidding contractors. As these projects are often reasonably small, a bid period of 4 weeks or 20 working days is the norm. With such a short tendering window, most contractors will wish to use unit rate estimations, and then some sort of quantities must be established. Most estimators will resort to using *builder's quantities* to facilitate the pricing of these works. These typically fall short in terms of detail as compared to bills prepared in accordance with, say, the NRM2, and will not be used outside the organisation concerned (apart from domestic subcontractors) but they do provide a reasonable base from which estimators can price projects.

Design and build

In design and build, contractors will not be issued with any measured works items. Most often tender returns will simply include elemental cost plans and a series of documents called 'contractors proposals'. Unlike with the traditional procurement route, JCT Practice Note 6 (JCT, 2012) identifies that the tender documents issued to contractors include:

- Invitation to tender (ITT);
- Instructions to tenders – date and time for return, site visit details, *et al.*;
- Tender evaluation criteria;
- Confirmation of receipt form;
- Conditions of proposed contract;
- Employer's requirements;
- Contractor's proposal documents;
- Pre-construction health and safety information;
- Form of tender and return envelope;
- Other appendix documents e.g. asbestos survey *et al.*

From the perspective of contractors, it is highly likely they will use some sort of unit rate estimations. As a result, it is important that as drawings and contractors' proposals are developed, reasonably detailed sets of quantities are established to allow unit rate pricing. To achieve this, most design and build contractors outsource development of bills of quantities to specialist quantity surveying practices, who as experts in quantification provide contractors with full sets of quantities; thus estimating departments may concentrate on pricing items based on detailed unit rates.

1.5.2 Unit rates – pricing measured works

Regardless of how projects are procured, and indeed who takes responsibility for quantifying works, project bids need to be developed. Given the dominance of lump sum, fixed price contracting via either traditional or design and build procurement routes (NBS, 2015), the risk of pricing invariably falls on contractors. Responsibility for ensuring projects are fully evaluated and appropriately priced during bidding phases lies initially with estimators in contracting companies. There are a number of high profile cases in the public domain involving contractors declaring millions of pounds of losses because they under-estimated risks and under-priced projects. These projects stand as testament to the need for accurate project bidding. Whilst bid development from a commercial management viewpoint is considered in chapter 4, this section of chapter 1 will focus on the importance of accurate measurement, quantification and pricing of *measured works* sections of contractors' bids.

When estimators are pricing measured works sections of projects, they are required to make forecasts of how much the final cost of projects will be if they are successful in winning the work. Some estimators suggest pricing is in part a tactical process, whereby they try to identify areas where they think variations will be required, and then price items higher than they might normally do, to make post-contract cost control from the viewpoint of contractors more commercially rewarding. In order that highly priced items do not unduly inflate bids and make them uncompetitive, contractors may compensate by pricing other items at lower prices. Estimators must still ensure they adequately forecast final costs for the measured work components of bids. This process requires estimators to develop unit rate costs for measured items. As depicted in Figure 1.5, the unit rate is derived from several items that are brought together for each item of work.

The constituent elements of the unit rate are considered below, although more attention is paid to costing the labour element of unit rates, as quantity surveyors or construction managers have most control over this resource. Materials and plant prices will be typically sourced from specialists within supply chains and will be explored in chapter 4.

Labour costs

Control of labour stems from the post-contract management of labour within firms, by construction managers who allocate labour resources to programmes to ensure the right labour is equipped with the right plant and materials to work in the most effective and efficient way possible. However, this post-contract financial management will also involve site managers in detailed record keeping. These records will include 'time sheets' that require site managers to monitor labour productivity. For example site managers feed data back to estimators about how long various tasks are taking to ensure forecasts for future projects

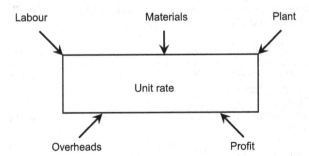

Figure 1.5 Unit rate constituent parts.

are either increased or decreased to improve estimating accuracy; this data is known as labour productivity constants. Whilst estimators should have their library of production constants, based on publicly available information, their own and company data, they need to make professional judgements on each project about potential adjustments to constants. For example, a factory project with a 2.4m-high perimeter long straight-facing brick walls, standard bucket-handle jointing, programmed for build in the summer period may have a 'fast' labour productivity constant for facing bricks. Alternatively, brickwork facades with many window openings, narrow pillars, many plumbing points, recessed pointing constructed at high level in winter months may have a 'slow' productivity constant.

When costing the labour element of the unit rate, the estimator will consider two elements:

- Labour productivity constants;
- The cost of various labour types per hour.

Table 1.19 illustrates a series of labour productivity constants for a variety of different trades sourced from several commercially available price books. As can be seen in Table 1.19, the labour constants used by estimators are expressed in terms of hours and decimal hours per unit of production. For example, the estimating constant for a joiner to fix a trussed rafter on a new house is 1.30 hours, so

Table 1.19 Labour and plant productivity constants

Item	Unit of measure	Plant production constant (hours)	Labour production constant (hours)
Excavating topsoil to be preserved, average depth 0.15m	m²	CAT D5 0.01hrs/m²	0.01hrs/m²
Excavating topsoil to be preserved, by hand. Average depth 0.15m	m²	–	0.30hrs/m²
Excavating foundation trenches, not exceed 0.30m wide and 1.0m deep	m³	JCB 0.27hrs/m³	0.27hrs/m³
Plain concrete beds/Slabs 150mm thick	m³	–	1.50hrs/m³
Proprietary roof trusses, Fink Pattern, 35 degree pitch, spanning 7.0m	nr	–	1.30hrs/each
Wall and Partition members, 50 x 75mm Sawn Softwood	m	–	0.24hrs/m
Plasterboard to walls, 12.5mm wallboard, height 2.4–2.7m	m	–	1.19hrs/m

Source: BCIS SMM7 Major Work Estimating Prices Database (2016).

assuming the house is a large 4 bedroomed detached that is 12 metres wide, with trusses at 0.60m centres (21 trusses) it will take the joiner:

21nr trusses x 1.30hrs/per truss = 27.30 hours

Since this data will not be specific to one project, it is usually stored and manipulated using a series of databases and spreadsheets.

Exercise 1.6

An estimator has been tasked with calculating the duration for a bricklayer to construct a boundary wall around the rear gardens of a show house complex. The wall will be 75m in length, and 1.6m in height, facing brickwork to both sides (225mm thick) and finished with a brick on edge coping. Assume there are 59 bricks per metre squared, and a bricklayer can lay 45 bricks per hour. How long will it take one bricklayer to complete the wall?

The second element of labour cost is hourly rates for labour. Table 1.20 illustrates a typical 'all-in rate' build-up for contractors using directly employed labour. In the construction industry, directly employed labour will have continuous contracts of employment with their employer. When setting the terms of conditions stipulated in this contract of employment, employers will normally follow the Construction Industry Council Joint Working Rule Agreement. The Working Rule Agreement is a negotiated settlement relating to employment terms and conditions between employers, represented by trade bodies such as the National Federation of Master Buildings, and the UK Contracting Group and employees represented

Table 1.20 Labour all-in rate for a general operative

	Quantity (hours/days)	Rate (£)	Total (£)
Flat time	1893.80 hours	8.52	16.135.18
Non-productive overtime	65.5 hours	8.52	558.08
Public holidays	63.00 hours	8.52	536.76
Holiday pay allowance	176.00 hours	8.52	1,499.52
Sick pay	5.00 days	23.00	115.00
Sub-total			**18,844.54**
National insurance employer contribution			1,488.26
CITB levy (0.5% of PAYE)			94.22
Retirement benefit			520.00
Death benefit			72.28
Severance pay (2%)			420.39
Employer's liability insurance (2%)			420.39
Cost for 1,856 productive hours			**21,868.46**
Cost per hour for labour			**11.78**

Source: BCIS Rates Database (2016).

Note: The rows in bold correspond to group elements as identified in NRM1. As a result, the costs against these items are sub-totals from the constituent elements.

Table 1.21 Labour classifications in CICJ Working Rule Agreement

Labour classification	Description
General Operative	General operatives are employed to undertake basic work, they are known as 'labourers' on construction sites.
Skilled Operative Rate 4	General operatives who carry out skilled work. For example driving a small dumper (less than 10 tonne) or a site banksman helping with the management of on-site traffic movements.
Skilled Operative Rate 3	General operatives who carry out skilled work. For example the driver of a telehandler (forklift truck) on a construction site or a traffic management operative putting out cones on a motorway.
Skilled Operative Rate 2	General operatives who carry out skilled work. At this level, examples would include piling rig operators, tower crane operators *et al.*
Skilled Operative Rate 1	General operatives who carry out skilled work on-site. Examples at this level include steel erectors or operators of extremely large plant, weighting over 50 tonnes such as that used on major civil engineering projects.
Craft Operative	Fully qualified craft operatives are typically 'time served' meaning they have completed a three-year site-based apprenticeship. Examples include joiners, bricklayers, plasterers, plumbers, heating engineers *et al.*

Source: Construction Industry Publications (CIP, 2013).

by trade unions such as UNITE and UCATT. The agreement includes five levels of labour operative, as outlined in Table 1.21. Schedule 1 of the agreement stipulates six bands for construction operatives; these range from general operatives (lowest band) to craft operatives (highest band). Each pay band is designed to reflect the increased skills and qualifications needed to complete the associated job function. An example from each level in the schedule has been included in Table 1.21. As a result of the different skill levels, each class of labour will need to have its own 'all-in rate' developed, similar to the example for a general operative shown in Table 1.20.

When using directly employed labour, contractors face additional costs over and above the base rate of pay that must be reflected in the 'all-in rate'. The most significant of these additional costs is employers national insurance contribution. The current percentage payment that employers must make to the government is 13.80% of wages. In addition contractors must also pay overtime rates, usually 1.5 hours' pay for 1 hour's work above the standard 40-hour working week, and 2.0 hours' pay for 1 hour's work on Sunday, and holiday pay for public and personal holidays, the current statutory entitlement is 5.6 weeks per year (Advisory, Conciliation and Arbitration Service, 2016). If operatives are on sick leave, contractors will also need to include monies to cover for payments made. Finally contractors will also need to make pension contributions if their employees contribute to the government's auto-enrolment scheme. At the moment the rates of contribution are being phased in as follows:

- 1.0% until April 2018;
- 2.0% until April 2019;
- 3.0% thereafter.

Contractors are also required to pay the Construction Industry Training Board (CITB) a levy on their wage bill, which the Board then uses to fund industry training schemes, and for grants to contractors who employ apprentices. However, from April 2017, the national apprentice levy will be payable by all businesses with a payroll exceeding £3m. This will be charged by HMRC at 0.5% of the total payroll (staff and operatives). The government has confirmed CITB will retain responsibility for the operation of the levy in the construction sector. Consequently a new levy system will operate from April 2017 requiring employers to pay a levy of 0.5% of the company's PAYE payments plus 1.25% of funds paid to net (taxed) Construction Industry Scheme subcontractors. The Construction Industry Scheme (CIS) is designed to reduce tax evasion by requiring employers to deduct tax at source from labour only subcontractors (HM Revenue and Customs, 2015). CITB will provide a levy exemption to businesses expending less than £80,000 on PAYE and CIS subcontractor labour only payments. There will also be a small business reduction of 50% for companies with combined labour costs of less than £300,000.

Finally, the all-in rate will include additional allowances for the cost of employer liability insurance and potential redundancy payments (severance pay). All these cost items are included in calculations to determine all-in rates. Chapter 6 includes an appraisal of using direct labour as opposed to indirect self-employed labour.

Material costs

Materials are a relatively simple item to include in unit rates. Prices are normally based on units of material including the cost of delivery, unloading and storage on-site before use. In developing material costs, estimators seek quotations from builders merchants or suppliers directly. These are evaluated and assumptions made about potential inflation, the period the quotation remains open for acceptance and the possible lead-in period required to source and deliver.

Before the material costs are added into unit rate, estimators need to consider allowances for waste of all kinds. Some waste, such as cutting waste can be unavoidable. Other types of waste can be minimised by good site management e.g. damage during handling and storage, theft, vandalism. Finally adequate allowance must be made for the cost of sundry associated materials or fixings (e.g. mortar in masonry walls, adhesives, nails or screws *et al*).

Plant costs

The inclusion of plant costs into the unit rate includes two major types of plant:

1   Non-mechanical and small items of plant;
2   Large mechanical plant.

Both of these types of plant need to be included in unit rates, or in the preliminaries section of contractors' estimates. Chapter 3 includes further appraisal of the arguments for hiring or buying plant.

Non-mechanical plant

Non-mechanical plant includes basic tools and small items plant required to complete projects such as wheelbarrows, hosepipes, mixers, small powered hand tools *et al.* Most often, contractors will include lump sums of monies in preliminaries sections of bills of quantities. Some of these items have a limited life span, and are purchased and written off at the end of projects; longer lasting non-mechanical plant items may be hired.

Mechanical plant

For most construction projects, mechanical plant is supplied by tier 2 or below subcontractors, who may use specialist items of plant on a daily basis. On some occasions, tier 1 contractors may hire in items of plant, and assign it to the preliminaries section of the contract. However, to give a comprehensive understanding of plant costing, and to demonstrate how plant can be integrated into unit rate costs for project, it has been assumed at this point that a contractor will be hiring or purchasing a JCB 3CX.

Hiring plant removes most of the significant risks from contractors including:

•   High up-front capital cost and depreciation;
•   Plant suitability for the project;
•   Down time due to breakdowns and servicing;
•   Extensive transport costs if projects are some distance from plant yards.

As a result, most general construction contractors will hire plant that is specifically designed for the projects they are bidding for, rather than owning their own. Assuming the plant is hired, a weekly cost will be calculated using the approach illustrated in Table 1.22.

The second option available to the contractor is to purchase items of plant they need. Given the problems with plant ownership identified above, this option is likely to be adopted by more specialist contractors who make extensive use of the same types of plant. For example, for groundworks or demolition contractors, who need the same types of heavy plant on a daily basis, sourcing plant from hire companies would be less efficient than owning their own. In this situation companies may employ plant managers, who would provide specialist advice on the plant charge-out rates that should be included in estimates. Although it is beyond

Table 1.22 Weekly rate for hired plant

	Quantity (hours/days)	Rate (£)	Total (£)
Labour			
General Operative (Skill Rate 2)	30 hours	14.21	426.30
Plant			
Fuel for mechanical plant	3.86l/hr x 30hrs	0.92	106.54
JCB 3CX Excavator	1 week	390.00	390.00
Cost for 30 productive hours			**922.84**
Cost per hour			**30.76**

Source: BCIS Rates Database (2016).

Note: The rows in bold correspond to group elements as identified in NRM1. As a result, the costs against these items are sub-totals from the constituent elements.

the scope of this text to evidence how these calculations are to be achieved, the CIOB *Code of Estimating Practice* (Flanagan and Jewell, 2016) suggests the following factors would need to be reflected in the internal hire rate:

* Capital sum based on the purchase price and expected economic life (this will depend on company accounting policy);
* Assessment of the cost of finance;
* Return required in capital invested;
* Grants and financial assistance available when purchasing plant;
* Administration and plant depot charges;
* Costs of insurances and road fund licences;
* Maintenance time and costs along with cost of stocks needed for maintenance purposes.

Discussion point 1.12: Why do you think most plant used on UK construction projects is hired rather than owned by main contractors?

Developing the final unit rate price

The final stage in the development of unit rates is for estimators to pull together the various rates they have established for labour, plant and materials for each item of work they are pricing in the measured works section of tenders. Estimators need to establish the exact amount of each resource needed relative to the item of work in question. This process has been illustrated in Table 1.23(a) with regards to pricing the construction of the internal leaf of a cavity wall. The build-up of the rate is further broken down in Table 1.23(b) showing the labour cost build-up using all-in rates, and Table 1.23(c) showing the materials build-up. In the example it has been assumed the work will be completed by a bricklaying gang. It is normal in the construction industry for bricklayers to work in a 2+1 gang.

However, in the case of laying heavy concrete blocks, Table 1.23(b) illustrates that three general operatives are needed to keep four bricklayers working.

The way in which the cost is calculated and included in the tender price depends on the tender documentation and how the price is to be expressed. If bills of quantities are provided for pricing, this is done by starting from the supply cost, adding the appropriate labour and plant resources, to arrive at a cost per unit (m, m², m³, tonne or nr). Subsequently this is applied to the total quantity in the tender. During this process, estimators will work through each item of work in turn, establishing an appropriate built-up rate and apply this to the given quantity. At the end of this process they will have developed a complete estimate.

Table 1.23(a) Unit rate for 100 thick blockwork internal leaf of a cavity wall

Rate for 1m² of blockwork	*Quantity*	*Rate (£)*	*Total (£)*
Bill Item			
Precast concrete blocks, BS 6073, furnace clinker aggregates, compressive strength 3.5 N/mm²; in gauged mortar (1:2:9)			
Walls: Solid blocks, Thickness: 100mm			
Labour			
Blockwork Gang	0.70 hours	24.58	17.20
Materials			
Gauged mortar (1:2:9)	0.01m³	127.11	0.64
100mm precast concrete blocks	1m²	7.59	7.59
Cost for 1m²			**25.43**

Source: BCIS Rates Database (2016).

Note: The rows in bold correspond to group elements as identified in NRM1. As a result, the costs against these items are sub-totals from the constituent elements.

Table 1.23(b) Labour element build-up

Rate for 1m² of blockwork	*Quantity*	*Rate (£)*	*Total (£)*
Labour			
Bricklayer	4.00 hours	15.74	62.96
General operative	3.00 hours	11.78	35.34
Sub-total	4.00 hours		**98.30**
Cost per hour	(£98.30/4.00hrs)		**24.58**

Source: BCIS Rates Database (2016).

Note: The rows in bold correspond to group elements as identified in NRM1. As a result, the costs against these items are sub-totals from the constituent elements.

Table 1.23(c) Material build-up for unit rate

Rate for 1m² of blockwork	*Quantity*	*Rate (£)*	*Total (£)*	*Quantity*	*Total (£)*
Cement					
Labour					
General operative	1.00hr	11.78	11.78		

Rate for 1m² of blockwork	Quantity	Rate (£)	Total (£)	Quantity	Total (£)
Materials					
Cement (bagged)	1.00 tonne	143.05	143.05		
Unit price			154.05		
Waste	5%		7.74		
Cement (bagged)			162.57	**0.16 tonne**	26.01
Hydrated Lime					
Labour					
General operative	1.00hr	11.78	11.78		
Materials					
Hydrated lime (bagged)	1.00 tonne	219.73	219.73		
Unit price			231.51		
Waste	5%		11.58		
Hydrated lime (bagged)			243.08	**0.16 tonne**	38.89
Building Sand					
Materials					
Building Sand	1.00 tonne	23.25	23.25		
Unit price			23.25		
Waste	10%		2.33		
Building Sand			25.58	**1.98 tonne**	50.65
Sub-total					115.55
Waste				10%	11.56
Total					127.11

Source: BCIS Rates Database (2016).

Note: The rows in bold correspond to group elements as identified in NRM1. As a result, the costs against these items are sub-totals from the constituent elements.

Exercise 1.7

Prepare a unit rate price for the replacement of windows and doors to 900 social housing units as part of a major planned maintenance scheme. The estimator has tasked you with developing a unit rate for window Type W78B. The work requires the contractor to remove existing windows and install a new double-glazed internally breaded white PVCu window. To assist you with the task the following information has been provided:

- Window Type W78B is 1170 x 1200mm, with 1 fixed light, 2 side hung and 3 panes beaded internally;
- Labour – 2 nr joiners for 2 hours;
- Materials – window frame £380.00, sundry fixings, sealants and other materials £10.00.

Exercise 1.8

Prepare a unit rate price for a fixing truss rafter on a new build semi-detached property. To assist you with the task the following information has been provided:

- Fink trusses, 450mm overhang both sides, 35 degree pitch, 8,000mm span;
- Labour – joiner – 1.40 hours, all-in hourly rate £16.75/hour;

- Plant – mobile crane – 0.02 hour; daily hire rate £285.00;
- Materials – Truss Rafter 65.03/each. Assume zero waste.

1.6 Chapter summary

Chapter 1 has provided a review of pre-contract financial management techniques used by quantity surveyors, project managers, construction managers and estimators to manage the early phases of projects. Pre-contract financial management tools such as option appraisal, cost planning, cost control and finally unit rate estimating that are used to develop increasingly detailed forecasts of the price of projects have been explained and illustrated. It is hoped that this chapter has demonstrated the importance of this pre-contract phase, and the need for accuracy in the cost plans developed alongside the need to implement proactive project risk identification and control strategies to ensure risk has been comprehensively reflected in cost plans (these concepts will be discussed in more detail in chapter 4). Towards the end of this chapter, three of the most widely used procurement routes have been briefly introduced, with some discussion of the main pricing documents provided. This provides an introduction to the next chapter, which explores procurement and the role of construction professionals in more detail.

1.7 Model answers to discussion points

Discussion point 1.1: How could the findings from a detailed option appraisal influence the employer's non-financial decisions about the project?

In addition to the potential financial viability of the scheme, the option appraisal will allow the employer to consider the potential benefits of either 'stepping back' the quality of the scheme or increasing the budget to provide a more comprehensive scheme. For example, if the employer asked the quantity surveyor to appraise the potential refurbishment of an existing building, the option appraisal may reveal significant cost increases if the level of the refurbishment moved from an update to current demands, towards a level of refurbishment that future proofed the building. Whilst the latter would be attractive, it may not safeguard against obsolescence whilst also not enhancing the usability (function) of the building. Clearly in these circumstances option appraisal allows the employer to make informed decisions about its investment to ensure demands are achieved.

Discussion point 1.2: Why you think it is critical for quantity surveyors to determine a cost limit for the project?

The cost limit not only determines the fit of the project against the funds available to the employer, it also provides reassurance to the employer that the building is affordable and will allow the employer to seek support from lenders and other financial institutions with a knowledge that the maximum funds required have been determined.

Discussion point 1.3: Why do you think the quantity surveyor reports initial budgets as a cost range rather than a single figure?

It is essentially a question of risk, and given the lack of information available at this stage in the project life cycle it is difficult to determine to a reasonable degree of accuracy exactly how much the building will cost. By providing a range, the quantity surveyor can be more certain the likely cost of the employer's scheme will fit within the range provided; the range is called a 'confidence interval'. As a result the probability of the final cost falling in a range is more certain. Since the employer is likely to remember the first estimate provided, it is essential this cost is reflective of the scheme. Without drawings or information relating to the specification documents, it is a challenge for the surveyor to accurately determine what this cost will actually be.

Discussion point 1.4: Identify the potential limitations of using a functional unit cost to determine the cost target for the proposed hotel.

The technique is highly unreliable, as it is based on a very variable proxy unit. At best the unit cost should be used only at the strategic business case, when the surveyor has no other options due to the lack of design information.

Discussion point 1.5: What are the advantages of using a GIFA rate (£/m² gross floor area) when compared to the unit rate approach?

There are numerous advantages of using GIFA-based rates as opposed to unit rate. The GIFA provides a far more realistic view of the cost of the building as the rate will be related to the floor area, not just a generalised proxy value such as area of the building divided by the number of occupants. However neither technique adequately deals with the design attributes of the building, and for this reason NRM identifies both techniques as a last resort.

Discussion point 1.6: Identify the various types of information you would need, and therefore expect your employer to provide at the 'business case' stage of projects.

NRM1 section 2.3.1 (RICS, 2012a, p. 20) provides a clear list of information it is expected that employers provide in order to provide them with the initial 'order of cost estimate'. It is important for quantity surveyors to read and understand this information. In furtherance of this, section 2.3.2 of NRM1 (RICS, 2012a, p. 21) outlines the core information architects are required to provide.

Discussion point 1.7: When would you use: group elemental, elemental, sub-element and component level data in an order of cost estimate or formal cost plan?

Remember NRM1 recommends quantity surveyors should work to the lowest level of measurement possible. However, it is also acknowledged that the level of design information provided to quantity surveyors will be highly influential.

Group elemental – Used when only limited design information is available. It would be considered suitable for the order of cost estimate if the quantity surveyor felt the elemental approach would be appropriate.

Elemental – Commonly used for early cost plans. Formal cost plan 1 is likely to only be developed to elemental level.

Sub-elemental – Once the cost consultant/quantity surveyor is able to develop elemental units, they will proceed to produce more detailed cost plans, using sub-elements such as roof structure, roof coverings, eaves and verge treatments, etc.

Component level – The final cost plan, developed using approximate quantities, will always be produced to the lowest level provided within the NRM1 tabulated measurement rules. With designs almost fully developed, quantity surveyors will be able to comprehensively break budgets down to component level and beyond.

Discussion point 1.8: How would you distinguish between an elemental cost analysis and an elemental cost plan?

The simple way to make this distinction is to consider the source data used to derive cost figures. The elemental cost analysis is sourced from BCIS. To ensure this data is usable by subscribers, and that useful comparisons can be drawn, subscribers loading cost data are asked to ensure the data is presented in a standard format, following the rules outlined in the *Standard Form of Cost Analysis*, 2nd edition. The elemental cost plan, on the other hand, is the document the quantity surveyor produces for the project using various sources of price data, including cost analysis reports sourced from BCIS. The elemental cost plan is produced in accordance with NRM1.

Discussion point 1.9: Why would quantity surveyors not adopt a national measure of inflation to make these adjustments such as the retail price index (RPI) or consumer price index (CPI)? Why do you think it is important for quantity surveyors to also consider regional variance?

This is a question of suitability. Both the RPI and CPI are based on price movements of a basket of products that are not proportional to construction costs e.g. domestic householder items. The selection of a more specialist index, such as the Tender Price Index (TPI) produced by BCIS, provides a more specific and specialist review of price movements in relation to the price of construction.

Regional variations are also essential, as the TPI or other specialist index provides a nation view of tender price movements. Certain parts of the country will have far more buoyant construction markets than others, and prices relate to supply and demand. For example it is expected that prices will be higher in London than in other parts of the UK. By using regional adjustment the quantity surveyor can ensure these differences are reflected in the estimated budget cost for the project.

Discussion point 1.10: In what circumstances would you say it is appropriate for quantity surveyors to use element unit quantities and element unit rates to build up element costs?

NRM1 identifies that the quantity surveyor should actively seek to use elemental quantities and rates wherever possible. As a result, if the surveyor is provided with a reasonable level of information, such as scalable drawings and specification details it is possible to develop elemental unit quantities and using elemental unit rates sought from commercially available or historical project cost data.

Discussion point 1.11: How would you differentiate between elemental units of measurement and approximate quantities?

Approximate quantities are in effect at the same level of detailed measurement to that used for a bill of quantities. Using such precise measurement, following

NRM2 rather than NRM1, requires the quantity surveyor to have access to almost complete drawings. On the other hand, elemental units of measurement are a less precise approach to measurement that is reflective of the level of detail available to the quantity surveyor at the earlier stages of design development.

> **Discussion point 1.12:** Why do you think most plant used on UK construction projects is hired rather than owned by main contractors?

This is really a question of economic efficiency. For most organisations the ownership of plant can be very expensive, especially if that plant is not in continuous operation. As a result, most construction organisations will either subcontract specialist civil engineering elements of projects such as groundworks and other external works, or they will hire in the items of plant they need for the specific durations required. Although some would argue this approach, especially for more heavily used plant, such as telehandlers, which are a common sight on most projects would be uneconomical. However, once lost time is considered due to maintenance and repair, agreeing a long-term rental arrangement with specialist plant hire organisations who will maintain and repair the plant, or if needed exchange it whilst the maintenance work is undertaken, presents an equally attractive proposition.

1.8 Model answers to exercises

Exercise 1.1

An employer is considering the development of a new 20,000m² retail warehouse on a retail park close to the M1 in Nottingham. It is expected the project will be procured using a design and build contract with work commencing on-site in third quarter (3Q):2019. Regardless of the year to attempt this exercise, your cost analysis should be based on 4Q:2018. Prepare an order of cost estimate for the project using both the data provided in Table 1.16 and listed below.

- Professional fees are assumed at 10%;
- Risk is to be allowed for at a rate of 5%;
- Facilitating works – allow £100,000.

Table 1.16 BCIS cost data for retail warehouses

Building function	£/M²						
Retail warehouses generally	Mean	Lowest	Lower quartile	Median	Upper quartile	Highest	Sample size
£/m² GIFA	751	381	578	669		773	2,254	51

Source: BCIS (2016).

SOLUTION

This exercise requires you to produce an order of cost estimate for the retail warehouse project. You should adopt the template provided in the textbook example. The table below illustrates one solution based on the assumption that this retail unit will be mid-range in terms of complexity and quality. For that reason the rates used are the lowest to the median for this type of development. Finally it has been assumed that construction will commence in 3Q:2019 and will take eight months to complete. Construction inflation has therefore been computed to 2Q:2020, with tender inflation computed until 4Q:2019.

Exercise 1.2

Using the data below, adjust the price data provided in Table 1.16 for movement in price and regional differences.

To calculate the tender and locational adjustment for the scheme you should use the formula (Index for project – Historical data index/historic data index)*100.

Tender price adjustment = (307 – 276/276)*100 = 11.23%
Regional adjustment = (96 – 100/100)*100 = –4%
Overall price movement adjustment = 11.23% – 4% = 7.23%

Table 1.24 Example order of cost estimate for retail warehouse

Constituent	%	Minimum (£)	Maximum (£)
Facilitating works		100,000	100,000
Building works estimate		3,048,000.00	5,352,000
Contractor preliminaries	Included	0	0
External works (additional allowance)	Say	50,000	80,000
Sub-total		**3,198,000**	**5,532,000**
Main contractors overheads and profit	Included	0	0
Works cost estimate		**3,198,000**	**5,532,000**
Project/design team fees	10%	319,800	553,200
Other development/Project costs		0	0
Base cost estimate		**3,517,800**	**6,085,200**
Risk allowance estimate			
Design development risk estimate	5%	175,890	304,260
Construction risk estimate	5%	175,890	304,260
Employer change risk estimate	5%	175,890	304,260
Employer other risk estimate	5%	175,890	304,260
Cost limit (excluding inflation)		**4,221,360**	**7,302,240**
Tender inflation estimate	2.61%	110,177	190,588
Cost limit (inc. tender inflation)		**4,331,537**	**7,492,828**
Construction inflation estimate	6.67%	288,914	499,772
Cost limit (inc. construction inflation)		**4,620,451**	**7,992,600**
VAT assessment		Excluded	Excluded

Note: The rows in bold correspond to group elements as identified in NRM1. As a result, the costs against these items are sub-totals from the constituent elements.

Table 1.17 BCIS tender and regional indices data

	BCIS data	*New project*
Tender Price Index	276	319 (F)
Locational Index	100	96

Exercise 1.3

Discuss the statement: 'The preparation of orders of cost estimate (OCE) is fraught with difficulties for cost consultants, making the OCE worthless; can the impact of such difficulties be minimised?'

The order of cost estimates presents a significant range of challenges for the quantity surveyor, which must be overcome or their possible effects at least minimised to ensure the estimate presented to the employer is robust and as accurate as possible. Some of the challenges presented to the employer can include:

- Lack of information about the building (meet with the team and obtain as much data as you possibly can);
- Minimal design development (will need discussion with architect to ensure QS understands design concept);
- Reliability of the single rate estimating techniques (ensure data used for OCE is as reliable as possible by ensuring projects are similar and allowance made for anticipated complexity and quality levels);
- Accuracy demands information, we do not have the information needed (use cost ranges in early estimates and refine as information becomes available);
- Lack of similarities between proposed project and available cost data (QS should use professional judgement to make appropriate adjustments for quality *et al.*, and will need to capture and price risk effectively in risk register).

Exercise 1.4

Using the project data complete the excerpt from an elemental cost plan provided below (filling in the gaps). You need to decide whether elemental unit quantities (EUQ) and elemental unit rates (EUR) data should be used, and compute the total cost and cost per square metre for the project.

DATA

- Project information: GIFA 1,200m²
- 4-storey office building
- Footprint area 300m²
- Element data:

  ○ 2.5 External Walls 2,876m² £468,898.32
  ○ 2.6 Windows 142nr £63,900

Table 1.25 Completed excerpt from the elemental cost plan

Element	Total cost	Cost per m² GIFA	Elemental unit quantity	Elemental unit rate
2.1 Frame	£446,556	£372.13	–	–
2.2 Upper floors	£64,143	£53.45	900m²	£71.27
2.3 Roof	£37,763	£31.47	318m²	£118.75
2.4 Stairs	£81,285	£67.74	12nr	£6,773.75
2.5 External walls	£468,899	£390.45	2,876m²	£163.04
2.6 Windows and external doors	£63,900	£53.35	142nr	£450
2.7 Internal walls and partitions	£80,544	£67.12	587m²	£137.21
2.8 Internal doors	£6,300	£5.25	35nr	£180
2 Superstructure	**£1,249,390**	**£1,041**	–	–

- ∘ 2.7 Internal walls 587m² £80,544
- ∘ 2.8 Internal doors 35nr £180 each

Exercise 1.5

Using your answer from Exercise 1.1, for the development of a new 8,000m² retail development, transform the order of cost estimates into an elemental cost plan (formal cost plan 1) to provide a budget breakdown for the design of the scheme. You should also download cost analysis 28177 Trade and Retail Warehouses, Cardinal Point.

SOLUTION

The comparative elemental cost plan below outlines the suggested answer, rebased to 4Q:2018. The cost plan is at the lower end of the cost range prepared in Exercise 1.4.

Table 1.26 Elemental cost plan

Element		Total cost	Cost per m²	EUQ	EUR
0	**Facilitating works**	**£100,000**	**£12.50**		
1	**Substructure**	**£776,000**	**£97**		
2	**Superstructure**	**£2,480,000**	**£310**		
2.1	Frame	£584,000	£73		
2.2	Upper floors	£0	£0		
2.3	Roof	£1,368,000	£171		
2.4	Stairs and ramp	£0	£0		
2.5	External walls	£200,000	£25		
2.6	Windows and external doors	£208,000	£26		
2.7	Internal walls and partitions	£120,000	£15		
2.8	Internal doors	£0	£0		

(Continued)

Table 1.26 (Continued)

	Element	Total cost	Cost per m²	EUQ	EUR
3	**Internal Finishes**	**£48,000**	**£6**		
3.1	Wall finishes	£48,000	£6		
3.2	Floor finishes	£0	See 3.1		
3.3	Ceiling finishes	£0	£0		
4	**Fittings, Fixtures and Equipment**	**£0**	**£0**		
4.1	Fittings, fixtures and equipment	£0	£0		
5	**Services**	**£192,000**	**£24**		
5.1	Sanitary installations	£0	0		
5.2	Services Equipment	£0	0		
5.3	Disposal Installations	£0	0		
5.4	Water Installations	£0	0		
5.5	Heat Source	£0	0		
5.6	Space heating and air conditioning	£72,000	£9		
5.7	Ventilation	£0	0		
5.8	Electrical installations	£120,000	15		
5.9	Fuel installations	£0	£0		
5.1	Lift and conveyor installations	£0	£0		
5.11	Fire and lightening protection	£0	£0		
5.12	Communications and security installations	£0	0		
5.13	Specialist installations	£0	£0		
5.14	Builders work in connection with services	£0	£0		
6	**Complete buildings and building units**	**£0**	**£0**		
7	**Work to existing buildings**	**£0**	**£0**		
	Sub-total	**£3,404,000**	**£425.50**		
8	**External Works**	**£1,272,000**	**£159**		
8.1	Site preparation works	£552,000	£69		
8.2	Roads, paths and pavings	£0	0		
8.3	Soft landscaping, planting and irrigation systems	£0	0		
8.4	Fencing, railings and walls	£0	0		
8.5	External fixtures	£0	See 8.1		
8.6	External drainage	£576,000	£72		
8.7	External services	£144,000	£18		
8.8	Minor building works and ancillary buildings	£0	£0		
	BUILDING WORKS ESTIMATE	**£4,676,000**	**£584.50**		
9	**Main contractors preliminaries**	**£200,000**	**£25**		0
10	**Main Contractors Overheads and Profit**	**£0**	**£0**		
10.1	Main contractor's overheads	£0	£0		
10.2	Main contractor's profit	£0	£0		
	TOTAL CONTRACT SUM (excluding risks and design fees)	**£4,876,000**	**£610**		0

Note: The rows in bold correspond to group elements as identified in NRM1. As a result, the costs against these items are sub-totals from the constituent elements.

Exercise 1.6

An estimator has been tasked with calculating the duration for a bricklayer to construct a boundary wall around the rear gardens of a show house complex. The wall will be 75m in length, and 1.6m in height, facing brickwork to both sides (225mm thick) and finished with a brick on edge coping. Assume there are 59 bricks per metre squared and the bricklayer can lay 45 bricks per hour. How long will it take one bricklayer to complete the wall?

SUGGESTED ANSWER

- Size of wall = 75m x 1.6m = 120m^2 of brickwork
- Amount of bricks in the wall = 120m^2 x 59 nr bricks/m^2 x 2 leaves = 14,160nr bricks
- Bricklaying rate = 45 bricks/hr
- Duration to lay bricks = 14,160 / 45 = 314.67 hours or 8 weeks

For one bricklayer that would mean eight weeks of work, however, bricklayers tend to work as a gang, with two bricklayers supported by a labourer, who will ensure the materials are in place to keep bricklayers working as efficiently as possible. As a result the rate needs to be adjusted.

- Duration of work = 8 weeks/2 bricklayers so it will actually take approximately 4 weeks for a 2 + 1 gang to complete the work

Exercise 1.7

Prepare a unit rate price for the replacement of windows and doors to 900 social housing units as part of a major planned maintenance scheme. The estimator has tasked you with developing a unit rate for window Type W78B. The work requires the contractor to remove the existing window and install a new double-glazed internally breaded white PVCu window. To assist you with the task the following information has been provided:

- Window Type W78B is 1170 x 1200, with 1 fixed light, 2 side hung and 3 panes beaded internally;
- Labour – 2 nr joiners for 2 hours;
- Materials – window frame £380.00, sundries £10.00.

Table 1.27 Solution for Exercise 1.7

Rate for 1 window	Quantity	Rate (£)	Total (£)
Labour			
Window fitters (joiners)	2 hours	33.50	67.00
Each joiner – £16.75			
Materials			
Window	1nr	380.00	380.00
Sundries	1nr	10.00	10.00
Cost for 1m^2			**457.00**

Note: The rows in bold correspond to group elements as identified in NRM1. As a result, the costs against these items are sub-totals from the constituent elements.

Exercise 1.8

Prepare a unit rate price for the fixing of a truss rafter roof on a new build semi-detached property. To assist you with the task the following information has been provided:

*   Fink trusses, 450mm overhang both sides, 35 degree pitch, 8,000mm span;
*   Labour – joiner – 1.40 hours, all-in hourly rate £16.75/hour;
*   Plant – mobile crane – 0.02 hour; daily hire rate £285.00;
*   Materials – Truss Rafter 65.03/each. Assume zero waste.

Table 1.28 Solution for Exercise 1.8

Rate per truss	Quantity	Rate (£)	Total (£)
Labour			
Joiner	1.46hrs	16.75	24.46
Materials			
Trussed Rafter Fink/W pattern; 450mm	1nr	65.03	65.03
overhang, 35 degree pitch, span 8,000mm			
Plant			
Mobile Crane @ £285/day	0.02hr	40.71	0.81
£285/7hrs = £40.71			
Cost for 1m^2			**£90.30**

Note: The rows in bold correspond to group elements as identified in NRM1. As a result, the costs against these items are sub-totals from the constituent elements.

References

Advisory, Conciliation and Arbitration Service (2016) Holidays and Holiday Pay. Available at: www.acas.org.uk/media/pdf/5/h/Acas-guide-Holidays-and-holiday-pay.pdf Accessed 23.05.16.

BCIS (2012) *Elemental Standard Form of Cost Analysis.* 4th Edition. London: Building Cost Information Service.

BCIS (2016) BCIS Independent Data for the Built Environment. Available by subscription at: service.bcis.co.uk/BCISOnline/ Accessed 02.05.16.

Benge, D.P. (2014) *NRM1 Cost Management Handbook.* Oxon: Routledge.

CIP (2013) Working Rule Agreement for the Construction Industry. Construction Industry Publications. Available at: www.ucatt.org.uk/files/publications/2013cijcagreement.pdf Accessed 02.05.16.

Flanagan, R. and Jewell, C. (2016) *CIOB Code of Estimating Practice.* 8th Edition. Oxford: Blackwell Science.

HM Revenue and Customs (2015) Construction Industry Scheme: Guide for Contractors and Sub-Contractors. Available at: www.gov.uk/government/uploads/system/uploads/attachment_data/file/448756/CIS340_06_15_V1_0.pdf Accessed 23.05.16.

JCT (2011) SBC/Q2011 Standard Form of Building Contract. The Joint Contracts Tribunal.

JCT (2012) Identifying Tenderers and Tender Documentation. Practice Note 6. Available at: www.isurv.com/pdf/cost_estimating__tendering.pdf Accessed 02.05.16.

NBS (2015) National Construction Contracts and Survey. National Building Specification. RIBA Enterprises. Available at: www.thenbs.com/knowledge/nbs-national-construction-contracts-and-law-survey-2015-finds-disputes-continue-to-blight-construction-industry Accessed 22.03.16.

Ramus, J., Birchall, S. and Griffiths, P. (2011) *Contract Practice for Surveyors.* 4th Edition. Oxon: Spon Press.

RIBA (2013) RIBA Plan of Work. Royal Institute of British Architects. Available at: www.ribaplanofwork.com/Download.aspx Accessed 02.05.16.

RICS (2007) *Code of Measuring Practice RICS Guidance Note: A Guide for Property Professionals.* 6th Edition. Available at: www.isurv.com/site/scripts/download.aspx?type=downloads&fileID=167 Accessed 13.05.16.

RICS (2012a) *NRM 1 – Order of Cost Estimating and Cost Planning for Capital Building Works1.* Coventry: Royal Institution of Chartered Surveyors.

RICS (2012b) *NRM 2 – Detailed Measurement for Building Works.* Coventry: Royal Institution of Chartered Surveyors.

2 Procurement systems

2.1 Introduction

Substitute the word 'procurement' with 'buy'. When we buy, we accept risks or pay to pass them to others. As consumers on the high street, the way we buy products such as motor cars or electrical goods influences the degree of risk that we take, and the degree of risk that manufacturers take. For example how long do guarantees last, and are we prepared to pay an extra sum of money to extend guarantees? Car manufacturers often provide guarantees for three years; some for seven. As a buyer, are you prepared to take the risk of defects with a car in years four to seven?

When members of the public 'buy' perhaps an extension to their home from the construction industry, in any price agreed, does it include for the possibility of the ground conditions being worse than envisaged? Which party will pay if final costs of projects are higher than estimates, or if they take longer than planned, or if there are problems with the quality of the completed work? Do homeowners employ designers and constructors separately, or do they ask one company to design and construct their extension? In manufacturing, when we agree a contract to buy a new motor car, the car will usually be designed and built by the same company. Construction employers often engage designers and constructors separately, though modern thinking is that just one contract agreed for design and build of projects can be better.

Procurement extends down the whole supply chain. Figure 2.1 illustrates the construction supply chain from employers or clients to raw material producers. At each of these interfaces, contracts are agreed or orders are placed that allocate risks between parties.

A report by Glenigan (2015) found that only 69% of projects were completed to, or lower than the budget; even worse, only 40% achieved time targets. Put another way, it is like employers agreeing contracts, usually with contractors, with both parties knowing there is only a 7 in 10 chance of completing to cost and 4 in 10 chance of completing to time. That would seem like high risks for all. When developers commit to spend, it is arguable that the construction process should be able to assure cost and time predictability, otherwise the funding models that

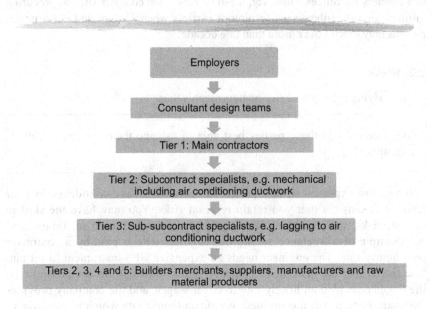

Figure 2.1 The construction supply chain.

determine project success can fail. Many would assert that all parties in the supply chain, starting with employers, should take their share of blame. Selection of procurement method is noted in literature to be the most crucial factor influencing whether projects are successful – if the wrong method is chosen at outset, projects may be destined to miss all employer objectives.

National Building Specification (NBS, 2015) conducted the 'National Construction Contracts and Law Survey' of employers consultants and contractors over the period 2014/15. Based upon 981 responses, it reported that the most popular contracts between employers and contractors are the Joint Contract Tribunal (JCT) suite comprising 39% and the New Engineering Contract version 3 (NEC3) comprising 30% Bespoke and others made up the remaining contracts. Bespoke contracts are written by employers and contractors, in their dealings with others lower down the supply chain. There are also standard JCT and NEC forms of appointment for use between employers and consultants, and standard contracts for use between contractors and subcontractors. Whilst standard contracts are designed to apportion risks sensibly and fairly between the parties, bespoke contracts may seek to pass more risks to others. Later sections in this chapter detail potential structures between parties in the construction process, depending on the method of procurement selected.

Procurement is not just for construction; it is in all spheres of all businesses internationally. In some cases it is relatively simple, procuring perhaps bulk items of stationery for offices. However, it can be very often complex such as procuring military aircraft, submarines or aircraft carriers where design and construction periods may extend over more than one decade.

2.2 Risks

An underlying principle of risk management is thus:

> Allocate risks to those parties best able to manage them; do not merely 'dump' risks.

If you are experienced in a particular sector and have confidence in your team, it is sensible that you retain relevant risks. You may have the skill to manage risks, such that things that could go wrong, do not do so. Take a simple example of a freelance setting out engineer who is paid by a contractor on a hourly rate. The engineer needs an expensive GPS instrument to set out. The risk is that the instrument could be stolen. Two options are possible: (i) the engineer is paid an hourly rate to cover wages, and the company provides the instrument, or (ii) the engineer is paid an hourly rate which is inclusive of wages and the engineer providing the instrument. Which party is in the best position to manage the risk of theft? In both situations, no doubt the engineer will be careful to keep the instrument safe, but the motivation to do so will be enhanced if the engineer owns the instrument. Therefore perhaps option (ii) is best, and importantly the construction process does not suffer the theft or an additional cost, since the risk is best managed, with not much greater effort, by the person best able to do so.

In a procurement context, consider an employer that is a regular procurer of construction work. A new project arises that will be similar in design to previous ones. Things could go wrong, but the employer decides to retain those risks, since it can manage them carefully. If however risks are identified with which an employer has no experience, and which it does not have the skill sets to manage, it is sensible that those risks are passed to others who do have the expertise. This latter instance may be the case if employers are inexperienced in construction, and one-off procurers of construction projects.

It is always to be mindful that if risks are passed to others, they would normally be priced, and the payee will pay more. However, there are on occasions that markets are 'distorted', and there is not a lot of work for contractors. Desperate for work, contractors may accept risks without pricing them, and keep their 'fingers crossed' that they do not come to fruition. They may be fearful that if they include a sum of money to cover risk, it might just be that sum that converts a potential winning bid into one that is lost, because the overall tender becomes second lowest. There are many examples in the construction press where contractors have not considered or priced risks correctly, and they have then lost huge amounts of money on projects. If contractors are losing money, they may try to fabricate claims for extra payments. Relationships between all can then be very fraught, and employers too may suffer. It may be argued that some employers merely 'dump' risks; just give the risk of design, bad ground, inflation *et al.* to contractors. If markets are buoyant, contractors may include high prices to cover risks, or may refuse to bid unless risks are rebalanced. Employers may then need to revert to procurement methods that allocate risks differently, since a method normally preferred may be just too expensive.

Lower down the supply chain, contractors frequently make decisions about whether to subcontract work or not. Many contractors are experienced at employing their own direct workforces for groundworks, steel bar reinforcement, formwork, concrete, brickwork, carpentry and joinery. On some projects given a certain set of circumstances, contractors may choose to do work in these trades themselves, thus retaining risks. On other projects, given a different set of circumstances, contractors may procure the services of subcontractors to undertake work, thus passing on risks to others.

2.3 Employer objectives

Employers may have objectives that are bespoke to each individual project. Each individual party in a supply chain will have its objectives too. Often quoted are the iron triangle of cost, time and quality. Employers and the construction industry, to their credit, are becoming more sophisticated. Society is rightly more demanding. Therefore objectives also include health and safety, and environmental issues or sustainability, since most employers and members of the supply chain are very keen to ensure that adverse incidents do not happen that could cause reputational damage. Projects should be 'hassle free'; that is they are enjoyable for people that work on them. Repeat work can be a key objective for all; that will only be attained if objectives of parties higher up the chain have been met. It can be argued that if you have not made as much money as you would have liked, this is not of huge consequence, since perhaps you can regain ground on the next repeat project, or the next one after. New objectives arise in the UK from the Public Services (Social Value) Act, which requires that in procurement systems, public organisations and their supply chains add value back into local communities; that may be by employing apprentices or supporting schools *et al.*

One scourge of construction is that selection is often based on price alone; in bidding situations, lowest price wins. However, more sophisticated employers may use a range of measures when selecting bidders, that are weighted; for example cost 40%, quality 15%, health and safety 15%, environmental performance 15% and social value 15%. Figure 2.2 illustrates a range of objectives that may be applicable to projects.

2.3.1 Cost objectives

All parties in the process have their cost and profit objectives. Often the headline issue is whether a project is completed to an employer's budget. That budget may be just the sum of money paid by an employer to a contractor. Employers

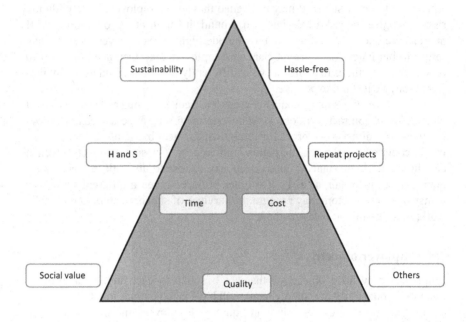

Figure 2.2 Project objectives.

will also be mindful of other parts of the budget such as design team fees, which may be as much as 10% (or 15% for smaller complex projects) of the amount of money paid to the contractor. On very large projects such as the 2012 Olympic Games, or the proposed HS2 rail link, the headline figures are those debated in Parliament, and in the case of the former included the cost of running the games and security *et al.*

The budget for design teams is the fees they receive from employers. Designers incur many costs in completing designs; the largest is usually staff salaries, but also a host of others connected with running businesses such as accommodation, utility services, information technology provision *et al.* Often reported in the public domain is the annual profit or loss amounts for consultant companies; on some occasions on flagship projects, it will be reported whether consultants have lost or made money.

Whether contractors and others lower down the supply make or lose money on individual projects is very often in the public domain. The overall annual performance of contractors is also reported in the construction press.

In the context that employer and contractor cost objectives are most often reported, it may be that one party will do really well, and the other really poorly. Using the construction of Wembley stadium as an example, it was reported in the national press that the project went well over budget; however the party that suffered most was the contractor, since its final cost far exceeded the money that it was paid by the employer. Whilst the employer did pay the contractor more than the original sum, it was not as large as the contractor's losses.

No doubt on most large projects, some parties will meet and others exceed cost objectives. A really difficult project for an employer may go well over its budget; however, on that project, the contractor or specialists may be required to complete extra work, and they may make far more money than they would normally expect – arguably an excellent project for them. However, it could be argued this is only a short-term win, since clients will be less inclined to work with the same teams on future projects. The best outcomes to ensure repeat business relationships are 'win-win' for all.

2.4 Four pragmatic high risks in procurement systems

2.4.1 Exceptionally adverse weather

When contractors bid for projects, they are usually required to make predictions about what weather will be like during the construction phase. There are lots of variables to consider: at what time of the year will groundworks be constructed; are projects in coastal areas or multi-storey developments subject to high winds. In the UK, take a benchmark figure of 8% that contractors would allow in their programmes. Both JCT and NEC contract standard clauses allow contractors to be given extra time to complete projects if the actual weather is more than benchmark figures. The contractor would only be allowed the extra time, such that 12% exceptionally adverse weather would give 4% extra time allowance. An actual benchmark may be set for each project by the contract, stipulating that records at the nearest weather centre will be used. Extra time may only be allowed if the actual weather is worse than the mean of the last 10 years, or more strictly, worse than the worst of the last 10 years.

2.4.2 Late availability of information

In traditional procurement systems, ideally all drawings and specifications should be 100% complete before contractors are invited to bid for projects; that is rarely the case. Designers may be working on superstructure details, whilst work is progressing in the ground. Contractors are required to produce programmes, and identify dates on which information is required. These dates can often be long before work on individual trades start on-site, since there may be long lead-in periods required to obtain materials; in a busy market, even commonly used materials such as bricks and blocks may need to have orders placed many months before delivery. Information should be available by the date stipulated by contractors. But in practice, it may not arrive in time. Designers may be very busy, working on many projects, and they are just unable to cope. Alternatively, employers may have not decided precisely what they want, because they are in a rapidly changing market place that might influence their choices. Contractors are often delayed because they do not have sufficient information to build; in such circumstances, the method of procurement will be an indicator of who will pay for delay costs.

2.4.3 Ground conditions

All projects should have site investigation reports. These reports should include desk top studies and results from borehole investigations to identity ground-bearing capacity and whether there are contaminants present. Before construction work commences, all parties should be able to reasonably predict what ground conditions will be encountered. However, unexpected things may still arise. Employers may be deterred from commissioning many boreholes because they are relatively expensive. Unluckily, a borehole may not have been taken on one part of a site where there is a pocket of bad ground. Determining ground condition on longitudinal civil engineering projects (roads, railways) can be more difficult than single plot construction sites. It may be the case that more excavation is required than originally envisaged to reach load-bearing ground. If the excavation is for a foundation to a structure, likely backfill will be concrete, or if the excavation is for a road surface, there will a requirement for extra compacted hardcore. The position of existing underground statutory services, mineshafts, drains, sewers and natural water courses can also be difficult to predict; on some occasions some of these things 'appear' totally unexpectedly.

2.4.4 Variations arising other than from ground conditions; changes to drawings, specifications and bills of quantities as work progresses

Ideally there should be no changes required as work progresses. However this is rarely the case in practice. Employers are often criticised for making changes, perhaps because they simply change their minds; designers too are criticised since there may be conflict between drawings and specifications, or drawings are not sufficiently developed at project bid stages, and more detailing is required during the construction phase. Contractors may argue that as new details are made available they are variations and warrant extra payment; designers may argue that new details are merely design development. It must be recognised that there is often a political or market need to get started on projects as soon as possible. For large projects in particular, it can be very difficult to think through precise requirements about finishing details before projects start on-site. Therefore, it is arguably inevitable that changes will arise.

Variations and changes are nothing new. Latham (1994) argued that an employer's brief should be signed-off before work commences. The signing-off process should make employers aware that any changes they subsequently make, may cost time and money. It is also worth being mindful that when prices are agreed with contractors for variations, employers have lost the advantage of competition since work has started. Therefore in some cases there is the possibility that agreed prices may be higher than would otherwise be the case, or to put it another way, employers 'pay through the nose'. Morgan (2010) stated that intelligent employers should spend a lot of time in the design period defining exactly what they want, and then get out of the way and let contractors build.

Compare briefly manufacturing and construction. In manufacturing one prototype is built, and then thousands more follow with no design changes. In construction there are no prototypes (except perhaps in housing to some extent), and there can be thousands of changes to design as work proceeds. This demonstrates the 'one-off' bespoke nature of the construction process.

2.5 Costs that arise if projects are completed late; who pays?

When unexpected things arise, procurement systems and the contracts that support them are designed to allocate resultant costs between the parties. If there is bad ground, it is not just the cost of the extra excavation, taking surplus material to tips, concrete and hardcore; there is the possibility that the contractor will need extra time on the programme. Two key cost areas need to be considered when projects finish later than the stipulated contract completion date:

Cost item 1. Employers lose the opportunity to earn profits; supermarkets for example may want a new store open in good time for Christmas trade; football teams may want a new stadium completed for the start of a new season; private developers want rental income; schools need classrooms open for the start of a new term. Generally speaking, if late completion is the employer's fault, the employer will have to withstand itself its lost profits, or withstand any costs incurred making alternative arrangements. However, if late completion is the contractor's fault, the contractor will pay the employer a sum of money in compensation. This is known as 'liquidated damages' in JCT contracts and 'delay damages' in NEC contracts. The amount is stipulated in tender documentation, so that bidders know the consequences of late completion. Some bidders may include a sum of money in their tender to cover potential late completion; other bidders may decline to bid because damages are relatively high. The amount of damages should be a reasonable assessment of likely losses to an employer; the amount should not be so high that it may be considered a penalty, and thus not admissible in law. Damages may be set on a scale, such that amounts in early weeks are higher than those in later weeks. For example, a college that may need to mobilise and make arrangements for alternative accommodation for students may incur its highest costs in that initial mobilisation process. Damages may be set at say £10k per week for the first two weeks, and £5k per week thereafter.

Cost item 2. Contractors need to stay on-site longer than they had planned, and need to pay their preliminary costs; this is known as 'prolongation'. These costs may be defined as 'all those costs incurred on-site that are not associated directly with the measured work or trades'. Typically preliminary costs may account for 10% of contract value, with perhaps a range between 6 and 15% depending on project size and complexity. The largest part of preliminary costs is usually site staff salaries; for complex projects such as hospitals, more supervisory and planning staff will be required if compared to less complex, but still large value, warehouses. The New Rules of Measurement 2 (RICS, 2010, pp. 50–119) lists preliminary type items. Other high-value costs may be temporary accommodation, temporary fencing, temporary roads, scaffolding, tower cranes *et al.* Relatively smaller-value items are perhaps personal protective equipment (PPE) for tradespeople, connection charges for temporary services, computers *et al.*

There are lots of other ancillary costs that arise too when projects are completed late. Contractors may argue that because they are delayed on one project, they are unable to take their staff away and make profits on other projects. Generally speaking, contracts are written such that if fault for delay lies with employers, then they will have to suffer their own losses that arise in cost item 1 above, and will additionally have to pay contractors' preliminary costs in cost item 2 above. If fault for delay lies with contractors, they will have to pay employers damages, and suffer losses arising from their own preliminary costs.

The risk of bad ground is usually given to employers; therefore extra monies are paid to contractors for extra excavation and for contractors' preliminaries. Employers withstand their loss of profits and contractors secure an extension of time, which relieves them of the responsibility to pay damages. However, consider the situation of an employer that has a fixed amount of funding; perhaps a charity that obtained a grant of some kind. It needs a facility built and has £1,000,000 (£1m), but absolutely no more. The charity may receive two bids:

Option 1: £950,000 (£950k) bid, with the employer to take the risk of the ground. If the ground is excellent, the contractor will be paid £950k. If the ground is poor, the contractor will be paid £1.05m and the project completed later than scheduled.

Option 2: £1m bid, with contractor to take the risk of the ground. If the ground is excellent, the contractor is paid £1m, and if the ground is poor, the contractor will be paid £1m and the project completed on schedule.

This employer may think that option 2 and the extra £50k is good value; like an insurance policy. If the ground is excellent, the contractor will be paid £1m for doing £0.95m of work, but if it is poor it will be required to do £1.05m of work for £1.00m; or somewhere in between. The clauses in the standard contracts would need to be struck out to make it clear to bidding contractors that they are required to take the risk of the ground.

Often it is the case that delays arise due to the fault of neither party. Which party pays the cost of delays, if there is exceptional weather, since it cannot be controlled? Apportionment of costs between parties will vary dependent on the method of procurement and the terms of the contract. The JCT Standard Forms of Contract are written such it is a shared risk, 'half/half'; employers withstand their loss of profits and contractors withstand their preliminary costs.

On some projects, employers may be in a position where they just cannot give extra time to complete; the school, for example, has got to be open for the start of the new term. In standard contract forms and tender documentation, employers may strike out clauses that permit an extension of time because of bad weather. Bidding contractors should include in their prices for the risks that this may give them and the extra costs they may incur; if they were to finish late they will be required to pay to the employer damages. The latter is clearly not palatable, so they must try to control the weather or more accurately the consequences of the weather. If there is wet weather, frost, hot spells or high humidity, contractors can use temporary shelters, temporary heating, dehumidifiers, work evenings, week-ends and nights, employ more supervisors, employ more labour and plant even though inefficient working may increase unit cost of production. All these things would cost contractors extra money; ideally there should be a sum of money in tenders to cover the possibility that they will be required.

2.6 Other risks

Standard forms of contract in construction are designed to apportion risks fairly and equitably. There are often options in these contracts, such that for some projects, employers may take certain risks, but on others it may be contractors. Many risks arise from market conditions and the consequential financial repercussions.

Discussion points 2.1, 2.2, 2.3 and 2.4: Which party would pay or benefit in the following scenarios?

2.1 The cost of labour, plant or materials are higher during the construction phase than was envisaged when tendering; or perhaps costs are lower.
2.2 An order of £1m is placed by a contractor for external cladding to a multi-storey building; the cladding is to be imported into the UK from Europe. At the time of the order £1 = 1.20 Euros, and the order therefore has an equivalent value of £1.2m Euros. The value of the UK £ weakens between the date of order and the date at which payment is due, such that £1 = 1.1 Euros. Should the cladding company be paid £1m, equivalent to 1.1m Euros, or £1.09m equivalent to 1.2m

Euros? Perhaps the currency movements are such that the £ strengthens, and £1 = 1.30 Euros. In such a circumstance £0.923m = 1.2m Euros.

2.3 A mechanical engineering specialist goes into liquidation part way through completion of its work. A new specialist is appointed who demands a higher price. There are costs in correcting work by the first specialist, which is found to be defective. There is a long delay appointing the new specialist, and overtime working is required to ensure the project is completed on time.

2.4 As part of a main contractor's bid, six respected electrical specialists submit prices. The contractor is equally happy to select any one of the specialists since they are all reputable, and decides that lowest price should be the only selection criteria. The lowest price of £1m is included as part of the main contractor's bid to the employer. This may be called 'a bid for a bid', since the specialist is giving the main contractor a bid to help with the main contractor's bid. The main contractor's bid, that includes all trades and work is £10m, and the contractor wins the project. The main contractor starts foundation works, and takes a short time to place an order with the specialist; in the meantime, the market changes and there is a lot more work around. The specialist revises its price to £1.1m. Other specialists advise they are now too busy to take the project on, and the contractor has no choice but to place an order in the value of £1.1m.

Table 2.1 illustrates some events that may occur in construction and cause delays. The JCT Standard Form of Building Contract is used, assuming a traditional form of procurement, to indicate which party pays for employers' lost profits, and which pays for contractors' preliminaries.

Table 2.1 Which party pays in the event of delays?

			Who pays – employer or contractor?	
A project in a remote location is delayed due to an outbreak of foot and mouth disease.	Extension of time clause JCT 2011–2.29 2.29.14	Loss and expense clause JCT 2011 2.29 –	Employer's lost profits Employer absorbs	Contractor's preliminaries Contractor
A bomb hoax closes a site.	2.29.11	–	Ditto	Contractor
There is a shortage of skilled bricklayers.	2.29.12	–	Ditto	Contractor
There is a very long winter frost.	2.29.9	–	Ditto	Contractor
The architect provides drainage drawings to the contractor two weeks after the date requested.	2.29.7	4.24.5	Ditto	Employer

			Who pays – employer or contractor?	
The design depth for the strip foundation is 900mm below ground level. The employer's quantity surveyor measures 900mm as the firm quantity, but also includes for 100mm extra depth as a provisional quantity. The actual depth averages 1,400mm across the site.	2.29.1	4.24.1	Ditto	Employer
There is a strike by drivers delivering fuel and diesel – material deliveries to site are delayed.	2.29.12	–	Ditto	Contractor
The carpenters go on strike.	2.29.12	–	Ditto	Contractor
There is a long hot summer – bricklaying and concreting operations are affected.	2.29.9	–	Ditto	Contractor
The installation of the site underground electrical supply is delayed by the statutory undertaking.	2.29.8	–	Ditto	Contractor
The employer suspends work on the project for one month because of sudden organisational problems and world market uncertainties.	2.29.7	4.24.5	Ditto	Employer
There are several hundred minor variations on the project.	2.29.1	4.24.1	Ditto	Employer
There is extensive flooding.	2.29.14	–	Ditto	Contractor

2.7 Integrated or separated teams

The reports by Latham (1994) and Egan (1998 and 2002) called for procurement systems to be integrated not separated; 'integrate the supply chain'. Integrated teams involve parties, including contractors, working together on designs at early stages; separate teams usually involve consultant teams putting designs together without contractor involvement, and at a later stage contractors bid and subsequently build. A plethora of reports since Latham and Egan (e.g. Wolstenholme, 2009; Government Construction Strategy 2025, BIS, 2013) have recommended, with limited success, that integrated teams are the best way for employers to achieve cost, time and quality objectives. The report by the Innovation and Growth Team (BIS, 2010) suggested that the way to get lower CO_2 emissions during the construction of projects, and over their life cycle up to demolition and recycle, is to take on board at early stages the best ideas of those lower down the supply chain. True integration requires employers at the top of the supply chain to have open communication channels with suppliers and raw material producers at the bottom.

The problem then arises with an often reported cultural barrier to integration: 'the UK is culturally wedded to the ethos of competition'. There are many benefits of competition recognised in the UK, the European Union and internationally. It is competition that often drives improvements in many spheres of business and society. However, whilst competition can have its place in integrated supply chains, it cannot be on the basis of six contractors or suppliers competing for the same order.

If procurement systems are separated, employers can get true competition when they invite bids for work. Traditionally, based upon the briefs of employers, consultants design and specify projects, often bills of quantities are produced, and six contractors may submit bids. In the compilation of their bids, contractors ask subcontractors and suppliers to submit their bids in competition with others. Employers can be reasonably assured that they have the lowest price possible, and the project has been 'market tested'. There are examples of where competition is taken to the extreme, with employers using e-tendering portals. In these systems, drawings and specifications for projects are fully developed, with or without bills of quantities. Contractors are invited to pre-qualify to bid for projects, but the criteria to bid are easily met by many potential bidders. In effect 'anyone', though not literally, can bid. Bidders are given user names and passwords to get electronic access to the bid documentation. For one project perhaps 50 companies may ask to bid and perhaps 20 may actually bid. Employers award to the lowest bidder. In public bodies, politicians and executives can argue that prices obtained have been truly market tested. However, it is arguably very undesirable that if there are 20 bidders, 19 have expended resource that comes to nothing.

The problem that may then also arise is that upon start of the work, contractors who are the experts in buildability point out elements of the project cannot be built the way that they are drawn or that there is conflict between drawings, specifications and bills of quantities. Specialist subcontractors report that materials that have been specified are not suitable for their intended use. Variations are required, and as Latham noted in the report *Never Waste a Good Crisis* (Wolstenholme, 2009):

> *"If lowest price is demanded by the client, the tender price will not be the actual financial outturn at the end of the project, because the supply side will be looking for claims and variations to make up for what was not in the tender".*

Additionally, because of variations, a delay in completion of projects may result.

Alternatively, a procurement method can be chosen that truly 'integrates the team'. At the early stages of a project as design progresses, the supply chain is brought together to put their best ideas on the table; clearly, for example, structural steel specialists have better ideas about structural steel innovation than consultant structural engineers. Similarly, there are specialists in smart building technologies that are likely to know more about heating and electronic control systems than consultant mechanical and electrical engineers. Using integrated teams, certainty in design is better assured, and construction may commence with, it is hoped, few variations and delays as work progresses. Indeed there may be cost and time benefits, since teamwork between all may be able to identify savings. The one huge cultural problem in this integrated team model is that at some point a price or budget needs to be agreed. If there has only been one specialist advising, when it comes to agree a price with that one specialist, the fear from an employer's viewpoint is that the price will be a high price that has not been marketed tested. Perhaps employers need to be brave at this point. Specialists will not spend time and money bringing their best ideas to the table, if they are then required to bid as one of six companies; only a one-in-six chance of winning the project.

A compromise position was used by the Olympic Delivery Authority (ODA) in its role to procure the London 2012 Olympic Games. Its system was called 'collaborative procurement' whereby just two or three specialists were brought to the table to contribute towards the design, and then when a price was invited there was a one-in-two or -three chance for suppliers to win the work. Whilst the ODA wished to negotiate, it sought to ensure that it took advantage of being a huge employer with immense buying power. It needed to use some element of competition to assure taxpayers that it had attained best value. Similar problems may exist, as the use of Building Information Modelling (BIM) takes hold in the UK construction industry. The supply chain will only be willing to contribute its best ideas to the model, if there is statistically a high chance that it will win the work. The survey by the NBS (2015) indicates that there is real hope that BIM level 3 will drive integrated procurement systems to the fore, and cost and time predictability objectives *et al.* for all will be more likely achieved. In a further development, Maqbool (2016) explains the function of a NEC3 additional clause introduced in January 2016 that deals with contractor costs for pre-commence-ment (stage 1) and project on-site (stage 2). If a contractor does not receive an instruction to proceed to stage 2, the employer will 'likely need to pay the con-tractor for its stage 1 costs'. This clause opens up the possibility for contractors to compete at stage 1, without fear of losing all their bidding costs if unsuccessful. Employers need to take the view that monies they will pay at stage 1 to unsuccess-ful contractors will be more than recovered through innovative solutions realised in stage 2 and beyond.

Table 2.2 illustrates simplistic hypothetical outcomes for a project procured using competition and integrated teams, assuming the development is to be let on a long lease. It illustrates potential cost and time savings resulting from integrated working.

The question asked may be 'whilst recognising that competition is often good, why do employers still use separate systems when most authoritative sources, including government, argue that integrated teams are best?' Work by Challender *et al.* (2016) found that often, it is chief executives with an accountancy back-ground who select procurement methods. They are responsible to their non-exec-utives and shareholders. They find safety in 'full-on' competition, and danger in integrated teams. It is not so easy to argue the hypothetical position in Table 2.2, since there is a statistical possibility that projects procured by competition will indeed complete on budget and on time. Also, there is a possibility that projects procured by integrated teams do not achieve savings. If a competitively procured project were to go badly wrong, there is an easy defence in saying 'it was market tested'. However, if a project was procured with limited competition or indeed by negotiating a price with just one contractor, the blame for the failure may be placed upon the chief executive. It is also argued that PLC companies that are

Table 2.2 Hypothetical outcomes for a project procured using competition or integrated teams

Procured by competition		Procured with integrated teams	
Bid amount	£10m	Negotiated amount	£10.25m
Additional cost due to variations	£0.50m	Savings in cost due to team working	£0.50m
Final account or final amount paid	£10.50m	Final account or final amount paid	£9.75m
Contract period	52 weeks	Contract period	52 weeks
Delay due to variations	4 weeks	Time saving due to team working	4 weeks
Actual contract completion	56 weeks	Actual contract completion	48 weeks
Loss of rental income due to delayed completion	£0.10m	Additional rental income arising from early completion	£0.10m
Total cost	**£10.60m**		**£9.65m**

accountable on the London Stock Exchange to shareholders feel the need to satisfy their investors that projects have been market tested by an appropriate level of competition.

2.8 Methods of agreeing prices; inviting bids and e-tendering

When employers place agree contracts or place orders with the supply chain, at one extreme, it is possible to negotiate with one contractor, and at the other it is possible to have completely open lists where many parties may bid. Two key documents that have traditionally guided tendering procedures in construction are the National Joint Consultative Committee (NJCC, 1996a and 1996b) publications 'The Code of Procedure for Single Stage Selective Tendering' and 'The Code of Procedure for Two Stage Selective Tendering'. Even though NJCC was disbanded in 1996, the principles of the codes have remained as part of custom and practice for many years. The publications are now replaced by the JCT (2012b) Tendering Practice Note. There is no standard method of inviting bids; these are only codes and practice notes. Employers may do as they wish.

Traditionally, some private sector employers may have a relatively small (or long) list of contractors who they regularly ask to bid for their projects. Selection for each project may be by rotation, so that all get a chance to bid for work. The NJCC codes suggest that six should be the maximum number of bidders, because that number allows appropriate competition, and since tendering can be expensive, it limits the number of contractors who bid unsuccessfully. Some employers, especially in the public sector, may prefer more bidders to assure themselves and the taxpayers to whom they are responsible that projects have been market tested.

For large projects, since bidding costs can be high, contractors may refuse to bid in a competitive environment, if statistically there is only a one-in-six chance of winning the tender. Therefore to obtain bids, a two-stage process may be used. The first stage with perhaps six bidders is where contractors may provide outline proposals and attend for interview with employers. Two or three bidders are then selected to go forward to the second stage where detailed bids are submitted.

Paper-based systems require that bids are submitted by a set deadline; perhaps noon on Fridays. Electronic systems or e-tendering can be used to provide bidding windows, and a 'reverse auction' that drives prices down. A typical arrangement may be that documents are made available to bidders on an electronic portal. Bidders will have a reasonable period to work on and formulate their bids; perhaps four weeks for medium-sized projects. Then an electronic bidding window is established, perhaps between 9:00 am on Monday and 4:00 pm on Friday. All bidders can see the amount that has been bid by competitors A, B and C, etc., though anonymity is preserved. Consider the case of a contractor who in a traditional bid situation may have bid £10m; on Monday morning, the contractor starts at £11m, and then having observed the bids of competitors over coming days, lowers its bid gradually to £10.10m by Thursday morning. However, by Thursday evening a competitor has submitted a bid of £9.90m. The contractor really wants to win the project, therefore it telephones around its supply chain – subcontractors and suppliers – to see if they can provide it with lower prices for their elements of the works. Perhaps on Friday the contractor can submit a price lower than £9.90m? Using this system, employers will hope they have received bids which are truly market competitive. However, projects will not have the advantages of integrated teams, and may run the risk of many problems on-site if it transpires that the winning contractor's bid is too low. Worth emphasis, is one view of project managers thus:

> Projects run well in the construction phase, if contractors are making money.

Consequently, project managers and employers should want contractors to make money, since then the whole process is much more likely to be enjoyable and hassle-free with projects completed to quality and on time.

The RICS (2010) has published its RICS practice standards; e-tendering guidance note that puts forward good practice in this area.

2.8.1 Covers and bid-rigging

Both covers and bid-rigging are illegal and fraudulent; bid-rigging is the more serious. They were brought to the attention of the media by an investigation by the UK Office of Fair Trading (OFT, 2009), where substantive fines were imposed upon many contractors.

Covers unfortunately have a long tradition in construction. They arise when an employer asks a contractor to bid, and the contractor decides that perhaps its estimating team is too busy, or if it were to win the project it does not have the time or resources to deliver it, or perhaps it is judged the project contains too much risk. The contractor should politely decline to bid, but there is often a cultural feeling that to do so would mean a contractor would lose its place on tender lists for future projects. Therefore, contractors submit a high bid rather than no bid at all. To make a judgement about what that bid should be, they will speak to competitors who are also known to be bidding. The request will be 'please indicate a sum of money that I can submit that will ensure that I am not the lowest bidder'. The OFT found in its investigation one example of where a charity-based employer obtained six bids for a project, five of which were covers. The charity had been misled into thinking the price it paid was based upon market prices in a competitive environment; this is clearly morally unacceptable.

Bid-rigging arises when contractors or suppliers meet before tenders are due to decide what prices they will submit, and which party shall win. Projects are shared between each other, and clearly the prices are higher than would have been the case if competition had been present. Employers may fear that in collaborative procurement models, where few bidders are present, even though it is illegal, there may be enhanced possibility of bid-rigging.

2.9 Procurement categorisation

Categorisation of procurement systems is difficult – categories vary between authoritative sources in the literature – that is prominent authors and professional institutions. One key authoritative source has been the Royal Institution of Chartered Surveyors (RICS) 'Contracts in Use' survey. Whilst it has been published regularly since 1985, its final and last output was in 2012. The current authoritative surveys of a similar type have been conducted by NBS, the latest being in 2015 (NBS, 2015). The RICS classifications are illustrated in Tables 2.3 and 2.4, together with the volume of use by number and value in the UK in 1985, 2004, 2007 and 2010. Trends can be observed; in some periods for example bills of quantities have been less popular than in others. Whilst the RICS classify lump sum, target cost, re-measurement and prime cost (cost re-imbursement) as procurement systems, NBS call these 'pricing mechanisms'. Other labels that are often used include 'fast-track', 'package deal', 'turnkey'; these are features that

Table 2.3 Trend in methods of procurement – by number of contracts

	Procurement method	1985 %	2004 %	2007 %	2010 %
1	Lump sum – firm bills of quantities	42.8	31.1	20.0	24.5
2	Lump sum – specification and drawings	47.1	42.7	47.2	52.1
3	Lump sum – design and build	3.6	13.3	21.9	17.5
4	Target cost		6.0	4.5	3.7
5	Re-measurement – approximate bills of quantities	2.7	2.0	1.7	0.3
6	Prime cost plus fixed fee	2.1	0.2	0.5	0.6
7	Management contract	1.7	0.2	0.7	0
8	Construction management		0.9	1.1	0.3
9	Partnering agreements		2.7	2.4	1.0
Total		**100**	**100**	**100**	**100**

Source: RICS (2012a).

Table 2.4 Trend in methods of procurement – by value of contract

	Procurement method	1985 %	2004 %	2007 %	2010 %
1	Lump sum – firm B of Q	59.3	23.6	13.2	18.8
2	Lump sum – specification and drawings	10.2	10.7	18.2	22.6
3	Lump sum – design and build	8.0	43.2	32.6	39.2
4	Target cost	0	11.6	7.6	17.1
5	Re-measurement – approximate B of Q	5.4	2.5	2.0	0.7
6	Prime cost plus fixed fee	2.7	0.1	0.2	0.6
7	Management contract	14.4	0.8	1.0	0
8	Construction management	0	0.9	9.6	0.1
9	Partnering agreements	0	6.6	15.6	0.9
Total		**100**	**100**	**100**	**100**

Source: RICS (2012a).

can be included in methods of procurement – they are not necessarily methods of procurement in their own right. Whilst this can add to the 'confusion' it is just the way it is and reflects custom and practice whereby there are no standard procurement systems and employers will pick and integrate methods as they like, mindful of their objectives.

When categorisation does take place, there is a tendency in the literature to put individual projects into either one category (box) or another. It is to be mindful that many projects will have features drawn from two or more procurement methods. A management contract for example may have an element of specification and drawings included, or design and build.

All forms of procurement may, or may not, have a project manager to represent the employer. Some employers engage their own project managers in-house. A separate person is sometimes used to oversee the quality of the works *et al.* On JCT contracts this may be a clerk of works, or on NEC contracts the job title used is 'supervisor'. This person may be engaged directly by the employer, or one of the design team or may be engaged independently. In all methods of procurement, to comply with the CDM Regulations (2015), there needs to be formal appointment of an Employer or Client, Principal Designer and Principal Contractor.

A question often asked is 'what is the best form of procurement?' There is no answer – it depends on the circumstances of individual projects – there are lots of 'variables at large' on each project and the method of procurement should be selected carefully to match these variables. For example, it may be that complex projects with possibilities for repeat work may be best suited to partnering; whilst simple industrial one-off buildings may best involve some competition or design and build. It is definitely 'horses for courses'. Another question to employers may be 'if you are using partnering successfully on many of your projects, why do you not use it on all of them'. The answer here is that employers like to have a toolbox with many tools. Partnering is one of them; it is not to throw away other methods of procurement since one day the market may fundamentally change, and there may be a need to use another method. It is definitely not to 'put all your eggs in one procurement basket'.

Alongside the procurement method is the selection of the contract to be used – there are many variables to consider such as risk apportionment, extent to which the design is well defined, size of project, market conditions and type of project (new build/refurbishment/maintenance). Table 2.5 indicates standard contracts that are often used in various methods of procurement.

Table 2.5 Standard contracts used on methods of procurement

	Source or user	Procurement method/ label	Possible form of contract between employer and contractor; note – there are many possible options
1	RICS (2012a) Contracts in Use classifications	Lump sum – firm bills of quantities	JCT Standard Building Contract with quantities – SBC/Q

Source or user	Procurement method/ label	Possible form of contract between employer and contractor; note – there are many possible options
2	Lump sum – specification and drawings	JCT Minor Works Building Contract – MW
3	Lump sum – design and build	JCT Design and Build Contract – DB
4	Target cost	NEC3 Option C: Target contract with activity schedule: Target contract with bill of quantities
5	Re-measurement – approximate bills of quantities	JCT Standard Building Contract without quantities – SBC/XQ
6	Prime cost plus fixed fee (also called 'cost reimbursement')	JCT Prime Cost Building Contract – PCC
7	Management contract	JCT Management Building Contract – MC
8	Construction Management	JCT Construction Management Agreement – CM
9	Partnering agreements	JCT Constructing Excellence or PPC 2000 (by Association of Consultant Architects) or NEC3
10	Prime contracting, Private Finance Initiative (PFI or PF2), or Public Private Partnership (PPP)	Government agencies	Employer's bespoke, adapted from industry standard forms, where possible

2.10 Lump sum methods of procurement

2.10.1 Introduction

The RICS procurement categories preface three methods of procurement with the phrase 'lump sum': traditional bills of quantities, specification and drawings and design and build. The intention of these methods is that projects are built at an agreed price, and those prices can only be changed in four circumstances, thus: (i) provisional sums; (ii) prime cost sums (PC sums); (iii) contingencies and (iv) variations.

2.10.2 Traditional bills of quantities

The contractual arrangements for this method of procurement are shown in Figure 2.3. This is often referred to as the traditional method of procurement since it

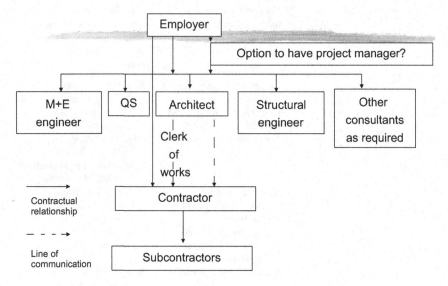

Figure 2.3 Traditional procurement structure.

has been done this way for many years. Employers' QS (often known as PQS or professional quantity surveyor) produce bills of quantities to assist bidders in the compilation of their prices. Whilst called 'lump sum', it is often the case by custom and practice that three key elements of work are re-measured: substructure, drainage and external works. Contractors are usually paid on the actual amount of work completed, and not on the quantities included in bills of quantities. This approach may result in savings or additional costs for employers.

A key feature of traditional methods of procurement is separation of design from construction. Employers appoint design consultants, and enter into agreements by forms of appointments. Architects often take the role of lead designer and also 'manage' contractors, although specialist project managers may be appointed to coordinate the work of designers and contractors. Architects are supported by quantity surveyors, structural engineers, mechanical and electrical engineers. Other specialist designers may also be needed on some projects e.g. BREEAM assessors, acoustics, landscaping. Employers, architects and quantity surveyors invite bids from contractors by a competitive process. If project managers are appointed, they may be asked to take responsibility for all consultants so that the employer has single-point responsibility for design work. Some project managers may enter into formal contracts with these other consultants, but some companies have multi-disciplinary consultancy structures such that many disciplines work directly for one employer. Textbooks say that the design is complete before bills are produced, but this is probably rarely the case. Often the lowest bid wins, and the successful contractor enters into a JCT contract with the employer. Some projects will be very successful using this method of procurement, but there are many examples of dissatisfied employers – projects over budget, finishing late, adversarial atmospheres. Of the four major risk items identified in 2.4 (weather, late receipt of information, ground conditions and variations), weather is a shared risk, and the employer carries the remaining three. Alongside paying for any extra work, employers will have to withstand their own loss of profits; and contractors will be awarded an 'extension of time', which will relieve them of the responsibility to pay damages. Contractors will also be awarded loss and expense to reimburse them for the extra preliminary costs they incur *et al.*

2.10.3 Specification and drawings

By number of projects procured, this is by far the most popular method of procurement used, accounting in the RICS (2012a) survey for 52% of projects placed by number, and 22% by value. Many of its features, including contractual arrangements and the apportionment of risk, are similar to the traditional bills of quantities method. It is not often written about in the weekly construction press, since it is not used on national or international major flagship projects. It is most often used on smaller or medium-sized projects that involve refurbishment, alteration or adaption; or as a small part of bigger projects. The contractual relationships are the same as when a bills of quantities is used.

Employers' PQS do not prepare bills of quantities, although elements of work can be itemised in the specification e.g. one description to embrace forming a new ramp for wheelchair access, that includes the multiplicity of trades involved, thus:

> *Cast new concrete ramp and landings comprising removal of existing paving and excavation, backfill with well consolidated limestone hardcore, blind top surface of filling with 25mm clean sand to receive 2000 gauge visqueen damp proof membrane taken up along length of external wall to 150mm minimum, overlaid with minimum 75mm thick concrete with a lightly tamped finish to the surface and smooth 100mm border with maximum slope of 1 in 15 as indicated on the drawing, plan surface area 3.2m².*
>
> *Item*

The specification will also include a preliminaries part to make it clear to bidding contractors that they should provide all items such as supervision, temporary power, security, insurance, cleaning *et al.* Just one item is indicated thus:

> *The works and any areas disturbed are to be left in a clean and tidy condition at the end of each working day. All debris resulting from the work is to be removed from site on a daily basis.*
>
> *Item*

The time and cost for contractors to bid may be slightly higher than if a bills of quantities were produced, since they all need to establish their own quantities for each element that is itemised, to enable prices to be determined. This may not be a huge deterrent for contractors to price small projects, but it can be for larger projects. When contractors do take quantities off they may not do so at the level of detail required by the RICS New Rules of Measurement (NRM). For example NRM requires that when measuring plastering works, any areas in narrow widths (less than 300mm wide) such as reveals to windows or beam sides and soffits are measured separately; also plaster angle beads must be measured in linear meters. For 'builders quantities' take-offs, these types of measured items may be ignored and measures are taken of overall area of plaster. Also they may not deduct the areas of openings for windows and doors. Then, when using a rate such that area multiplied by unit rate = amount, a higher rate is used to compensate for the items that have not been measured.

An hypothetical example of how using a higher rate can compensate for items that are not measured, is illustrated thus:

Use BCIS (2016) plastering rate:

13mm two coat plaster work to brick or block surface exceeding 300mm wide

Labour 0.39 hours/m² x £24.58	= £9.46/m²
Material	= £2.48/m²
Total	= £11.94/m²

Assume 100m²; total estimate = £1,194.00

Ditto not exceeding 300mm wide

Labour 0.77 hours x £24.58	= £18.92/m²
Material	= £2.48/m²
Total	=£21.40/m²

Assume 10m²; total estimate = £119.40

Expamet internal beads fixed with plaster dabs ref 550

Labour 0.13 hours/m x £24.58	= £3.07/m
Material	= £0.92/m
Total	= £3.99/m

Assume 50m; total estimate = £199.50

Overall estimate = £1194.00 + £119.40 + £199.50 = £1512.90

Use builders quantities and one composite rate:

Assume builders quantity measured approximately = perhaps 120m²
A builder may use a rate from experience of say £13.00, that includes for an element of narrow widths and angle beads = 120m² x £13.00 = £1,560.00. Assume this sum is sufficiently accurate for an estimate.

Whilst project documentation may include drawings and a separate A4 type written specification, alternatively the documentation may be drawings only with specification notes written on drawings. Such specification notes are often not comprehensive, and leave much to bidding contractors to interpret.

The concept of specification and drawing may be included as part (or indeed all) of other methods of procurement such as the lump sum bills of quantities, design and build, management contracting or construction management. In all of these, for example, electrical and mechanical work may be merely itemised and not taken off in accordance with NRM. Specialists who bid for these works need to make their own estimates of quantities of materials. In trades such as plastering or painting and decoration, if bills of quantities have not been prepared by the employer's side, this can present difficulties for contractors who need to obtain prices for these elements of work. A busy estimator working for one of these specialists, with many projects to price, may just select the easiest ones on which to bid; those with bills of quantities provided. It can be a time-consuming process to take off quantities on a large scheme, with only perhaps statistically a one-in-six chance of winning it. In these circumstances, to get prices, a main contractor may need to take off the quantities, and provide the details to specialists.

One downside of specification and drawings is that when variations occur, there may be no similar items in the specification that can be used as a basis for agreeing prices. This is not however a deterrent to using this method of procurement.

2.10.4 Design and build

Introduction

Design and build is the first step towards integrated team working. Rather than employers engage separate design and contracting teams, and sign a contract with each of those parties, there is just one contract signed with a design and build contractor. In the event that something goes wrong, employers have single-point responsibility from one party; it does not have to get involved in potential arguments about whether a fault lies with the designer or the contractor. For large projects, design and build contractors most often engage designers who would otherwise have worked for employers as illustrated in Figure 2.4, or alternatively may have their own in-house design team as shown in Figure 2.5.

When it was first used, it was not with the benefit of integrated team working in mind; it was to gain single-point responsibility and pass risks. This may be viewed negatively, if employers merely want to dump risks; alternatively it can be seen positively, where employers pass risks to parties best able to manage them. For those risks that contractors are confident about managing, it does not follow that contractors may increase their prices. However, some risks that are passed may be difficult to manage, and contractors will usually add money to their tenders. Employers may be happy to pay higher prices to sensibly apportion risks to contractors. In buoyant markets, contractors will be looking to win jobs with low risk; if contractors perceive risks are high they may decline to bid or add large sums of money to tenders to cover risks. Employers may then take the view that they are better to retain some risks themselves by changing method of procurement,

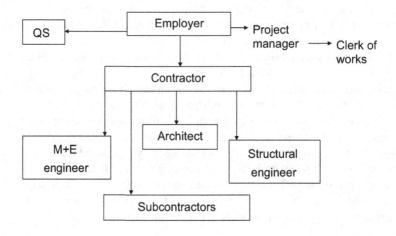

Figure 2.4 Design and build structure; contractors using consultants.

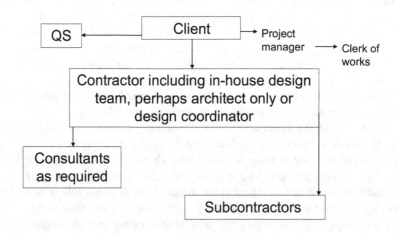

Figure 2.5 Design and build structure; in-house.

since design and build can become too expensive. Alternatively, in markets where little work is available, in order to secure turnover, contractors may be willing to take many risks and add little or no money; just hoping that problems will not arise. Contractors may also hope they can make some profits during the construction phase from employer variations. Over many decades, there are many large contractors that have reported significant losses in their annual reports, or have gone into liquidation, because they have not included sufficient money in tenders for risk.

Single-point responsibility for receipt of information and variations

Of the four major risks in section 2.4 that cause delays, design and build systems can pass to contractors the frequently occurring risk of late availability of information. There are however two reasons why contractors may receive information late: (i) because employers have not made up their minds about what they want, or (ii) because design teams have not had time to produce detailed drawings. Clearly, in design and build the risk of the former remains with the employer. If employers change their minds, they must pay. It is the latter risk which frequently arises in traditional methods of procurement. In design and build, the contractor must now manage information flows from its own in-house designers, or its consultants.

It is similarly the case for variations. Some variations may arise because employers change their mind, and in these cases the risks and costs must again lie with employers. But in many cases, variations occur because there are mistakes in designs, or discrepancies between drawings and specifications. In traditional systems employers also carry these latter risks. Although employers may ask for reimbursement from designers, in practice this seldom happens. This may be because it is rarely the case that professional mistakes are 'black or white' issues. In the design and build method, such variations now become the responsibility of design and build contractors. If consultant designers make mistakes, there is arguably a high possibility that contractors will seek to recover any costs from designers.

Whilst the employer has now passed two of the four major risks to the design and build contractor, it 'might as well', if market conditions suit, pass the other two risks also: weather and bad ground.

An often used tactic of contractors in traditional systems is to exploit delays due to late receipt of information from designers, and variations caused by mistakes of designers. Contractors may have their own problems on projects which they know will cause a delay to a project; perhaps they have planned things badly or site management has not been as good as it should. Perhaps a project has had several key managers leave, and this has caused disruption. All of these things are the contractor's responsibility, and if delayed completion of the works does occur, the contractor would be required to pay the employer liquidated damages. However, contractors may mask their failings by arguing that delays are due to late receipt of information. Contractors are required to issue a formal schedule of information requirements that is linked to the master programme. However, if a contractor is in delay because of its own fault, it may not pursue the information from design teams vigorously; it will fulfil its obligations under the contract to forewarn about potential delays, but it may do so politely, and hopes information does not arrive. The contractor is happy to let a delay occur, ask for extra time to complete, and within this time, complete works on which they had their own problems. Importantly, rather than pay damages, it will receive payment for its preliminary costs. Figure 2.6 illustrates how in traditional systems of procurement, employers may pay for faults of the designers.

Employers can become quite aggrieved at this, since they have done nothing wrong. It is the designers who are late providing information or require variations due to their mistakes. Employers receive their buildings later than scheduled, and have to pay contractors extra money. Indeed employers become so aggrieved that they look for a method of procurement where there is a party to the contract that is better able to manage the risk of information being supplied late by design teams. That method is design and build. Since the design and build contractor is now responsible for providing the information, whether the designer be in-house or an external consultant, the information is now requested assertively and not politely. If information provision is slow, the contractor using its own in-house designers can ask people to work overtime, or recruit extra people; the contractor is in control. If external consultants are being used, the contract between the two parties can be written such that consultants pay liquidatedand ascertained damages that may arise due to late issue of information.

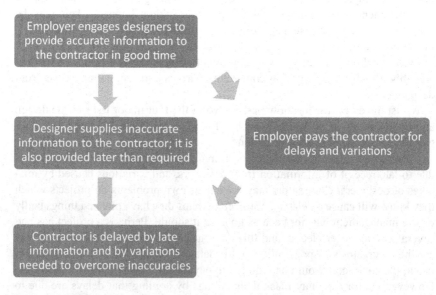

Figure 2.6 The employer pays.

In these circumstances, external designers will be highly motivated to ensure information is supplied in good time. Importantly in design and build, the risk of late information is passed to the party best able to manage it; that is a design and build contractor. Design and build contractors will also work hard to minimise the possibility of variations caused by design mistakes. Since the risk is managed well, delays that sometime occur on traditional projects do not arise on design and build projects.

Single-point responsibility for defects

A second very important feature that makes design and build attractive to employers is that they secure 'single-point responsibility' for any potential defects. In traditional systems, there are split responsibilities to employers, designers responsible for design and contractors responsible for materials and workmanship. There are many flagship examples of where problems have arisen during construction or during the occupancy period where there is a dispute between designers and contractors about whether a defect is a design or materials/workmanship problem. Some defects may be so serious to prevent the use of buildings such as foundation settlement, floor screed failures or substantial water penetration through wall cladding or roofs. If a major defect does arise in a traditionally procured project, a contractor may refuse to undertake corrective work without payment, since it argues the defect is a design issue; long delays can occur, whilst dialogue takes place. Employers may be very uncomfortable to find themselves in the middle of these arguments. In design and build, the issue is simple. The design and build contractor must attend to the defect immediately; it is left to the design and build contractor to deal internally or with a consultant designer where the fault lies.

How much design should be complete before it is passed to design and build contractors? Novation.

Employers may employ architects initially to develop briefs for projects and then invite several design and build contractors to build-up some proposals and submit bids. In such cases, employers need to make it very clear to bidders about the criteria they will use to select winning bids; will they pick best designs, which may be highest price, or will they pick simpler designs at lower prices?

Alternatively, employers' architects may produce some outline drawings; perhaps some 1:200 scale plans, elevations, sections and external works layouts. That would still leave design and build contractors the opportunity to be innovative in the remaining parts of the design, and certainly leave the successful design and build contractor to develop working drawings. Design and build contractors may be left with lots of choices around specifications for materials.

Some employers however may ask their architects to develop drawings and specifications in some detail, leaving only design and build contractors to complete working drawings. Perhaps the project information is as much as 90% complete at tender stage, leaving the design and build contractor almost nothing to design. Employers then ask contractors to complete the remaining 10% of the design by directly employing the same design team that did the initial 90% design. This is known as 'novation' or switching. The designers switch responsibility for designs from the employer to the design and build contractor. Why would employers appoint a design and build contractor when 90% of the design is already complete? The answer is simple; the employer gets single-point responsibility. The design and build contractor is responsible for (i) overseeing there are no delays in completing the remaining parts of the design, (ii) dealing with variations that may arise due to mistakes in the design and (iii) defects that may arise during construction or occupation.

Another advantage of seeking substantial completion of designs before novation is that if designs are not well advanced, contractors have a reputation for substituting good specification materials with those of lower cost. If employers have well developed specifications, this will not be possible. Also, in design and build, contractors may steer towards less complex designs of lower cost; if employers use architects to produce detailed drawings, aesthetically pleasing and ornate designs are more likely to be achieved.

2.11 Management contracts

Figures 2.7 and 2.8 illustrate the structure for management contracting and construction management. Both these methods involve employers carrying more risks, but there are opportunities to share them with management contractors or construction managers. They are only used for very large projects, often £10m plus. The role of the management contractor or construction manager is one of coordination, organisation, planning, trouble shooting and financial control. The tender process is likely to be two stage, where the first stage may involve perhaps six bidders putting together some outline programmes, method statements and preliminary budgets. A proposed management fee will also be detailed, to include costs of staff and the amount required for head office overheads and profit. The second stage may involve two or three bidders being selected to put together more detailed bids. It is unlikely that a bills of quantities will be used for building up prices; it's more the case that budgets will be established for each individual trade, based upon some approximate quantities and elemental or unit rates.

Many main contractors take the role of management contractors or construction managers; some companies traditionally known as consultant designers take the role too. The intention is to put management contractors or construction managers in a professional capacity alongside the rest of the design team, where they are able to provide management skills and practical buildability knowledge in an integrated team, for a fee. Management contracts facilitate early starts on sites, since design detailing may progress as the works are being built.

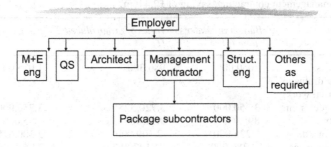

Figure 2.7 Management contracting structure.

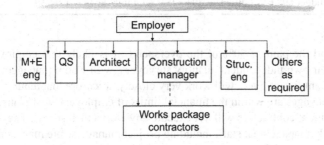

Figure 2.8 Construction management structure.

2.11.1 *Management contracting*

As work proceeds, the successful management contractor will seek bids and place subcontract orders with each trade, hopefully at or below the budgets set at tender stage. If a budget for one trade proves to be low, the management contractor may seek to secure savings elsewhere in the project, in order that the overall budget is not exceeded. Employers will pay the actual amount that is paid to each subcontract specialist. Each trade will have its budget at tender stage, an order value and a final account, as illustrated briefly in Table 2.6. Whilst savings were achieved between the tender budget and the value of orders placed (circa £98.65m − £95.20 = circa £3.45m), the final saving was only £2.45m. This may reflect adjustments for any variations and other difficulties that may have occurred during the progress of the works. Overall, it is hoped the employer is pleased with the £2.45m saving achieved.

Management contractors may be incentivised to secure savings on projects. If the total budget for a project is £100m, there may be a gain share/pain share arrangement. If the final cost is £95m the management contractor may get 50% (or some other percentage) of the gain; however, if it is £105m it may be required

Table 2.6 Tender figure, value of orders and final accounts for a project procured using management contracting; saving of £2.446m

Budget item	Budget at tender stage (£)	Value of order placed (£)	Final account (£)
Groundworks, including drainage and external works	1,050,000	980,300	1,025,000
Structural steel frame	3,650,000	3,720,525	3,750,000
Precast floors	895,000	895,000	895,000
Brickwork facade	2,225,000	2,100,000	2,300,625
All other trades not detailed	73,380,000	74,127,857	74,557,000
Soft floor finishes	450,000	380,000	375,000
Preliminaries	10,500,000	10,500,000	10,800,500
Management fee	2,500,000	2,500,000	2,500,000
Totals	**98,650,000**	**95,203,682**	**96,203,125**

Saving on budget £98,650 000 − £96,203,125 = £2,446,875 = say £2.45m

to withstand the pain of 50% of the overspend. It is likely that employers will appoint their own independent quantity surveyors, with whom management contractors' quantity surveyors will work very closely, to ensure that tenders for subcontract packages are within the financial limits of employers' cost plans.

Management contractors will not execute any works themselves. They employ directly only management staff, including project managers, site managers, quantity surveyors, planning surveyors; some at a senior level and some people in training too. Also perhaps some general operatives to undertake cleaning works *et al*. All trades are subcontracted. Much of the work required for the site preliminaries or the temporary site set-up are also subcontracted, such as temporary roads, temporary fencing, temporary electrical supply, scaffolding, catering, security *et al*. Management contractors may employ setting out engineers and retain control for setting out of base lines and establishing datums, since mistakes by specialists can be potentially very costly. These engineers may also check the setting out and levels of subcontractors to minimise the risk of errors. It would for example be catastrophic for all, if there is a requirement to break out large quantities of reinforced concrete because of specialist engineers' mistakes.

Management contractors enter into subcontract agreements with subcontractors. Payments are made by employers for all subcontract trades, normally on a monthly basis to management contractors; management contractors pass that money to subcontractors.

Management contracting has fallen into disrepute with some, since on some occasions management contractors do not pay subcontractors promptly. Management contractors use projects as a vehicle to generate positive cashflows to be used elsewhere in their businesses. Also, they may unreasonably impose 'set-off'. Set-off is when management contractors withhold part of payments due to subcontractors, since it is alleged subcontractors have failed to comply with subcontracts in some way. It may be for relatively trivial items, such as not cleaning up after completing sections of work; or it could be for defective work or failure to keep to programmes. Sometimes the reasons for set-off and the amount of money set-off may be reasonable. On other occasions the reasons for set-off may be tenuous, and/or the amount of money taken may be grossly inflated. Late payment and unreasonable set-off taken together can leave subcontractors very disgruntled, and unwilling to participate in this method of procurement. Some of the very large specialists, such as mechanical, electrical, structural steel *et al.*, may have balance sheets larger than management contractors, and they may complain vehemently to employers and even refuse to work on such projects.

The JCT2005 Management Building Contract (MC) does suggest a formula for how management contractors will gain an extra fee should project final accounts be different to the contract cost plans. This formula is:

$$\text{ACPMF} = \text{CPMF} \times \frac{100 + \text{or} - (D - T)}{100}$$

Where:

ACPMF is the Adjusted Construction Period Management Fee.

CPMF is the Construction Period Fee as stated in the Appendix.

D is the increase or decrease of the total Prime Cost when compared with the Contract Cost Plan Total expressed as a percentage of the Contract Cost Plan Total.

T is 5 or such other number as is stated in the Appendix under reference to clause 4.10.2 and 4.10.3.

+ shall be plus if the Prime Cost exceeds the Contract Cost Plan Total.

− shall be minus if the Prime Cost is less than the Contract Cost Plan Total.

For example:
£30 million project, overruns by £3 million, construction period fee = £600,000.

$$\text{ACPMF} = 6000,000 \times \frac{100 + (10 - 5)}{100}$$

therefore ACPMF = £630,000

However, this formula is subject to negotiation, or may indeed be struck out by employers who wish to include a gain/pain arrangement, or include a guaranteed maximum price.

2.11.2 Construction management

Construction management is similar to management contracting, but the way contracts are arranged overcomes the problems of subcontract specialists being subject to unreasonable set-off and late payment. The construction manager takes on many of the roles detailed in management contracting, but when forms of contract are signed for trades, those contracts are between the employer and the specialists. Thus, there are no subcontractors, since the construction manager is not a middle party. Specialists are now called 'works package contractors'. The works package contractors will make monthly payment applications to the construction manager, who with the oversight of the employer's quantity surveyor will approve or adjust them. Payment will be made directly from the employer to the works package contractors, it is hoped on time. If there is a good reason to set off some money from works package contractors, the construction manager is not incentivised to make recommendations to do so unreasonably.

2.12 Partnering and frameworks

Partnering has been described as a structured management approach which facilitates team working across contractual boundaries by integrating project teams and smoothing the supply chain. The three fundamental characteristics are:

- Formalised mutual objectives (which may or may not be binding) of improved performance at reduced cost;
- The active search for continuous measurable improvement, which is perhaps measured against industry key performance indicators (KPIs);
- An agreed common approach to problem resolution.

It was brought to prominence by Latham (1994) and Egan (1998 and 2002). There is also the seminal publication by Bennett and Jayes (1998) 'The Seven Pillars of Partnering: A Guide to Second Generation Partnering'. All parties in the process should work together, mindful of the need to meet their objectives, including profitability. In many other procurement systems, the processes are adversarial, and at the end of projects, a judgement can be made that one side or the other has won. It clearly should not be that way. Latham in 1994 cited the words of Dodo in Alice's Adventures in Wonderland, '*Everybody has won and all must have prizes*'.

There are many employers who have huge procurement programmes extending over many years, for example the utility industries, supermarkets, Highways England and airport authorities. In traditional systems, there are different parties coming together on each project. That can make it difficult to develop trusting relationships, and therefore increased costs and delays arise from misunderstandings and parties relentlessly pursuing their own self-interests in disputes. There is no standard method of partnering; it may be for one or for many projects. An employer may simply go to a contractor that it trusts and negotiate a price or budget for a single project, or for several projects over a long period. Prices may be based on a lump sum where the contractor takes risks, or it may be based upon a budget with a management contracting or construction management reimbursement arrangement, so that employers take the risk of final accounts being higher or lower than budget. There could also be a design and build element included.

If many projects are to be constructed, to introduce an element of competition, employers may establish frameworks over perhaps five years. A selection process will take place; expressions of interest are invited and after due preliminary consideration, perhaps 12 contractors will be invited to apply for a position on the framework. They will be required to submit detailed documentation about their companies, and give presentations to employers. There may be 6 or 8 successful contractors. The criteria used to reduce the number of contractors from 12 to 8 or 6 may be around previous experience, reputation, health and safety, sustainability, CVs of professional staff employed, strength of balance sheets, *et al.* Contractors may also be required to include some estimates of likely costs for professional staff, or percentages that successful bidders will require for overheads and profit. When the framework has been established, as individual projects are to be built, employers may negotiate prices in rotation with each member on the framework; again on lump sum or management style arrangements. Alternatively, to ensure competition, they may negotiate with several contractors on the framework about each project, or indeed they may invite bids for some projects. In traditional bidding situations, the number of bidders can be six; in a bid for one project on a framework, the number of bidders is likely to be less than that; perhaps between two and four. Some contractors have complained they have not secured any work whilst on frameworks.

Workshops are thought to be a key to successful partnering, and there needs to be a cultural shift in the way that professionals work. The opportunity should be taken to ensure the partnering ethos extends fully down the supply chain such that it is truly integrated. Collaboration should focus on building trust between teams, and use that trust as a driver to change traditional adversarial behaviours. One criticism often levelled at partnering is that employers partner with contractors and then contractors enter into traditional adversarial relationships with subcontractors. It is claimed that projects that are procured using partnering methods are far more likely to meet employer objectives, including completion to budget and on time.

2.13 Prime contracting and the private finance initiative (PFI or PF2)

In the design and build system in section 2.10.4, it was suggested that employers may pass to contractors the four keys risks of weather, ground conditions, variations and late availability of information. In prime contracting, responsibility is also passed for caring for projects for their life cycle, though perhaps agreements are initially just for 25 years. Prime contracting and PFI are used mostly in public procurement systems. Designers, contractors and those who will be responsible for maintenance may come together to form a consortium for one large project. Some contractors have a division which specialises in maintenance, and there may therefore be just two members in the consortium. The maintenance may include soft aspects of facilities management, such as daily cleaning, catering and landscaping. Built into maintenance schedules may be periodic replacement due to fair wear and tear of elements such as floor finishes, perhaps after every 5 or 10 years. The consortia will also be responsible for defects that may arise during the 25-year period, perhaps due to failings in workmanship or materials. The ethos of taking responsibility for facilities for a long period can focus the minds of designers and others in consortiums; should cheaper materials with short life expectancies be specified, or should it be those with initial higher capital costs and a longer life?

One key difficulty in taking responsibility over such long periods is how much will things cost in the future. In 20 years, what will be the cost of employing grounds maintenance workers and buying equipment; how will technological aspects of intelligent building systems have developed, and how much will they cost? Since there are many unknowns, when consortiums put together their estimates and bids, they need to include sums of money that cover all potential risks; employers may end up paying more money than if they had retained the risks. It must also be acknowledged however, that on another project, the reverse may apply; that may leave consortiums with losses that they may predict to run for many years. In prime contracting, the consortia will be paid in the conventional way for the capital build cost, and then regular payments to an agreed schedule over 25 years to cover facilities management and maintenance.

In the private finance procurement systems, one more risk is passed to consortia; that of providing the money. In civil engineering projects it is sometimes called DBFO – design, build, finance and operate. The operation side may be more than just maintenance and facilities management; for example if the facility were a prison, the consortium may take responsibility for providing all security staff. Financiers, as members of the consortium, fund the capital cost of developments. Clearly employers have to pay eventually; this is by regular fixed payments over the life of agreements, again often 25 years. Payments made include sufficient to pay off capital costs of facilities and also importantly, interest charges. There are analogies to leasing facilities or to taking mortgages. As individuals we may take occupation of a dwelling, and fund that occupation by a mortgage. The owner is actually the lender, but when the term of a mortgage expires, ownership reverts to the individual mortgagee. Table 2.7 illustrates how a £100k mortgage would be repaid over 25 years, assuming for convenience an interest rate of 7%. Since private finance projects are usually very large, perhaps add six zeros, so the capital cost is £100m not £100k. It will be noted that since money changes its value over time, the actual amount paid for a £100m facility is actually £214m.

On some PFI projects, employers may ask a consortium to take the risk of how much a facility is used, such that payments are not fixed amounts, but are linked to income. This could be for conventional roads where sensors are laid to detect the number of vehicles passing, or toll roads, or prisons where payments are based upon the number of prisoners in occupation.

PFI has been the focus of much criticism. There are examples where consortia have made huge or windfall profits, but also those where heavy losses have been incurred. Some public bodies such as the NHS have made cuts to front line services, since they have found it difficult to pay amounts due to consortia. Consortia are dissatisfied with high bidding costs, especially if they are unsuccessful. If perhaps four

Table 2.7 Repayment schedule for £100k loan, over 25 years at 7% – all figures per year. For PFI or PF2 projects, multiply by 10^6

Period in years	Payment (£)	Principal (£)	Interest (£)	Balance (£)
1	8,581	1,581	7,000	98,418
2	8,581	1,691	6,889	96,727
3	8,581	1,810	6,770	94,917
4	8,581	1,936	6,644	92,980
5	8,581	2,072	6,508	90,907
6	8,581	2,217	6,363	88,690
7	8,581	2,372	6,208	86,317
8	8,581	2,538	6,042	83,778
9	8,581	2,716	5,864	81,062
10	8,581	2,906	5,674	78,155
11	8,581	3,110	5,470	75,045
12	8,581	3,327	5,253	71,717
13	8,581	3,560	5,020	68,156
14	8,581	3,810	4,770	64,346
15	8,581	4,076	4,504	60,269
16	8,581	4,362	4,218	55,907
17	8,581	4,667	3,913	51,240
18	8,581	4,994	3,586	46,245
19	8,581	5,343	3,237	40,902
20	8,581	5,717	2,863	35,184
21	8,581	6,118	2,462	29,065
22	8,581	6,546	2,034	22,519
23	8,581	7,004	1,576	15,514
24	8,581	7,495	1,086	8,019
25	8,581	8,019	561	0
Totals	**214,526**	**100,000**	**114,526**	

consortia bid, statistical chances of success are not good enough. Since these are public projects, huge profits are seen to be at the expense of taxpayers. To mitigate this, a system of public private partnership (PPP) has been used on some projects, where the public sector provides a percentage of the capital cost (perhaps 51% so that it retains control), and thus is the recipient of 51% of profits. In 2010, the government introduced private finance 2 (PF2) to replace PFI (HM Treasury, 2010). Some of the aims of the new system were to (i) curb the ability of the private sector to make windfall gains, (ii) support the public sector in retaining many risks which the private sector may otherwise price highly and (iii) establish a board for each project and require an annual report such that if windfall profits are being made, they are transparent to all. Facilities management contracts within PF2 may be established over 5-year cycles rather than 25 years, such that new financial arrangements can be agreed mindful of market conditions prevailing at the time.

2.14 Chapter summary

Procurement systems are designed to allocate and control risks. Whilst risks should be given to those parties best able to manage them, it may be the case that employers try to merely dump risks. A key issue is to what extent design and build is separated or integrated. Many authoritative sources argue that the best way is to integrate teams, and to incorporate some form of negotiation or collaboration in the processes required to determine prices. It is also argued that the selection criteria for winning bids should not be based on lowest price alone. All parties to contracts should be able to meet their objectives, including but not limited to profitability. Supply chains may only be willing to contribute their best ideas to Building Information Models if they are part of integrated teams, and if they have only one or two competitors. There is a desire, especially in public procurement systems, to make best use of buying power and to have some element of competition present in bidding systems. Selecting the correct procurement method is seen as the key decision in determining whether projects will meet their objectives. Research indicates that 60% of projects are completed late, and 31% exceed employer budgets. Table 2.8 illustrates key differences between methods of procurement.

Table 2.8 Indicative similarities and differences in methods of procurement

Procurement method/label	The extent to which design and construction are separated or integrated	The basis on which the contractor is selected	Maintenance in use – say 25 years	Lump sum payment or 'lease' – the issue of ownership by the constructor	Balance of risks – employer or constructor
1 Lump sum – firm bills of quantities	Separate	Competition	No	Lump sum	Both
2 Lump sum – spec and drawings	Separate		No		Both
3 Lump sum – design and build	Integrated if contractor involved at an early stage		No		Most with constructor
4 Management contracts; management contracting or construction management	Integrated	Limited competition	No		Most with employer
5 Partnering/ frameworks	Integrated		To be agreed; usually no		To be agreed
6 Prime Contracting PFI/PPP	Integrated		Yes	Lease	Most with constructor

2.15 Model answers to discussion points

Discussion points 2.1, 2.2, 2.3 and 2.4: Which party would pay or benefit in the following scenarios?

2.1 The cost of labour, plant or materials are higher during the construction phase than was envisaged when tendering; or perhaps costs are lower.

This depends upon which party has been asked in contracts to accept the risk of fluctuating prices. In recent years, it has often been the case that contractors take the risk, especially for short-duration contracts in a low-inflation environment. However, many contractors have reported significant losses especially when they have accepted risks over longer-term projects and construction inflation has been much higher than inflation in the general economy. Mindful of this, there may be a trend developing where contractors are only prepared to bid on projects where employers take the risk of fluctuating prices.

2.2 An order of £1m is placed by a contractor for external cladding to a multi-storey building; the cladding is to be imported into the UK from Europe. At the time of the order £1 = 1.20 Euros, and the order therefore has an equivalent value of £1.2m Euros. The value of the UK £ weakens between the date of order and the date at which payment is due, such that £1 = 1.1 Euros. Should the cladding company be paid £1m, equivalent to 1.1m Euros, or £1.09m equivalent to 1.2m Euros? Perhaps the currency movements are such that the £ strengthens, and £1 = 1.30 Euros. In such a circumstance £0.923m = 1.2m Euros.

The circumstances in this situation are similar to that in discussion point 2.1. In some market conditions employers may be able to secure prices whereby the cladding specialist takes the risk of currency movements. It is possible however that the specialist may add monies to its bid to cover this risk, and then currency movement does not occur! It is for the employer to decide. If the employer asks the main contractor to take the risk, it is likely this will be passed to the specialist.

2.3 A mechanical engineering specialist goes into liquidation part way through completion of its work. A new specialist is appointed who demands a higher price. There are costs in correcting work by the first specialist, which is found to be defective. There is a long delay appointing the new specialist, and overtime working is required to ensure the project is completed on time.

If the specialist were nominated (see chapter 3.8), many of these costs will fall to the employer. However, assuming the specialist is a domestic subcontractor, selected by the main contractor, all these costs fall to the main contractor;

substitute the word 'contractor' for 'risk taker'. The contractor can only hope that the losses it may incur as a consequence of this liquidation will be made up for by gains (net gains; see chapter 6.9) made on other elements of work.

> *2.4 As part of a main contractor's bid, six respected electrical specialists submit prices. The contractor is equally happy to select any one of the specialists since they are all reputable, and decides that lowest price should be the only selection criteria. The lowest price of £1m is included as part of the main contractor's bid to the employer. This may be called 'a bid for a bid', since the specialist is giving the main contractor a bid to help with the main contractor's bid. The main contractor's bid, that includes all trades and work, is £10m, and the contractor wins the project. The main contractor starts foundation works, and takes a short time to place an order with the specialist; in the meantime, the market changes and there is a lot more work around. The specialist revises its price to £1.1m. Other specialists advise they are now too busy to take the project on, and the contractor has no choice but to place an order in the value of £1.1m.*

This risk falls to the main contractor. Many contractors have reported large losses across portfolios of their work because of this type of problem. There is a difficult balance to be struck between including monies in tenders to cover risks, and trying to submit competitive bids; including money may move a contractor from being potentially the lowest bidder to be second lowest and therefore losing the project.

References

BCIS (2016) *Building Cost Information Service.* The Royal Institution of Chartered Surveyors. Available by subscription at: service.bcis.co.uk. Accessed 22.03.16.

Bennett, J. and Jayes, S. (1998) *The Seven Pillars of Partnering: A Guide to Second Generation Partnering.* Reading Construction Forum. Partnering Task Force. London: Telford. Available at: capitadiscovery.co.uk/bolton-ac/items/33997?query=The+Seven+Pillars+of+Partnering%3A+A+Guide+to+Second+Generation+Partnering&resultsUri=items%3Fquery%3DThe%2BSeven%2BPillars%2Bof%2BPartnering%253A%2BA%2BGuide%2Bto%2BSecond%2BGeneration%2BPartnering%26target%3Dcatalogue&target=catalogue Accessed 25.08.16.

BIS (2010) Low Carbon Construction: Innovation and Growth Team. Available at: www.gov.uk/government/uploads/system/uploads/attachment_data/file/31773/10–1266-low-carbon-construction-IGT-final-report.pdf Accessed 20.03.16.

BIS (2013) Construction 2025. Industrial Strategy: Government and Industry in Partnership. Available at: www.gov.uk/government/uploads/system/uploads/attachment_data/file/210099/bis-13–955-construction-2025-industrial-strategy.pdf Accessed 01.05.16.

CDM (2015) Construction (Design and Management) Regulations 2015. Available at: www.hse.gov.uk/pubns/books/l153.htm Accessed 23.03.16.

Challender, J., Farrell, P. and Sherratt, F. (2016) Effects of an economic downturn on construction partnering. *Proceedings of the Institution of Civil Engineers: Management Procurement and Law.* DOI: dx.doi.org/10.1680/jmapl.15.00033

Egan, J. (1998) *Rethinking Construction.* London: HMSO. Available at: constructing excellence.org.uk/wp-content/uploads/2014/10/rethinking_construction_report.pdf Accessed 09.09.15.

Egan, J. (2002) *Accelerating Change.* London. Available at: constructingexcellence.org. uk/key-industry-publications/ Accessed 09.09.15.

Glenigan (2015) *KPI Zone.* UK Industry Performance Report. Available at: www.glenigan. com/construction-market-analysis/news/2015-construction-kpis Accessed 23.09.15.

HM Treasury (2010) A New Approach to Public Private Partnerships. Available at: www.gov.uk/government/uploads/system/uploads/attachment_data/file/205112/pf2_ infrastructure_new_approach_to_public_private_parnerships_051212.pdf   Accessed 20.03.16.

JCT (2012a) SBC/Q2011 Standard Form of Building Contract. The Joint Contracts Tribunal.

JCT (2012b) Tendering Practice Note. The Joint Contracts Tribunal.

Latham, M. (1994) Constructing the Team. Final Report on the Joint Review of Procurement and Contractual Arrangements in the UK Construction Industry. July. London: HMSO. Available at: constructingexcellence.org.uk/wp-content/uploads/2014/10/ Constructing-the-team-The-Latham-Report.pdf Accessed 09.09.15.

Maqbool, A. (2016) Start at the beginning. *Building.* 11 March 2016, pp. 40–41.

Morgan, S. (2009) The right kind of bribe: BAA's Steven Morgan on project roles. *Building.* 9 October 2009. Available at: www.building.co.uk/analysis/the-right-kind-of-bribe-baas-steven-morgan-on-project-roles/3150411.article Accessed 25.08.16.

NBS (2015) National Construction Contracts and Survey. National Building Specification. RIBA Enterprises. Available at: www.thenbs.com/knowledge/nbs-national-construction-contracts-and-law-survey-2015-finds-disputes-continue-to-blight-construction-industry Accessed 22.03.16.

NJCC (1996a) Code of Procedure for Single Stage Selective Tendering. The National Joint Consultative Committee. Available at: nh.tt/cablebayfiles/Phase%203A%20Pre%20Bid% 20Meeting%2017_12_09/Appendix%20C%20-%20NJCC%20D&B%20Code.pdf Accessed 20.03.16.

NJCC (1996b) Code of Procedure for Two Stage Selective Tendering. The National Joint Consultative Committee.

OFT (2009) Bid rigging in the construction industry in England. Office of Fair Trading. Available at: webarchive.nationalarchives.gov.uk/20140402142426/http:/www.oft.gov. uk/OFTwork/competition-act-and-cartels/ca98/decisions/bid_rigging_construction Accessed 20.03.16.

RICS (2010) *RICS Practice Standards: E-Tendering Guidance Note.* 2nd Edition. Available at: www.rics.org/uk/knowledge/professional-guidance/black-book/e-tendering-black-book/ Members only download. Accessed 20.03.15.

RICS (2012a) Contracts in Use. A Survey of Building Contracts in Use during 2010. Available at: www.rics.org/Global/CONTRACTS%20IN%20USE_FINAL_%20Nov2012_% 20lteage_081112.pdf Accessed 20.03.16.

RICS (2012b) *NRM2: RICS New Rules of Measurement.* Coventry: Royal Institution of Chartered Surveyors.

Wolstenholme, A. (2009) Never Waste a Good Crisis: A Review of Progress Since Rethinking Construction and Thoughts for Our Future. Constructing Excellence. Available at: dspace.lboro.ac.uk/dspace-jspui/bitstream/2134/6040/1/Wolstenholme%20Report% 20Oct%202010.pdf Accessed 24.07.15.

3 Elements of a contractor's bid

3.1 The decision to bid

3.1.1 Initial inquiry

There are a number of approaches available to quantity surveyors when sending projects to tender or inviting bids from suitable contractors. These methods range from uncontrolled open advertisements in trade magazines or local newspapers, whereby any party interested in tendering can make an enquiry and perhaps for a small fee receive the bid documents, to the use of a tightly controlled range of potential contractors, selected from an existing database of pre-approved organisations, such as *Constructionline*. Equally, the contractors could be selected from an externally managed framework such as the *North West Construction Hub*, which serves public sector employers operating in the Northwest of England. Some employers have their own select list of contractors who are regularly invited to tender for projects, perhaps as part of a framework.

Discussion point 3.1: Why do you think it is so important for employers to use contractor pre-qualification as part of the tender process?

With the majority of employers now seeking to manage risk by instigating some form of pre-qualification, it is normal practice, prior to the issue of tender documents, for contractors to receive an initial enquiry from architects inviting them to express an interest in tendering for projects. The invitation, usually sent in the form of a letter, with a return form to indicate interest, will outline key information to allow the contractor's estimator or quantity surveyor to make an initial decision as to whether to proceed with the bid. The letter will typically outline:

- Tender process (single stage, two stage *et al.*);
- The tender award criteria (price, or a combination of price and quality, presentation and interview requirements, the relative weightings of the elements and scoring methodology);
- The approximate value of the project;

- The nature and location of the project;
- Overview of the employer;
- Key dates, such as document issue, tender return, construction commencement, project duration, etc.;
- Length of tender period (typically 4–6 weeks);
- Form of contract to be used;
- Contractors input into the design;
- OJEU procedure (if applicable);
- Form of procurement.

Once invitations are received, estimators or quantity surveyors in contracting organisations need to vet documents and provide quick overviews of projects and employers' requirements to management teams or commercial directors to make initial decisions as to whether or not to submit an expression of interest. This information is important, as contractors need to ascertain how projects would impact on their organisation. Primarily they need to decide whether they have sufficient resources and cashflow to complete projects, whilst also deciding whether the risks presented are acceptable and within the tolerance of the organisation. After all checks are completed and assuming projects fit well with organisational experience and there is required capacity in terms of labour and financial resources, contractors may respond positively to invitations.

Discussion point 3.2: How would you determine whether to proceed with tendering for a project if the employer invited you to submit a bid?

3.1.2 Bid documentation

Following acceptance of invitations to tender, contractors receive full sets of tender documentation in hard copy or electronic format; the latter if employers have e-tendering platforms (see chapter 2.8). Bid documents are likely to include:

- Notice to tenderer;
- Form of tender;
- General conditions of the contract to be used;
- Specifications and employers' requirements;
- Bills of quantities, work schedules or other pricing documents.

It is important documents are comprehensive, as this avoids later disputes or the possibility of claims due to a lack of or unreliable information. It is for estimators to check documents with a view to ensuring they:

- Relate to the project under which the contractor was invited to tender for;
- Align with the details provided in the invitation to tender;

- Are complete, and contain no missing pages;
- Are adequate for assessing costs;
- Allow sufficient time for the production of a robust and accurate bid for the works.

Discussion point 3.3: If you were invited to bid for a design and build project what bid documents would you expect to receive?

3.1.3 The decision

If the documentation received is acceptable, contractors will review and re-consider the decision they made during invitation stages. Decisions to tender are very significant commercial judgements. Bid processes can be very expensive, and usually there is no financial recompense for unsuccessful submissions. However, decisions to withdraw could jeopardise or damage relationships with employers or design teams, and a consequence may be to lose the opportunity to bid for future projects. If contractors decide to proceed with bids, they need to be aware of the balance between risk and reward and the opportunity cost of the capital they invest (could this have been used to facilitate alternative projects with a higher return or better profit?).

The bidding decision will be affected by a vast array of factors which must ultimately be evaluated by contractors before they proceed with the development of their bid. Flanagan and Jewell (2016) in the latest iteration of the *CIOB Code of Estimating Practice* have classified these using three main themes:

1 *External factors* – For example the level of market competiveness, the sector's ability to source debt funding if needed and the state of the wider economy (boom–bust cycle position).
2 *Internal factors* – Including the firm's current workload, availability of cashflow to fund the project, resourcing availability and capacity within the estimating department.
3 *Project related factors* – There are numerous considerations here, which can include size and scope of the works, complexity, ground conditions, contract conditions, employer, design team, cost of bidding, supply chain complexity, location, levels of information about the site, timeframe for completion, overall risk evaluation *et al.*

Assuming contractors are assured projects fit well into these three themes, authorisation is given to estimating departments to proceed with production of fully developed bids. Once raw cost forecasts are developed by estimators, there should be internal 'tender adjudication meetings'. For large projects, these may take place over many days. Key contracting personnel need to be involved, often including chief executives and directors; also estimators, planners, contracts managers, supply chain managers *et al.* One purpose of these meetings will be to scrutinise bids;

where can cost savings be achieved and alternatively are there areas that have been potentially under-priced? Perhaps there are significant recent movements (up or down) in wages paid to tradespeople? Adjustments will be made to estimators' initial figures as required. Key areas of risk will be examined in a risk register, and judgements made about the pricing of these risks. On some occasions, significant sums of money may be included to cover risk items, but on other occasions contractors may come to judgements that they will hold some risks at no costs, since if they were priced, that may make bids uncompetitive. Final decisions are about the level of mark-up or profit assigned, and the amount for company overheads. This will ultimately be a commercial decision, based on circumstances such as employers, design teams, probability of winning, level of competition and levels of demand within the market place. Contractors at this point are faced with a significant range of possible outcomes, which could include:

- Markets are in recession and companies urgently need work (income) to sustain turnover. Contractors may decide to submit 'suicide bids' with no profit or even negative margins (i.e. bid below cost), in the hope they can make some money through variations, subcontractor procurement or claims post-contract. The term 'suicide' is used, since on some occasions, contractors are unable to make money and even one single project with losses can lead to company failure;
- Projects have incomplete or conflicting information and contractors feel there is a commercial opportunity to make profit using post-contract change, so they submit low bids in the hope of later recovery through post-contract change;
- Contractors apply a normal profit margin to the bid to ensure profitability;
- Employers are regular developers and there is a wish to sustain on-going relationships so they submit bids with lower to medium profit margins.

When making bidding decisions, contractors need to manage the risk of 'winner's curse', which is a situation often encountered in competitive bidding markets with low margins such as in construction, and very tight tendering timeframes (4–6 weeks). As a result of these pressures, it is not uncommon for successful bidders to have made errors somewhere within bids. When this is realised, low or even negative margins are possible. It is important that estimates are carefully audited to ensure any errors are detected before bid submission.

The approach to tender preparation, pricing key documents, supply chain management, commercial management and decision making will be discussed further in other sections of this book.

Discussion point 3.4: An employer is concerned a contractor may have submitted a suicide bid that could lead to excessive claims post-contract. What actions should the employer take?

3.2 Performance bonds

Performance bonds are sometimes thought of as though insurance; they are similar in principle, but different in the way that they operate. Consider individuals who take out insurance, perhaps for homes or cars. If there is an untoward event, the individual who paid the insurance premium will make a claim and if that claim is successful, receive a payment in compensation. For bonds, the party that makes the payment will not be the beneficiary in the case of an untoward event, since it will often be (but not always) contractors who pay premiums and employers who receive payments in compensation.

Employers use bonds to protect themselves against the non-performance of contractors; or contractors to protect themselves from lower tiers. Most frequently non-performance is not that the progress or quality of works is unsatisfactory; more often it is because contractors go into liquidation. In cases of liquidation, employers will incur many additional costs to complete works. Employers may try to recover monies from the company that has failed, but it is often the case that anything received may be minimal. If the project was secured in a competitive bid situation, an employer may approach the contractor that was in second position. If that contractor's bid was 1% higher than the winning bid, it may be prepared to complete remaining work at that higher figure or even more. As the new contractor takes possession of the site, it may wish to undertake an array of tests on work already completed, and may find defects that need rectification. Some defects may only become apparent at the final test and commissioning stages, such as on underground drainage and mechanical and electrical works. The new contractor will need to be paid to rectify these defects. There can be a long period between a site being closed as a result of liquidation, and reopening with a new contractor. Most likely there will be a delay to the completion date, and the employer will lose the opportunity to make profits from the project for that period of time. Indeed the original contractor may have been behind programme as a result of its own problems on the project, and the employer will therefore lose the opportunity to claim damages.

To protect themselves against these potential extra costs, in tender documentation, employers state that the successful contractor will be required to provide a performance bond. Often the amount is 10% of the project value, such that a £1,000,000 (£1m) project will have a £100,000 (£100k) performance bond. The successful contractor will then go to the market and pay a premium for the bond. The amount of the premium will depend upon the strength of the contractor's balance sheet. Those with a strong balance sheet will pay a lower premium since they are less likely to go into liquidation than those with a balance sheet that is not so strong, and therefore constitute less risk to the bond provider. Perhaps a guideline premium figure may be 3% of the value of the bond, so on a £1m project and £100k bond that will be £3k. Contractors will include the cost of premiums in their bids, so therefore effectively, employers pay. If the winning contractor does fail, the bond provider will pay the employer for the losses incurred, up to the maximum value of the bond.

For employers who are regular procurers of construction work, it is recommended that bonds are not requested, since it is rarely the case that they are 'called in'; that is, contractors do not fail and employers do not need to ask bond providers for compensation. It is thought better to check carefully the balance sheets of contractors before they are invited to bid, so that the risk of liquidation is assessed. Commercial assessments of companies are available through on-line organisations such as Dun and Bradstreet; percentage scores are given to rate the financial security of contractors. Those contractors who are not strong ideally should not be on tender lists. Consider an employer with a portfolio of developments over many years; perhaps 50 projects on which they have paid premiums amounting to £150k. Perhaps just one contractor fails, and only £100k is recovered, therefore asking for bonds in such scenarios could be false economy. Also, it is argued that if employers have strong balance sheets themselves, they should be able to withstand the 'shock' of failures. However, bonds do have their place for employers who are occasional procurers of construction, especially if they are small companies, since a one-off project that fails could be very serious for the whole business.

Contractors may also ask for performance bonds from their subcontract specialists, in instances where there is large value and complex work. If mechanical or electrical tier 2 specialists were to go into liquidation part way through their works, the potential delays to programmes and costs to find a replacement are potentially high. There may also be risks from tier 3 contractor failures. However, it is unlikely that contractors would ask for bonds for lower value, or less complex trades, since contractors would not want specialists to include in their bids bond premiums; contactors will be happy to take the risk themselves.

There is a range of other bonds available on the market. An on-demand performance bond could be called if the employer is not happy with the performance of a contractor; perhaps progress is slow. On-demand bonds represent higher risk to contractors, since they do not want to have a bond called on them. It is like making a claim on an insurance policy; the premium for the next performance bond will be higher, or indeed it may not be possible at all to secure a bond for the next project. Unless there are potential high rewards on a project, contractors may decline to bid on projects where employers require on-demand bonds.

Retention bonds are possible between employers and contractors or contractors and specialists. The principles of retention are explained in more detail in chapter 5. Retention is a sum of money held from a payment, typically 3 or 5%; half is usually released on project completion and the other half after any defects have been certified as being corrected. One reason retention monies are held is so that there is a lever to ensure contractors or subcontractors return to correct their defects; if they do not return, another party may be asked to complete the works and be paid from retention monies held. However, retention damages cashflow of supply chains. Take for example a piling specialist who completes high value work in the early weeks of a project, with an overall contact period of two years. Under standard contract terms, the first half of retention would only be released after that two-year period. Whilst special arrangements can be made for the piling company, far simpler is that there is zero retention for that portion of the works or indeed the whole project. That would particularly benefit those lower down the supply chain. If there is zero retention the employer may require that the main contractor supplies a retention bond, such that if the contractor or a party lower down the supply chain fails to rectify any defects, the employer can call the bond, and use the money to pay someone else. The contractor will makes its judgement about which specialists it will require a retention bond from; again likely to be high value complex trades only. Some contracts may be written such that occupation of facilities may be taken subject to completion of some minor 'snagging' items. Snagging is attention to minor defects that do not affect immediately the use of facilities. In such circumstances, the retention may only be released after completion of the snagging.

Also available are advanced payment bonds. There may be trades that have long manufacturing periods for expensive materials, for perhaps large volume cladding on multi-storey structures. Subcontractors and suppliers may be concerned about the risk of parties higher up supply chains going into liquidation; perhaps contractors or employers. More likely there is a need for cash to fund manufacturing processes, since if those periods are lengthy, in traditional payments systems it may be many months before lower-tier subcontractors and suppliers receive monies. In such circumstances, employers may facilitate advanced payments. However, for security of those payments, in cases where subcontractors or suppliers go into liquidation or fail in some other way, employers may require bonds.

The National Joint Consultative Committee, noted in chapter 2 to have been disbanded in 1996, produced a code and template for a performance bond (NJCC, 1995). The Association of British Insurers (ABI, 2016) provides a model form of guarantee bond. However, it is often the case that bond wording is based upon bespoke writing of employers and insurance companies, on the advice of legal specialists.

Discussion point 3.5: A contractor submits the lowest tender for a project, but it exceeds the employer's budget. As part of the process to negotiate a reduction in the tender price, the contractor states that it has a strong 'balance sheet' and suggests that the employer omits the requirement for a bond. Advise the employer.

Discussion point 3.6: A contractor has three performance bonds lodged with its bank for different projects and also has a large overdraft facility. The contractor applies to the bank for a fourth bond, but there is concern that the bank may be taking on board too much risk. Advise the bank.

Discussion point 3.7: A national contractor undergoes substantive reorganisation. Each region in the UK is registered as a separate PLC. Advise an insurance company which is considering an application for a bond from the North West–based PLC.

3.3 Project insurance

Insurance provides a way of managing risks on construction projects. The liability of contractors can be substantial, it is therefore essential that they consider the full implications of project risks at the time of bidding, and consider whether incorporating additional insurances would be a suitable risk management approach. This section focuses on insurances required by various standard forms of building contract and insurances all businesses are required by law to hold. Contractors can take out non-mandatory types of insurance to protect themselves from other risks on projects; these are also explored as they often form a key consideration when managing project risks, and deciding whether or not the risks are acceptable.

Insurance forms a fundamental aspect of risk management on all construction projects. Procurement routes and associated contract strategies establish project culture and dictate how much risk is allocated to parties and how much risk will ultimately be distributed throughout construction supply chains. Despite some types of project insurance being mandatory under UK law, the use of insurance nevertheless provides all parties with opportunities to protect themselves against major risks inherent in construction processes. It also provides a means through which risk can be transferred to specialist third-party organisations in return for a fee (the insurance premium).

3.3.1 Overview of insurance in the UK construction industry

As with many other industries, insurance is a fundamental feature of the construction industry as it provides protection for employers, building users, designers and constructors. As a result, the purchase of insurance can represent a principal method for managing project-level risks. Insurance gives parties the option to transfer risks (events involving some uncertainty about whether they would occur or not) to other parties (insurance companies) who are better able to deal with the full financial effect of risk events should they actually be realised. In return for accepting risks, insurance companies charge a premium to parties taking out policies.

In the construction sector, there are two main types of insurance:

1 *Liability Insurance* – Provides financial cover for legal liabilities that employers, consultants and contractors owe to others. These payments could result from statutory, contractual or professional commitments. It covers compensation awarded by the courts or the legal expenses incurred defending a claim. Insurance would not, however, cover fines imposed by the court.
2 *Loss Insurance* – Provides cover for any losses falling directly on insured parties. This type of insurance can include:

   • *Protection against loss or damage to property*, including temporary works and work in progress, owned construction plant, hired-in plant *et al.*
   • *Protection against pecuniary loss*; insurance that deals with monetary loss in some form or other; examples could include the loss of a tender/

bid/mobilisation bond, employee theft of cash, ransom demands if operating in volatile countries *et al.*

How much of the risk parties ultimately decide to insure depends on their appetite for risk and the amount of money they wish to spend on insurance premiums. As a general rule of thumb, the higher the risk and its associated liabilities, the higher the premium insurance companies charge. The difficulty for most in the construction industry is achieving the correct balance. It is very expensive to over-insure projects, especially given historically low profit margins associated with construction contracting. However, it is equally very expensive to under-insure in the event that untoward incidents occur, and claims need to be made.

Discussion point 3.8: Why do you think insurance is so important to construction?

In deciding what the optimum balance of insurance is for companies, it must be remembered that not all risk should be insured. Some risks, such as inclement weather would simply be not worth insuring. It would be expected that most contractors operating in the UK would be well versed in managing the risk of rain. It is expected that they would allow for this eventuality in their tender and build the risk into their construction programme. For other risks, however, the impact of the event would be such that insurance becomes essential. For example the financial losses associated with a fire in buildings under construction could, if uninsured, have a crippling impact on contractors and potentially lead to insolvency. The impact of fires could also be disastrous for employers who, in the event of contractor insolvency, would be unlikely to recover any of the money paid in interim valuations for destroyed structures. As a result risks with potential severe consequences for projects, employers and contractors, if they were to occur, need to be mitigated through insurance. When making decisions about the balance of risk and what is to be insured, it is essential to seek the advice of insurance experts.

The UK insurance market contains a significant array of potential insurance policies. Building Design Wiki (Cantor, 2016) identifies 15 different types of insurance for someone operating in the UK construction sector. These range from employers' liability to directors' and officers' insurance. However to simplify this complex range of policies, Lock (2013) suggests insurances to protect those involved in construction projects can, generally speaking, be grouped into the four clusters illustrated in Table 3.1.

The CIOB Code of Estimating Practice (Flanagan and Jewell, 2016) identifies five main insurance policies construction contractors need, thus:

1   All-risks insurance;
2   Public liability insurance;
3   Employers liability insurance;

Table 3.1 Main insurance clusters for construction projects

Insurance clusters	Description
Legally required insurance	These are the insurances that any firm operating in the UK is required to have in place by law. Currently only two types of insurance are mandatory: (1) employers liability insurance (required under the Employers Liability (Compulsory Insurance) Act) and (2) motor vehicle insurance (a requirement of the Road Traffic Act).
Insurance required as part of construction contracts	Liabilities allocated under a construction contract can be substantial. Therefore most construction contracts will require parties to insure these risks, through the operation of several insurance clauses.
Insurance required as part of a professional or industry body certification scheme	A number of professional bodies such as the Royal Institution of Chartered Surveyors (RICS) or Chartered Institute of Building (CIOB) provide registration schemes for construction firms and consultancy practices. The professional bodies will often mandate the insurance policies they require firms to hold who are registered as part of the scheme.
Optional insurances	These are additional policies construction firms or consultancy practices can take out to provide additional protection from risk, assuming they are happy to pay insurance premiums.

4   Professional indemnity insurance;
5   Latent defects insurance.

3.3.2 Contractors all-risks insurance

All-risks insurance is a policy that provides coverage for the two main construction risks: damage to property and third-party injury or damage claims (Dunning, 2008). The cover is usually in operation or effective for the period of the works, from the date of commencement to the date of practical completion. If both parties agree, the cover can be extended to cover the period after practical completion to include 'defects liability periods' (Joint Contracts Tribunal, JCT) or 'defects correction periods' (New Engineering Contract, NEC3).

Contractors all-risks insurance or contract works insurance is usually taken out in joint names (contractors and employers) so that regardless of fault, funds will be available to cover risk events affecting projects. As such, each party will retain the right to file a claim against policies. However, all parties also have an obligation to inform insurers at the earliest opportunity of any injuries or damage that may lead to claims being made.

Insurance provisions, and therefore the types of insurance required, differ significantly between Association of Consultant Architects (ACA), JCT and NEC standard forms. Under JCT contracts the insurance clauses are standardised, although these can still sometimes be amended if employers want additional protection. Insurance is always arranged on the basis of joint names (contractor and employer), and policies cover 'all risks', which in JCT forms are deemed to include the works, materials, reasonable costs of removal and disposal of debris and shoring and propping up works. Employer stipulations about the level of cover and any limits defined must be fully outlined in the conditions and contract particulars. NEC3 and ACA (Project Partnering Contract, PPC 2000 and Term Partnering Contract, TPC 2005) insurance provisions are far less straightforward. Both contracts are written to include the concept of employer and contractor risks, so that parties to the contract choose the insurances they feel are relevant to projects, usually decided by considering risks identified in risk registers. Responsibility for taking out insurances is assigned to either employers or contractors and included in contracts. This could mean insurance is the sole responsibility of employers, contractors or both.

Discussion point 3.9: Why do you think insurances are covered by an express clause in standard forms of contract?

All-risks insurance under JCT

Under the JCT suite of contracts, insurance clauses are standardised. The Standard Form of Contract, and Design and Build Standard form, Section 6 (clauses 6.4–6.12) outlines the main aspects of the insurance requirements. In terms of all-risks insurance, JCT specifies:

> *Insurance in the joint names of the contractor and the employer, to cover injury or damage to property due to collapse, subsidence, vibrations, weakening or removal of support or lowering of groundwater attributable to the carrying out of the works.*
>
> *(Cl. 6.7 – Schedule 3 Insurance Option A)*

As a result, contractors are only deemed responsible for damage to property due to negligence, breach of statutory duty, omissions or default on their part.

As part of standard insurances, clauses in section 6 of JCT, Clause 6.7 makes provision for insuring the works by way of all-risks policies, providing employers with a number of choices, namely insurance options A, B and C. Full details of these policies are set out in schedule 3 at the rear of the contract:

- Option A (New Buildings) requires the **contractor** to take out and maintain an all-risks insurance policy for the works.
- Option B (New Buildings) requires the **employer** to take out and maintain an all-risks insurance policy for the works.

- Option C (Existing Structures) requires the **employer** to take out and maintain (a) insurance in respect of the existing structure and its contents and (b) an all-risks insurance policy for the works. This option is usually used for extensions (vertical or horizontal and alterations to buildings).

Employers are required to specify which option is to be selected.

> **Discussion point 3.10:** You are drafting tender documents for the refurbishment of 20 private terraced properties as part of a housing regeneration project. Each property will be fully occupied during the work. Which insurance option in JCT would be most appropriate?

All-risks insurance under NEC3

The NEC3 standard form is far less prescriptive when documenting insurance obligations of parties to contracts. In the NEC3 standard form, risk and insurance are covered in clause 8 (a core clause applicable to all NEC3 options). Whilst this combination can cause confusion to those who are not familiar to the NEC standard form, it does not necessarily mean that all risks associated with projects must be insured. That is ultimately a decision for employers and contractors when negotiating the terms of contracts. Despite this rather flexible approach to allocation, the NEC3 does try to provide some framework around which insurance can be arranged. Clause 80.1 identifies a series of employer risks:

- Unavoidable damage;
- Negligence of the employer;
- Fault in the employers' design;
- Plant and equipment until received and accepted by the contractor;
- War risks;
- Strike and riots, not confined to the contractor's employees;
- Loss, wear or damage to parts of the work taken over (expect when due to a defect);
- Loss, wear or damage to works/materials retained on-site by the employer in the event of termination other than due to the activities of the contractor;
- Risks stated in the contract data, which are additional and voluntarily assumed by the employer.

As a result, it would be expected that employers would insure these if they feel that is required, although clearly not all of the situations above would be insurable. Similarly clause 81.1 stipulates that contractors' risks are *"the risks which are not carried by the employer"* so it is safe to assume contractors would be responsible for insuring those (again if insurance is needed). The types of insurance required are explained in a simple insurance table, as illustrated in Table 3.2.

In relation to all-risks insurance in the NEC3 standard form, the contract is only slightly more complex than JCT. As with the JCT standard form, NEC3 stipulates that contractors all-risks insurance is required and that it must be taken out in joint names (cl.84.2). However, unlike JCT, the contract extends the period of insurance cover beyond practical completion; insurance under NEC3 must be in operation from the date the contract commences until the date of the defects certificate. The other substantive difference between the two standard forms is the scope of 'all-risks policies'; in JCT, policies cover the full range of works, however, under NEC, this has been restricted due to the separation of contractor and employer risks. As a result, the all-risks insurance policy is now only required for 'contractor risk events'. Employer risk events are insured separately, under a different form of insurance and not in joint names.

3.3.3 Public liability insurance

Public liability insurance covers the cost of claims made by third parties to the contract (Association of British Insurers, ABI, 2014a). In the case of contracts between employers and main contractors this would include members of the public and subcontractors. As a result public liability insurance policies typically insure employers and contractor against:

- Personal injuries;
- Loss of or damage to property;
- Death.

Table 3.2 NEC3 Insurance Table

Insurance	Minimum amount of cover or minimum limit of indemnity
Loss of or damage to the works, plant and materials.	The replacement cost, including the amount stated in the Contract Data for the replacement of any Plant and Materials provided by the employer.
Loss or damage to equipment.	The replacement cost.
Liability for loss of or damage to property (except the works, plant and materials, equipment) and liability for bodily injury to or death of a person (not an employee of the contractor) caused by activity in connection with this contract.	The amount stated in the Contract Data for any one event with cross liability so that the insurance applies to the Parties separately.
Liability for death of or bodily injury to employees of the contractor arising out of and in the course of their employment in connection with this contract.	The greater of the amount required by the applicable law and the amount stated in the Contract Data for any one event.

Source: NEC (2005).

Most standard forms of construction contract will include an express provision in insurance clauses making it the responsibility of contractors to ensure public liability insurance is in place from the commencement of projects. Examples of such clauses can be seen in both the JCT Standard Form of Building Contract (Cl. 6.4.1) and the NEC3 standard form (Insurance table in Clause 84).

3.3.4 Employers liability insurance

In addition to public liability insurance, all organisations operating in the UK have a legal obligation under the Employers Liability (Compulsory Insurance) Act 1969 to have insurance policies in place to cover the costs of compensating employees who are injured or become ill as a result of work they have undertaken, together with any legal fees incurred. This legal obligation is reinforced through a strict penalty system under which employers can be fined £2,500 for every day they are without appropriate insurance. Despite this statutory obligation most standard forms of construction contract will again reinforce the need for employers liability insurance. Examples of this can be seen in both the JCT Standard Form of Building Contract (Cl. 6.4.1) and the NEC3 standard form (Insurance table in Clause 84). As a result public liability insurance and employers liability insurance are usually offered in a single policy that may also cover office contents and buildings insurance requirements (Cantor, 2016).

3.3.5 Professional indemnity insurance

Professional indemnity (PI) insurance, as its name implies, is designed to provide cover for those who provide a professional service. The description of a 'professional' service would at first appear to exclude contractors, however the advent of design and build procurement systems has caused a blurring of liability in relation to professional services. Most contracting companies will now take some (under the contractors design supplement) if not all (under Design and Build) design liability as part of projects and so will need to ensure they have professional indemnity insurance in place.

Insurance will cover the cost of compensating employers for any loss or damage that has resulted from negligent services or advice they have been provided by a business or individual (Association of British Insurers, ABI, 2014b). Such claims could include negligence, misrepresentation, inaccurate advice or unfair dealing. Additionally, professional indemnity insurance will cover consultant or contractor defence costs associated with claims that turn out to be groundless; in which case the PI policy acts as a legal expenses policy in defending policy holders for alleged negligence.

It is important to point out that the rules of conduct published by professional bodies such as the Royal Institution of Chartered Surveyors (RICS) and the Chartered Institute of Building (CIOB) make express provision that all members or registered organisations hold professional indemnity insurance. Some, such as the RICS, also stipulate the level of cover, and the run-off period (length of protection after companies or individuals have stopped trading) for these policies.

Discussion point 3.11: What are the professional indemnity requirements for firms registered with the RICS (these are in the Rules of Conduct for Firms; see www.rics.org/us/regulation1/rules-of-conduct1/)?

3.3.6 Latent defects insurance

There is a common misconception amongst some in the construction industry that
responsibilities for latent defects are covered by express provisions of contracts.
As such they perceive the end of the defects liability period also signifies the end
of liability for buildings. In reality, the latter simply marks the stage when the contractual right to insist that the contractor rectifies defects ends. However, employers are still entitled to seek redress through the courts for damages due to breach
of contract or negligence. Given that most construction contracts are signed under
deed, the actual limitation period (period of responsibility) is 12 years. Whilst this
is often not seen as a significant issue, the risks this presents to contractors needs
to be evaluated.

For those contractors who wish to protect themselves against the risk of
claims during limitation periods, latent defects insurance (decennial insurance)
can be purchased. Typically lasting for 10 years from the original construction
of buildings, the insurance protects both the owners of buildings and contractors
against the cost of repairs to structures resulting from design or construction
defects.

3.3.7 Other insurance

Some contracts, such as ACA and PPC 2000 standard forms, make provision
for other, more obscure insurance policies that may need to be incorporated into
bids. For example, PPC 2000 identifies seven principal insurance requirements
including:

- Insurance of the works, goods, materials, equipment and existing structures
  (Cl.19.1);
- Public and employers liability insurance (Cl.19.1);
- Third-party liability insurance, covering injury/damage to any person and/or
  property (cl.19.3);
- Professional indemnity or product liability insurance (cl.19.4);
- Environmental risk insurance (cl.19.5);
- Latent defects insurance (cl.19.6);
- Whole project insurance (cl.19.7).

Some of these, such as professional indemnity insurance and all-risks insurance,
are similar to policies required by other standard forms. Others, such as whole

project insurance, are more specialist and have not yet been explored. This final section provides an overview of these more obscure policies.

Product liability insurance

This is very similar to professional indemnity insurance however it protects employers, contractors or subcontractors against liability for injury to people or damage to properties, resulting from products rather than services businesses provided. An example of this may be seen in relation to MEP (mechanical, electrical and plumbing) packages, where suppliers and installers of equipment such as lifts or escalators may be required to provide product liability insurance.

Environmental risk insurance

This insurance covers the cost of restoration after environmental accidents, such as pollution of land, water, air and biodiversity damage.

Third-party liability insurance

This insurance is very similar to public liability insurance, in that it aims to protect employers and contractors from third-party claims such as those received from members of the public. Unlike with public liability insurances, it is possible to purchase specific non-negligent third-party insurance policies such as policies covering potential damage to neighbouring buildings (Flanagan and Jewell, 2016). Construction works may involve working in close proximity to existing structures whilst undertaking for example piling, underpinning, excavating near existing foundations, works to listed buildings in a poor state of repair *et al*. The consequences of vibration that arises from such work may be difficult to judge. No matter how much care is taken, there is always the possibility that adjacent properties will suffer damage. In this situation no party has been negligent, but adjacent owners nevertheless suffer loss. Insurers will be mindful when setting premiums in difficult circumstances such as these of the possibility of claims being made.

Whole project insurance

Traditionally all participants in construction projects obtain their own insurance, or they purchase 'all-risks' insurance when stipulated in contracts, that cover two parties, to protect against risk of financial loss. In recent years, however, 'wrap-up' insurance programs have emerged as an alternative to traditional methods of risk management. In a wrap-up policy, project owners, in this case employers, purchase insurance policies that will cover all project participants throughout the supply chain.

The benefits of these policies have recently been championed in the UK government's *Government Construction Strategy (2011)* that made the case for reformed procurement practices that would affect behavioural and cultural change in the industry (Cabinet Office, 2011) whilst also seeking to drive further efficiency savings into capital projects undertaken in the public sector. As part of this policy, three new procurement routes were introduced; these new routes included one entitled *Integrated Project Insurance* which is loosely based on the whole project insurance option in the PPC 2000 standard form and the growing use of wrap-up insurance.

Exercise 3.1

Your employer is proposing to construct a new hotel complex. The value of the project is £8 million and the project duration is 15 months. Explain what the insurance provisions are within JCT Standard Building Contract 2011 for this project and advise the employer on which provisions they should use, including details of any appropriate financial limitations.

Exercise 3.2

In 2011 the government proposed a new procurement route called 'Integrated Project Insurance'. It is focused on the employer holding a competition to appoint an integrated project team to deliver the design and construction of the scheme. However, the novelty in this route is the use of a single insurance policy covering all the insurance policies that exist in other projects. The policy would protect the top slice of commercial risks covering any cost overruns in the project above and beyond the pain-share threshold set in the contract and apportioned between the employer, the contractor and its supply chain.

To what extent do you think this proposal will work? Would it remove the blame culture and adversarial nature of construction and make insurance claims and liability easier to proportion and resolve?

3.4 Collateral warranties in construction

Collateral warranties are neither an insurance policy nor are they a bond; they are a form of protection for employers (although they can be used by other stakeholders) should third parties involved in contracts, for example specialist subcontractors, fail to perform their contractual obligations.

A collateral warranty is an agreement that exists alongside and related to another contract, and its purpose is to create a contractual agreement between parties where one would not normally exist. To illustrate the point, consider a traditional lump sum procurement route based on a bills of quantities and using a JCT Standard Form of Contract. There are normally contractual links between employers, design teams, contractors and potentially building occupants. There are also contracts between main contractors and tier 2 subcontractors and so on down the supply chain as illustrated by the solid black lines in Figure 3.1. This

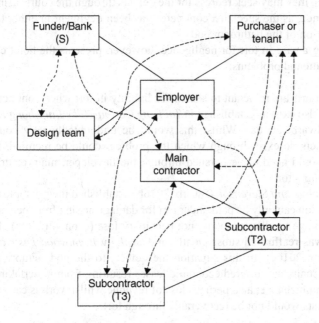

Figure 3.1 Contractual and collateral warranty links for a traditionally procured construction project.

situation would lend itself to the creation of additional protections for major parties, through the use of collateral warranties. Figure 3.1 illustrates potential collateral warranty links with dashed lines.

3.4.1 Need for collateral warranties

Returning to the hypothetical hotel project in chapter 1, and assuming that AZX Property Limited has reached a lease agreement with a national budget hotel chain, it is normal for the lease agreement to be something called an 'all repairing and insuring lease'. This transfers responsibility for maintenance, upkeep and insurance of the building to the tenant (the national hotel chain). So now consider a situation where, two years after taking possession, the budget hotel chain discovers that significant defects in the design and installation of pod bathrooms is causing them to fail. What action should the hotel chain take? Bear in mind that under the terms of the lease they hold responsibility for rectifying any defects that occur during the period of their tenancy.

Clearly the hotel chain would have to make the necessary corrective works, as that is their contractual obligation under the lease. However, as the defect has been caused by the specialist subcontractor who designed and installed the bathroom pods, the hotel chain would want to recover its costs. As they do not have a direct contractual link (that was between the subcontractor and the main contractor), they may seek redress for the defects through the courts using the tort of negligence, as the specialist contractor has been negligent in either the design or installation of the bathroom pods.

Pursuing a case in tort for negligence, however, presents the hotel chain with two fundamental problems:

- The courts are reluctant to search for liability in tort when contract relationships also exist as established in *Tai Hing Cotton Mill v Liu Chong Hing Bank Re* (Swarb, 2016a). Whilst this would be undoubtedly very complicated, contracts do exist through which the problem could be rectified. However, this would need to be pursued through the developer, main contractor and subcontractor.
- *Donoghue and Stevenson* (Swarb, 2016b) established the principle that compensation can only be paid under tort for damage arising from personal injury or damage to other property, not pecuniary loss (economic loss). This position was reaffirmed subsequently in *Murphy v Brentwood District Council* (Swarb, 2016c). In this situation the defects to the pods simply mean the hotel chain has suffered economic losses resulting from undertaking corrective maintenance; also perhaps loss of revenue whilst work is carried out. So the costs would not be recoverable through tort.

The obvious solution to this type of problem is to ensure adequate protections are put into place between the various parties involved in the project to protect themselves should a situation such as this arise. Had the hotel chain put a collateral warranty in place between itself and the various parties involved in the project, a direct contractual relationship would have been established between the hotel chain and the specialist bathroom pod subcontractor, allowing the hotel chain to take legal action to rectify the defect using the liabilities established in the collateral warranty. As a result, collateral warranties are now a mainstay of many construction projects. Standard forms of collateral warranty are available from organisations such as:

- The British Property Foundation (BPF);
- Royal Institute of British Architects (RIBA);
- Royal Institution of Chartered Surveyors (RICS);
- Joint Contracts Tribunal (JCT);
- Construction Industry Council (CIC).

In some cases, however, the need to impose collateral warranties throughout supply chains can be avoided if parties used the provisions in statute of the Contacts (Rights of Third Parties) Act 1999, that came into force in the UK on 11 May 2000. Section 7 of the JCT Standard Building Contract 2011 for example includes express provisions for the enforcement of third-party contractual rights, and these clauses give employers the choice of whether to use statutory provisions and create rights under the contract or obligate main contractors to enter into collateral warranties.

3.5 Pricing other risks

During the development of bids for projects, contractors will at some point need to make provision for risk. This will require estimators to determine:

- The extent of risk transfer to the contractor;
- Levels of risk sharing between employers and contractors;
- Levels of risk retained by employers.

From this analysis the estimating team will then need to evaluate and compute a risk allowance for projects. A risk allowance can be defined as

> *The amount added to the base cost estimate for items that cannot be precisely predicted to arrive at an allowance that reflects the potential risk.*
>
> (Flanagan and Jewell, 2016, p. 82)

In the case of bid development, these allowances could be computed as overall percentage additions that are applied to bids at the end of tender periods. In public projects, quantitative risk analyses (QRA) (see chapter 4) are used to generate risk allowances. This method focuses on an 80% confidence level of risk allowances not being exceeded. This is sometimes referred to as a P80 risk allowance. Implying that for 8 out of 10 projects, it is predicted calculated risk allowances will not be exceeded. Another method bid teams use to manage risks is to develop project risk registers, whereby risk events are identified during project team meetings, documented and their potential probability, impact and value established and included in bids. Risk allowances contractors include in their bids will consist of a combination of:

- Compensation for inaccuracies/errors in the estimating process;
- Cover for identified risks;
- Residual allowances decided upon by senior management teams based on their assessment of projects and needs of their businesses.

A comprehensive discussion of risk management is provided in chapter 4. It will explore risk management from the perspective of multiple stakeholders and provide explanations of risk management techniques such as those mentioned above. As part of the bidding process for complex projects, there will be a significant number of potential risk events that need to be considered. An overview of how estimators could develop a risk allowance for more common types of risk such as inclement weather and liquidated or delay damages is provided below.

Discussion point 3.12: What other types of common risk do you think contractors should make allowance for in their tenders?

3.5.1 Weather and climate risks

Weather and climate often have a considerable impact on construction projects, affecting not only time, but also productivity, safety, cost and quality. It is therefore particularly important during the tender phase of projects for estimators to consider the financial effects of both and make suitable provision within bids. Chan and Au (2008) make the point that whilst bids are developed based on tender documents received from the employer's consultants, weather and climate are not usually one of the bills of quantity items.

When considering the potential impact of weather events on construction projects, estimators need to consider both the weather that falls purely under the responsibility of the contractor (usually the normal weather conditions for the season and time of year). Also weather events under the form of contract will possibly lead to some level of risk sharing. For example clause 2.29 in the *JCT Standard Building Contract 2011* identifies a series of *relevant events* that are either employer risk events or shared risk events. Within the list, at sub-clause 2.29.9 the contract makes express provision for *exceptionally adverse weather conditions.* Consequently, the risk of time for certain types of weather activity is shared. Employers accept the time risk associated with weather by extending construction periods, whereas contractors accept the financial impact of the risk events' occurrence.

Discussion point 3.13: Assuming you are applying for an extension of time for exceptionally inclement weather, how would you prove the weather was exceptionally inclement?

Similarly, NEC3 makes provision for compensation events arising as a result of exceptional weather in clause 60.1(13) and 60.2. However, the clause provides a far more robust approach to weather analysis, based on a comparison of 1-in-10-year Met Office data at its regional weather stations. The weather station nearest to the project will be stipulated in contracts. Typical data is illustrated in Table 3.3. The data is calculated based on 30 years of monthly data, and the totals are tabulated from the most extreme to the lowest. The third most extreme value is then used as the 1-in-10-year value. It is this data that is used as a benchmark for compensation events in NEC3.

In addition to the 1-in-10-year data, used to analyse delays under the NEC3 Standard Form and therefore used by estimators to compute weather risk in the context of risk sharing, the Met Office also publishes long-term average data as illustrated in Table 3.4. This will allow estimators to gauge the levels of climatic and weather risk present on sites. From this data estimators can formulate a view on the sum to be added to works to allow for potential risk events. This process is illustrated in Figures 3.2 and 3.3, as a contractor is considering the risk allowance needed for brickwork in January, using both the Met Office data and a probability

Table 3.3 Met Office 1-in-10-year values (1971–2010) for Exeter, EX1 3PB

Month	Daily rainfall total (mm)	Days of rain > 5mm	Days of snow	Days with snow lying at 09:00 UTC	Days of freezing
January	149	11	5	4	2
February	137	10	5	3	1
March	100	7	3	1	0
April	93	7	2	0	0
May	104	7	0	0	0
June	97	7	0	0	0
July	79	6	0	0	0
August	102	7	0	0	0
September	114	8	0	0	0
October	141	10	0	0	0
November	135	10	1	0	0
December	136	11	2	2	0

Source: Met Office (N.D. b).

and impact assessment supported by expected monetary value (EMV) analysis (see chapter 4 for a full explanation of EMV analysis).

3.5.2 *Liquidated or delay damages risk*

Liquidated damages (JCT contracts) and 'delay damages' (NEC contracts) are also discussed in section 2.5. All construction contracts have fixed commencement and end dates; in law this type of contract is called a 'time is of the essence' agreement, as contractors are under pressure to achieve completion dates specified.

JCT (Clause 2.29 relevant events) and NEC (Clause 60.1 compensation events) provide mechanisms for adjusting dates for completion if employer-generated risk events (i.e. variations) or neutral risk events (i.e. exceptional weather) occur during construction periods. The contractor is often faced with the need to either complete the works on the scheduled completion date or pay a pre-determined sum of money to the employer to cover their loss, in the form of liquidated damages (Liquidated and ascertained damages).

Although this section of the book does not consider the full contractual and legal framework for the payment of these sums, it is important estimators make provision for possible risks to ensure contractors are protected from losses the enforcement of damage clauses would have. These risks can once again be considered using the EMV approach, as illustrated in Figure 3.4.

Exercise 3.3

Working as the estimator for a major contractor you have been tasked with evaluating the risk associated with the tight programme the employer has insisted upon

Table 3.4 Long-term averages (1981–2010) for Exeter, EX1 3PB

Month	Daily rainfall total (mm)	Days of rain > 5mm	Days of snow	Days with snow lying at 09:00 UTC	Days of freezing
January	85	6	2	1	0
February	68	5	3	1	0
March	61	4	1	0	0
April	57	4	0	0	0
May	59	4	0	0	0
June	50	3	0	0	0
July	46	3	0	0	0
August	55	3	0	0	0
September	59	4	0	0	0
October	88	6	0	0	0
November	88	6	0	0	0
December	94	6	2	1	0

Source: Met Office (N.D. a).

Time impact

Based on the data from the Met Office, January will have eight days of inclement weather due to rain or snow in a typical year. As a result, masonry work will be disrupted, causing delays and financial impact. It is assumed the bricklayers are 'cards in' employees of the company and no labour-only subcontractors. As a result the estimator needs to make an allowance for this risk event in the bid.

Time allowance: 8 days added to critical path on tender programme

Financial impact

Assuming the works consist of 380m^2 of facing brickwork, and the contractor's labour constant suggests a 2 + 1 gang will lay 800 bricks per day (800/59 bricks/m^2 = 13.5m^2). So let's say the work will take six weeks (30 days). However, based on the Met Office data, eight of those are likely to be non-productive due to weather. So the work will in actual fact take 38 days.

The contractors all-in rate is set at £28.00 per hour (without mark-up), so assuming eight days will be lost to weather and the contractor needs to pay non-productive time, and encounter additional preliminary costs:

Base estimate:

- (Masonry £45.75/m^2 x 380m^2)

£17,385

Risk allowance:

- Non-productive time (8 days x 8 hours x £28.00.hr) = £1,792
- Prelims allowance for extra programme time (say £300/day) = £2,400

£4,192

Figure 3.2 Risk allowance for weather – Method 1.

Risk allowance for Weather
Method 2 – Scientific approach using weather data
Probability

Seeking out weather data is both time-consuming and expensive, so the estimator may instead evaluate the risk using their professional judgement based on past projects. This resulted in the following probabilities:

- 20% chance of a 4-day delay due to weather
- 30% chance of an 8-day delay due to weather
- 30% chance of a 12-day delay due to weather
- 20% chance of a 16-day delay due to weather

Impact

The impact of the risk event is computed using Expected Monetary Value (EMV) as follows:

Outcome (i)	Probability $\{P(X_i)\}$	Time impact	$X_iP(X_i)$	Financial impact	$X_iP(X_i)$
1	0.20	4	0.8	£1,136	£227.20
2	0.30	8	2.4	£2,272	£681.60
3	0.30	12	3.6	£3,408	£1,022.40
4	0.20	16	3.2	£4,544	£908.80
		Time E(X)	**10 days**	**Financial E(X)**	**£2,840**

It is assumed the financial impact will be £224/day (8 hours x £28.00) for non-productive labour and £60 per day for prelim costs (supervision, etc.).

As a result of the Expected Monetary Value Analysis, the following outcomes are recorded.

Base estimate:

- (Masonry £45.75/m² x 380m²)

£17,385
Risk allowance:

- Non-productive time
- Prelims allowance for extra programme time

£2,840

Figure 3.3 Risk allowance for weather – Method 2.

Rate of LDs stipulated in the contract:

• £1,000 per week or part thereof

Probability of delay

The contractor's team will need to prepare a tender programme for the project to evaluate the anticipated period for completion, considering the full scope of project risk. From this analysis let's assume the following delay due to contractor owned risk:

• 10% chance of 1-week delay
• 30% chance of 2-week delay
• 50% chance of 1-month delay
• 10% chance of 6-week delay

Impact

The impact of the risk event is computed using Expected Monetary Value (EMV) as follows:

Outcome (i)	Probability $P(X_i)$	LDs (£)	$X_iP(X_i)$
1	0.10	£1,000	£100
2	0.30	£2,000	£600
3	0.50	£4,000	£2,000
4	0.10	£6,000	£600
Financial E(X)			**£3,300**

As a result of the analysis it is suggested the contractor allows for £3,300 of risk monies within its bid for the project.

Figure 3.4 Risk allowance for liquidated or delay damages.

in the tender documents. Using the following data determine the risk allowance to be included in the bid.

• LDs are set at £3,450 per week or part thereof
• 20% chance of 1-week delay
• 24% chance of 2-week delay
• 18% chance of 1-month delay
• 22% chance of 6-week delay
• 16% chance of 2-month delay

3.6 Contingency sums

Contingency sums are monies placed in tenders, with the hope from the perspective of employers they will not be spent. They will be used only if necessary, on items that arise that are 'unforeseen'. Contingencies can be a political football. Some employers, particularly public bodies who are accountable to taxpayers, do not like to be seen including contingencies. They may fear that if contingencies are available to professional design teams and contractors, they will spend them. If extra costs are on the horizon, with contingencies in place, design teams or contractors may just say 'no problem, full steam ahead, we have contingencies to pay for that'. Some contractors may fabricate claims, mindful that employers have contingencies to fund them. Alternatively, if there are no contingencies, construction teams would rightly have to work hard to minimise costs of unforeseen items that may arise, and also work hard to secure savings from other elements of projects to pay for them. However, unforeseen events will arise in construction work; high-risk areas are often groundwork, weather and variations, and therefore contingencies do have their place.

In both the private and public sector, headline figures are tender sums. Approval will be gained from boards of directors or committees based on tender sums. On completion of projects, final amounts paid (final accounts) will be compared to tender sums. If final accounts are lower than tender sums, 'everybody is happy'; however if they are higher, then 'heads may roll'. Design teams will clearly be in a more comfortable position if they have contingencies.

Two scenarios illustrate this point, thus:

(i) Bid for work described £100,000

  10% contingency   £10,000
  Total bid   £110,000
  Total approved by the Board or Committee: £110k
  Final account £105k; £5k saving
  Outcome: 'everybody is happy'.

(ii) Bid for work described £100,000

  0% contingency   nil
  Total bid   £100,000
  Total approved by the Board or Committee: £100k
  Final account £105k; £5k overspend
  Outcome: 'heads may roll'.

If it is acceptable for employers to include contingencies within contract sums, it is transparent in documents and to all parties, thus:

  'Allow the contingency sum of £100,000'.

There is a case of 'who controls the contingency'. Whilst contingencies are always 'owned' by employers, accountants within employer organisations may prefer to retain control by seeking approval from boards of a tender sum without a contingency sum, and then having separate approval documents for contingencies not known to professional teams. If unforeseen events do arise, professional teams have to work hard to make requests for additional funding. Employer structures may be such that project teams can authorise expenditure up to certain values, perhaps £50k. Executives can authorise up to perhaps £100k, and over that amount board approval is required.

Discussion point 3.14: It is the policy of an employer not to include contingencies in projects. A design team knows that whilst there are some uncertainties in the ground, there are many risks that the employer should best retain in the design of superstructure elements. How may the design team ensure it has some monies somewhere in the tender documentation to pay for unforeseens?

3.7 Provisional sums

For JCT contracts, provisional sums are defined in New Rules of Measurement 2 (NRM2, 2010, pp. 12, 21 and 25). Provisional sums are different to contingency sums; whilst it is hoped contingency sums will not be spent, the intention is that provisional sums will be. Provisional sums are money allocated for elements of work that are known to be required, but at the time of putting tender documents together, sufficient information is not available to allow bidders for projects to estimate likely cost. For example, an employer may know that it will require direction or location signs around a complex. These can likely be fixed on-site towards the end of the project, without affecting other work. The employer has still to make some key decisions about its own future organisational structure that may influence the number and type of signs required. These decisions are not urgent and may depend on potential market changes ahead. However, signs will be required, and it is desirable to ensure a sum of money is in the tender bid, so that approval for extra sums are not required at a later date. Tender documentation will read 'Allow the provisional sum of £x for directional signs'.

A provisional sum may be 'defined' or 'undefined'. For a defined provisional sum contractors are deemed to include in their programmes and preliminary elements of bids for executing work. Clearly design teams must 'define' using an appropriate narrative the work to be carried out so that contractors may make allowances.

An undefined provisional sum is used where it is not possible to define the work; it follows that contractors cannot include in their bids for something that is not defined. When contractors are given details of the work, they may include in any prices they negotiate, amounts of money, if appropriate, for consequences on programmes and preliminaries.

From the perspective of employers, perhaps provisional sums of high value are not good, particularly if projects are awarded on a competitive basis. The market may determine that prices obtained whilst in competition are lower (or better value) than those negotiated after main contract awards.

3.8 Prime cost or PC sums; nominated subcontractors and suppliers

3.8.1 Reasons for nomination

Prime cost (PC) sums are for works to be executed by nominated subcontractors or for materials to be provided by nominated suppliers. Nomination occurs when employers (perhaps on the advice of design teams) wish to ensure an element of the work is executed by, or materials supplied, by specific companies. Employers may prefer to have control over who is appointed, with important criteria being company reputation and strength of balance sheet. If contractors have responsibility for selection of specialists, the most important criteria may be lowest price, potentially leading to inappropriate appointments and problems later in the life of buildings. Therefore, employers through their design teams instruct main contractors to place orders with specialists of their choosing. Nomination was very popular in the 1970s to 1990s, but it has more recently fallen out of favour.

Nomination may occur in cases where:

1   Employers and design teams may need to negotiate with specialists before contractors are appointed. Agreements may need to be made with specialists who need to start their work at early stages in construction programmes and have a long lead-in time for reserving equipment or manufacturing materials e.g. piling (may be six months to reserve heavy piling rigs) or structural steelwork (may be six months to prepare working drawings and manufacture steel).
2   Employers wish to retain control for trades that contain some technical complexity or sensitivity, with potential for long-standing problems during use and maintenance of buildings if things go wrong e.g. mechanical, electrical, lift installations. These trades are fundamentally important to the successful operation of buildings over a period of decades.
3   Employers have long-standing relationships with specialists that give high confidence in the quality of product and service provided.
4   The knowledge of specialists is required in the development of specifications for products; an example of this, perhaps surprisingly, is door and window ironmongery suppliers where the range of products is wide and highly bespoke to individual companies. Given the complexity involved in the security and fire protection of doors and windows, and the multiplicity of situations that can arise, it would be unreasonable to expect non-specialist designers to

specify with accuracy correct products of suppliers. Designers may therefore require that ironmongery specialists commit time at a relatively early stage of projects to develop ironmongery schedules; it would seem unreasonable they would do this work without an assurance they would receive orders.

A key problem with nominated subcontractors and suppliers is that if they do not perform as would be reasonably expected by contractors, and they delay projects, employers become responsible. That may mean that contractors may ask for an extension of time with associated loss and expense. The costs involved may be extremely high, and it does not seem commercially sensible that with three parties involved – employer 'A', contractor 'B' and specialist 'C' – party C does not perform as reasonably expected, and party A pays party B for the consequences. The ability of party A to recover its costs from party C may be limited. Further party B is not incentivised to assertively and pro-actively pressurise party C to perform, particularly if it is the case that delays are occurring elsewhere on projects; party B may claim for an extension of time as a result of party C failure even though the project would have still been delayed by a failure of party B in another element of the project. Consequently, the ability to nominate was dropped from new JCT standard forms developed in 2005 and 2011. However, employers and consultants who still wish to nominate may simply amend their forms of contract to insert JCT 1998 or similar relevant clauses.

Subcontractors who are selected by contractors are known as 'domestic subcontractors'. Contractors may enter into agreements with domestic subcontractors using standard NEC or JCT forms, though it is often the case that contractors will invite subcontractors to sign up to sub-contracts that are bespoke or written by contractors' legal teams. In such circumstances, subcontractors need to be careful they do not agree to conditions which may unwittingly increase their risks.

Discussion point 3.15: A supermarket employer has an excellent long-standing supply and maintenance relationship with a specialist manufacturer of food refrigeration equipment. It has a new project which is put out to tender to six main contractors. How can the employer ensure its favoured specialist secures the work without using nomination provisions?

3.8.2 Including a PC sum

When nomination occurs, design teams will have been involved in negotiations with nominated subcontractors or suppliers, without the involvement of contractors. Those negotiations may have been around design and specification, and also by necessity, about price. There may have been just one subcontractor or supplier involved, in which case price will be negotiated, or there may have been more than one, and a price invited and agreed in competitive situations. Negotiations may be taking place whilst contractors are separately bidding for projects. To support contractors in compiling their tenders, if bills of quantities are used, an allowance for PC sums is presented thus:

Allow the PC sum of £100,000.00 for lift installation work by a specialist to be nominated by the employer inclusive of 2.5% main contractor's discount.

When design teams come to an agreement with nominated subcontractors or suppliers, instructions will be issued to contractors thus:

Omit:
PC sum of £100,000.00 for specialist lift installation
Add:
Place an order with XYZ Ltd for the installation of lifts in the value of £98,500.00 inclusive of contractor's discount of 2.5%.

Section 3.14 details how the discount is handled.

3.9 Daywork in tender bids

Daywork, termed the *price cost plus percentage addition*, is something of a 'marmite' concept for those working construction. Daywork is often favoured by contractors and loathed by employers and their professional advisors because invariably it will be far more expensive than measure and value alternatives. As a result, daywork is seen as a valuation method of last resort in most standard forms of contract, and when all other approaches have been ruled not possible. This is usually because the work cannot be quantified for whatever reason. During the course of construction projects, daywork items can arise surprisingly frequently, especially on refurbishment contracts, where it is almost impossible to anticipate every possible eventuality. As a result, when additional work needs to be executed contractors can be paid on the basis of the cost of materials, plant and labour, plus a percentage for overheads and profit.

To ensure daywork items are used and applied in an appropriate manner, the Royal Institution of Chartered Surveyors (RICS) and the Construction Confederation jointly publish detailed guidance on the use of daywork. The latest version is entitled *Definition of Prime Cost of Daywork Carried out under a Building Contract* (RICS, 2007). The guidance provides a definition and series of rules that determine what may or may not be claimed within a daywork application. A summary of the key rules is provided below.

3.9.1 Prime cost of labour

The time that can be included daywork sheets is the time spent by operatives directly engaged on daywork, including those operating mechanical plant and those delivering or erecting and dismantling other plant that comes within the scope of the works covered by the daywork application. However, no additional monies will be paid if the operatives are working overtime (unless this is explicitly requested by the contract administrator). Contractors are unable to claim for supervisory staff such as managers, foreman, 'non-working' gangers *et al.*

The prime cost of labour can be determined in two different ways under the *Definition of Prime Cost of Daywork Carried out under a Building Contract.* These are:

A   Option A – a percentage addition to agreed hourly rates, which allows for incidental costs, overheads and profit, to the prime cost of labour applicable at the time daywork is carried out.
B   Option B – an all-inclusive rate that includes not only the prime cost of labour, but also provides an allowance for incidental costs, overheads and profit. All-inclusive rates are deemed to be fixed for the period of the contract. However, where fluctuating price contracts are used, or where the rates in contracts are to be index-linked, the all-inclusive rates shall be adjusted by a suitable index in accordance with contract conditions.

For most contracts, construction professionals have continued to use option A. Using this method, the rates used for labour claimed under daywork are the standard hourly rates based on industry agreed Working Rule Agreements. Daywork rates do not take account of actual rates contractors may decide to pay staff due to local labour supply problems for instance. The current prime cost of daywork labour rates for the various operatives engaged on construction projects is shown in Table 3.5, although these are only rates for building operatives. Electricians, plumbers and heating engineers have their own daywork rates.

Table 3.6 illustrates how the rate of £10.67 for a general operative is established using the definition referred to earlier. It should be noted that this calculation is not the same as the process shown in chapter 1 which illustrates how the all-in rate for the same operative would be established. As you can see from Table 3.6, the prime cost of labour does not make any allowance for other labour costs such as:

- Travel;
- Accommodation.

Table 3.5 Prime cost of labour rates for building operatives

Operative skill level	Prime cost of labour
Building Craft Operative	£14.28
Building Skill level 1	£13.59
Building Skill level 2	£13.07
Building Skill level 3	£12.21
Building Skill level 4	£11.52
Building General Operative	£10.67

Source: BCIS (2016).

Table 3.6 Prime cost of labour rate calculation

	Quantity (hours/days)	Rate (£)	Total (£)
Hourly base rate (£19,220.50/1,802hrs)			10.67
Working hours per year (2,028 hours – 226 hours)	1,802.00 hours		
Standard hours (52wks * 39hrs)	2,028.00 hours		
Annual Holidays	163.00 hours		
Public Holidays	63.00 hours		
Annual labour costs			19,220.50
Sub-total			**17,276.86**
*Basic wages (46.2wks * £332.28)*			15,351.34
*Holidays with pay (226 hrs * £8.52)*			1,925.52
Extra payments			
Annual NI total *(£24.3266 * 52wks)*			**1,264.98**
Welfare benefit *(52wks * £11.39)*			592.28
CITB levy *(0.5% of £17,276.86)*			86.38
Calculations			
National insurance contributions on £332.28 per week			24.33
Band A – Up to £111.99	0.00%		0.00
Band B – £112 to £156	0.00%		0.00
Band C – £156.01 to £827.00	13.80%		24.33
Weekly wage (£8.52 * 39hrs)		332.28	
Working weeks per year (1,802hrs/39hours)		46.20wks	

Source: BCIS (2016).

- Non-employment costs such as:

  - Administration;
  - Supervision;
  - Materials, *et al.*

3.9.2 Prime cost of materials

Prime cost of materials is the cost of materials including any delivery charges paid to contractors at prime cost; that is the invoiced cost after trade discounts have been deducted. Any other discounts such as a prompt payment discount for example will not be deducted. However, this is capped at 5%, so if a 10% prompt payment discount is included in invoices, contractors will only be entitled to retain 5%; the other 5% will be deducted using the same approach applied to trade discounts. In the event contractors use materials left over from other projects, these can be charged under daywork, and contractors will be entitled to receive the current market value of the materials, not the invoiced cost (whether that is higher or lower). All payments are exclusive of VAT, whether or not that is identified on invoices.

3.9.3 Prime cost of plant

Plant is generally charged at the rates provided for in the contract documents. Most tender documents make allowance for the use of the *Schedule of Basic Plant Charges* published by the RICS.

3.9.4 Pricing daywork in the tender bid

Contractors bidding for projects have the opportunity to include an additional sum, known as the *percentage addition*, to allow them the opportunity to recover items of expenditure not covered in the prime cost of daywork such as labour supervision, administration *et al.* Provision is made in a section of bidding documents, called the 'schedule of daywork'; an example of this is shown in Figure 3.5. By including daywork in tender sums, employers are trying to discourage contractors from applying excessive percentage additions to prime sums, thereby reducing their financial exposure should daywork be required on projects.

For this reason contractors must make a commercial decision when pricing daywork percentage additions. Keeping percentage additions low will reduce tender sums. In the event that award decisions are driven solely by cost, this could make the difference between winning and losing. However, if estimators judge there will be an opportunity for extensive daywork items post-contract, increasing percentage additions could improve profitability. In the example shown Figure 3.5, it can be seen that the bidding contractor has decided to add a percentage addition of 120% to the prime cost of labour, 20% to the prime cost of materials and plant. Chapter 6 provides a further narrative about the administration of dayworks post-contract.

3.10 Fixed/firm price or fluctuation

During the construction phase of projects, costs predicted at tender stage will often rise, though occasionally, for some elements or trades they may decrease. When inviting bids, employers need to stipulate which party will take the risk of cost increases or decreases.

If employers prefer that contractors take the risk, employers will specify that a firm or fixed price is required, and contractors and their supply chains will need to make predictions of likely future costs for their portions of the works. If for example there is a prediction of a 3% increase over one year, and that work is to be executed over a full year, a contractor may add to its bid the average price increase of 1.5%. If however, the 3% predicted increase is for one element of work that it is known will be completed in one year, perhaps a subcontract finishing trade, that part of the bid will be increased by the full 3%. When inviting bids at tender

Dayworks

For the valuation of variations relating to the execution
of additional or substituted work which cannot be
properly be valued for measurement, the contractor
will be allowed to charge daywork in accordance
with the provisions of clause 5.7 of the conditions of
contract.

The basis of charging will therefore be the Prime Cost
of such calculated in accordance with the 'Definition
of Prime Cost of Daywork carried out under a
Building Contract' which was current at the Base
Date, together with percentage additions to each
section of the prime cost at the rates set out below by
the contractor.

Provide the following Provisional Sums and add thereto
the percentage addition required for the incidental
costs, overhead and profit.

For the Prime Cost of:

A	Labour		£2,500.00
	Add Percentage addition required	120%	£3,000.00
B	Materials		£1,000.00
	Add Percentage addition required	20%	£200.00
C	Plant		£1,000.00
	Add Percentage addition required	20%	£200.00

Note: In pricing the percentage addition required in
respect of Plant the contractor should note that the
rates allowed for the individual items of plant will be
either:

a.   a.   The rates contained in the 'Schedule of Basic
Plant Charges' for use in connection with Daywork
under a Building contract issued by the RICS and
which was current at Base Date; or

b.   b.   In those cases where, for any reason the latter
Schedule is not applicable at rates current at the
time the work is executed.

Carried to Collection

Figure 3.5 Example of a bill of quantities daywork section.

Source: Ramus *et al.* (2011, p. 129).

stage, if employers ask contractors for a firm or fixed price bid, contractors will ask their suppliers and subcontractors similarly. Contractors may stipulate their bids are open for acceptance for a limited period, perhaps three months, and any award after that time will be subject to increased prices. Predicting price movements in low inflationary environments is less risky than in high inflationary environments. Also, predicting price movements for short duration projects is less risky than for long duration projects over many years or even decades. Table 3.7 illustrates building cost movements by indices over the five-year period 1978 to

Table 3.7 Actual and forecast cost indices with percentage annual increase; P = provisional, F = forecast

Year ending december	Index	Percentage annual increase
1978	52.9	
1979	62.9	18.9
1980	73.5	16.9
1981	79.8	8.6
1982	85.7	7.4
2014	318.5	
2015	319.3P	0.3
2016	325.1F	1.8
2017	336.7F	3.6
2018	349.6F	3.8
2019	362.9F	3.8
2020	377.4F	4.0

Source: Building Cost Information Services (BCIS, 2016) Cost Indices.

1982 and forecast cost movements for the period 2014 to 2020 (BCIS, 2016), with percentage annual increases. Percentage increases are calculated arithmetically by expressing the increase in the index for a year as a percentage of the previous year e.g. 1978 to 1979: 62.9 minus 52.9 = 10.0; 10 divided by 52.9 multiplied by 100 = 18.9%.

It can be seen that in the late 1970s and early 1980s, forecasting costs just one year ahead was very difficult. In 1980, the two previous years had seen cost increases close to 20%, what lay ahead in 1981? Whilst there may have been forecasts in 1980 for lower inflation in 1981, it was unlikely that with such volatility the actual out-turn figure of 8.6% was accurate. In 1980, a contractor may have won a bid, forecasting increased costs based on an average of the two preceding years. That would have been 17.9%, and the contractor would therefore have been paid 9.3% (17.9 minus 8.6) more than it needed. Excellent for the contractor, but not good for the employer. Table 3.7 indicates a forecast cost increase of 4.0% in 2020; even the best forecasters would acknowledge that that could have a range of ±1%, which on large projects is a significant amount of money.

Another stark illustration of the consequences of incorrectly predicting inflation is the period in the UK 2011 to 2015, when in these latter years many large contractors were reporting significant losses in their annual reports. One reason frequently offered was that in 2011/12 they had signed agreements with employers for long-term projects to be constructed in the period 2013/15 on a firm or fixed price basis. In their bids, they made forecasts of likely cost increases; according to BCIS (2016) Tender Price Indices increased by 1.4 and 0.4% in 2011 and 2012, respectively, and forecasts for 2013/15 may have been in this range. However, there was a resurgence in market confidence in 2013/15. Contractors' supply chains increased their prices significantly as illustrated by increases in Tender Price Indices (TPIs) of

8.8%, 5.6% and 9.3% for 2012, 2013 and 2014, respectively. In the years 2012/15 therefore, many contractors were building at 2011/12 prices, plus perhaps only circa 1.4% or 0.4% for inflation, when actual prices were as much as 9.3% higher. To put this in perspective, a benchmark profit figure that contractors seek on turnover is 2.5%; many companies were therefore constructing at a loss.

Even in lower inflationary environments, it is the case that forecasts of cost movements can be the sole criteria in determining which contractor wins a project. In Table 3.8, contractor 'A' has worked particularly hard on a bid; it has thought carefully about how to best construct the project, using some innovative construction techniques. There is a detailed programme, method statements, health and safety analyses and environmental risk assessments. If contractor 'A' wins the project, its tender stage planning provides a good platform from which to start construction. Contractor 'A' has spent a lot of time and money on the bid, since it wanted to win it. Contractor 'A' adds 5% to its tender to cover its prediction of likely cost increases. Contractor 'B' has not planned the project in such a detailed way at tender stage; its technical solutions are not so innovative, and if it wins the project it is not so well positioned for a good start to construction. Contractor 'B' has not spent so much time and money on the bid. Contractor 'B' adds 2% to its tender to cover its prediction of likely cost increases. If selection is based on cost alone, Contractors 'B' wins the project. Clearly, this is not desirable. Contractors will argue they are in business to make profits through efficiency and effectiveness *et al.* in their construction practice; not by speculating on rates of inflation or deflation. Nor indeed as discussion point 2.2 in section 2.6 illustrates, do contractors seek to make money by speculating on international currency exchange rates.

As an alternative to firm or fixed prices, employers may specify that they will take the risk of inflation or deflation. Employers may not want to pay a premium in the case that contractors over-estimate inflation, and reputable employers will not want to see contractors suffer losses because of under-estimates, and potentially therefore be at risk of liquidation; also repeating from section 2.8 'projects run well in the construction phase, if contractors are making money'. Employers taking the risk of inflation is particularly relevant in the public sector, where there are very large projects that run over many years, making forecasting, particularly in volatile markets, difficult. To take the risk of inflation, in NEC contracts employers use the 'price adjustment for inflation' in its option X1; in JCT

Table 3.8 Illustration of how judgements of inflation forecasts can win or lose tenders

	'A'	'B'
Estimated cost of construction based on prices prevailing at the date of tender	£1,000,000	£1,025,000
Addition for forecast of likely cost increases ('A' = 5%; 'B' = 2%)	£50,000	£20,500
Tender bid	£1,050,000	£1,045,000

contracts employers specify in schedule 7, three 'fluctuation options', A, B and C thus:

JCT option 'A contributions, levy and tax' allows for increases due to government changes in legislation, that were not foreseeable at tender stage, such as training levies or employers national insurance contributions. For example, as at 2015/16, the employer's national insurance contribution is set at 13.80% (Gov.uk, 2016). If government were to decrease or increase this figure, appropriate adjustments would be made.

JCT option 'B labour and materials cost and tax fluctuations' permits that contractors claim increases, or reimburse if there are decreases, for labour and materials based upon the submission of detailed schedules. . . . These schedules detail differences between unit costs at the date of tender, and actual costs at the date of construction. For example, if the hourly rate for a bricklayer were £11.50 at the date of tender, and that rate increases to £11.73 at time of construction (2% increase), the contractor would claim £0.23 for each bricklayer hour worked. Also, if the rate for concrete per m^3 at date of tender was £97.80, and the rate at the time it was poured was £99.80 per m^3, the contractor would claim £2.00 per m^3. Using option 'B' would necessitate that in tenders, contractors include a list of material and labour unit prices that have been used in building-up prices. JCT option 'B' can be time-consuming to administer.

JCT option 'C' and NEC usually operate on the Price Adjustment Formula Indices (PAFI) system (RICS, 2016a), which can be administered quickly using appropriate software. JCT contracts use the 'Building Series 4' indices formerly known as NEDO after the disbanded National Economic Development Organisation or 'Osborne indices', after the chair of the committee who developed them. NEC contracts usually specify the 'Civil Engineering Series' sometimes called the 'Baxter indices', again after the development committee chair. Also available separately are indices for two further disciplines; specialist engineering and highways maintenance. The principles are that when bidding, contractors and their supply chains price projects at prices current at the 'base date'; that base date will usually be the month of the date of tender, at which point there are published (or to be published) indices. At each point during a project when contractors are paid, either monthly or by stage payments, calculations are performed for each trade to determine increases or decreases in price. Table 3.9 illustrates a sample calculation for the trade 'concrete precast' using the Building Indices. Similar principles in calculation methodologies apply for civil engineering, specialist engineering and highways maintenance work. The indices themselves, available on-line by subscription from the RICS, are based upon a basket of construction prices within each individual trade, in much the same way as the government calculates consumer price indices (CPIs) and retail price indices (RPIs). The RICS

Table 3.9 Sample calculation for fluctuations. price adjustment formulae indices (Building) Series 3, 3/09 concrete: precast. indices copyright of the RICS and reproduced with permission.

Valuation Date number		Column C. Amount of valuation £	Column D. Index at month of valuation	Column E. Percentage increase between base date and month of valuation. $E3 = D3 - 210/210 * 100$	Column F. Amount due including fluctuations. $F3 = E3/100 + (1) * C3$
1	December 2015	11,525	213	1.429	11,689.69
2	January 2016	18,352	213	1.429	18,614.25
3	February 2016	27,751	213	1.429	28,147.56
4	March 2016	6,222	213	1.429	6,310.91
	Totals	**63,850.00**			**64,762.41**

(2016b) publication 'PAFI List of Contents' details trade categories of both civil engineering and building.

3.11 Subcontractors

Subcontractor costs can be the major element of contractors' bids. Indeed, as chapter 6 explains, some main contractors do not employ any direct labour, nor indeed purchase materials. Therefore a contractor's bid may comprise in its entirety only subcontractors, preliminary costs, overheads and profit. Some contractors employ supply chain managers, who are tasked to support lower tiers to achieve optimum efficiency and effectiveness, and thus optimum prices. Contractors who are able to manage supply chains well are likely to be more competitive in their bids to employers. Other contractors may not nurture supply chains so well.

3.11.1 'Subby-bashing'

Some contractors may take a firm approach in the management of subcontractors; sometimes called colloquially 'subby-bashing'. In such cases extreme pressure can be placed upon subcontractors to submit lower prices. One tool to 'bash subbies' is the 'dutch auction', whereby prices go down, rather than in conventional auctions, where prices go up. Dutch auctions are frowned upon by many captains of industry, but they seem to be just a fact of business life that will not go away. In the housing market vendors (sellers) may decide they value a property at £90k, and decide to market it at £100k – that is a rate not too high to frighten away potential viewings. A sale is agreed at £95k. The vendors are happy since they receive £5k more than the value, and the buyer is happy since there is a perception of a 5% discount.

In construction, main contractors may enter into dutch auctions with subcontractors. Consider the example where a contractor is one of six main contractors bidding for a project. It obtains four quotations from plastering subcontractors, thus (rounded figures used for ease):

Subcontractor A	£50,000
Subcontractor B	£52,000
Subcontractor C	£54,000
Subcontractor D	£56,000

All less 2.5% main contractor's discount.

The main contractor, in its bid to the employer, will include the sum of £50k less 2.5% for plastering. The contractor will do similarly for all other subcontract trades e.g. groundworks, roofing, floor finishes.

If the main contractor wins the project from the employer, on the one hand it may simply place an order with the subcontract plasterer at £50k less 2.5%. More often, though, the main contractor may go back to the subcontractors and say something like 'when you gave me your first bid, that was only a "bid for a bid"; now as a main contractor I have won the project, and I would now like your "bid for an order" – now give me your best bid'. Subcontractors may now bid thus:

Subcontractor A	£49,000
Subcontractor B	£49,500
Subcontractor C	£50,000
Subcontractor D	£48,000

In protracted negotiations, a main contractor may find a new Subcontractor E and say something like 'I have a lowest bid of £47k (note it does not – the lowest is £48k). If you bid £46k, you can have the order'.

If the order is placed at £46k, from a main contractor's point of view this becomes a valuable £4k profit that it did not anticipate making at tender stage; a 'net gain'. However, if markets are extremely keen, and main contractors need to win work just to secure turnover, they may in this example take a risk when bidding, and knowing that it is often possible to negotiate subcontractors down, include only £46k or a similar figure in their bid. It is possible that the main contractor would make a loss if it could only negotiate a subcontractor down to £47k.

It is also important to be mindful that risks can turn the other way for main contractors. A contractor may receive bids from subcontractors, and submit its own bid to an employer, for a large project in January. The main contractor wins the project, and signs a contract and starts work in February. In July it looks to place an order for plastering works that are scheduled to commence in October. The market is booming; prices are rising. Subcontractor A reports back to the main contractor that whilst its price in January was £50k, now its price is £55k. The main contractor may be forced to take a loss of £5k.

Arguably, dutch auctions add cost to the process. There is extensive time spent by many parties re-bidding. If contractors are able to secure lower prices than in their tender, it must be remembered in most forms of procurement, employers pay the higher prices. High prices from contractors are a deterrent for employers to procure construction work.

'Subby-bashing' may continue after works starts on-site. To secure work, subcontractors may feel obliged to sign conditions of contract that are bespoke to contractors, rather than standard NEC or JCT forms. These contracts may impose more erroneous conditions on subcontractors than are placed on main contractors by employers, and they may have financial consequences. As work proceeds, there may be instances of unreasonable set-off, as described in chapter 2.11.1. Additionally, sub contractors may not receive payment on time. When it comes to agreeing final accounts, subcontractors may be pressured to accept a lower figure than they reasonably ought. For example, if a subcontractor has been paid £90k and the projected final account is £100k, a main contractor may offer to pay £98k totally with a cheque of £8k now, or the subcontractor be forced to haggle and wait for a long period before a contractor's assessment is made about whether £100k is fair. Included as part of the final account negotiations is retention. Perhaps this subcontractor will take £97.5k if retention is released immediately? Some subcontractors may 'never get round' to negotiating a final account, and the payment made on the last payment certificate is effectively the subcontract final account; half retention that should be released on project completion, and the other half after defects rectification is 'forgotten', and retained permanently by the main contractor. The final 'card' that contractors may play is that if subcontractors complain, or becomes contractual, they will not secure any more work. However, contractors who 'subby-bash' may need to be careful, mindful of the idiom 'every dog has its day'. When markets are tough, it may be possible to 'bash', but when markets are buoyant, subcontractors may decline work, or only work on projects with higher profitability. There are many cases in the public domain where contractors complain they have been 'bashed by subbies' into accepting higher prices than contractors have in their bids to employers.

3.11.2 *Working with subcontractors*

The alternative to 'subbie-bashing' is to treat subcontractors as partners, and to nurture improvements for all. Long-term trusting relationships can bring about technical and process improvements that support shorter project durations and lower costs. These lower costs can be real savings, and permit all in supply chains to make better profits that can be re-invested into seeking more technical and process improvements. Framework agreements between employers and contractors provide a strong basis for learning and improvements. Ideally, subcontractors should be part of these frameworks, so they are able to contribute. It does not have to be employer procurement systems that drive long-term relationships between contractors and lower tiers. It can also be part of more traditional competitive tendering systems. Contractors may deal with just one or two specialists at the stage that contractors are bidding, similar to the collaborative procurement process described in section 2.12. If few parties are able to discuss innovative ways of approaching projects, one or two bidders may be able to submit lower prices than if contractors were to invite perhaps six bidders. As noted in section 2.12, specialists bidding as one of six may not be prepared to spend too much resource seeking lower-cost innovative solutions with statistically a one-in-six chance of success; more likely the contractor will get the 'standard' higher bid. In return for long-term framework orders, the higher tier may quite reasonably demand lower prices as time progresses, brought about by improvements. Supply chain management requires that contractors get close to their supply chain, by regular discussions, visits to each other's premises and visits to manufacturing facilities. Contractors may also wish to meet and integrate with sub-subcontractors and suppliers. On large projects, integration may be served by sharing of temporary office accommodation.

3.11.3 Obtaining subcontractor and supplier prices at tender stage

The first stage in preparing a bid is to visit site. Contractors have standard check sheets that they complete to detail such things as site topography, access, existing underground and overhead utilities, adjacent water courses, railway lines, proximity of the public *et al.* Existing manholes within and without the curtledge of the site should be located. Site photographs can be useful. Judgements can be made about the potential for temporary site layouts for cabins, roads, cranes, materials storage, car parking and fencing. Details about all boundaries should be obtained, and assessments made about the risks of construction work causing damage to boundaries and perhaps to adjacent footpaths or roads. Judgements also need to be made about local markets, availability of materials and subcontractors, and likely risks that could accrue from security aspects.

Bidding contractors then need to assess project documentation, and if a bills of quantities is provided, allocate items in the bills to trade specialists. On some occasions whole pages of a bill may be for one specialist, but on other pages, there may be items that should be allocated to two or three different trades. Contractors will also identify those trades, if any, that it may wish to be responsible for itself; perhaps brickwork and carpentry? Contractors will always want to invite bids on a 'back-to-back' basis. That is all risks put on it by the employer must be passed to subcontractors. That means that subcontractors should have access to all relevant tender documentation, including drawings, specifications and preliminaries sections of bills of quantities. It would be foolhardy of a contractor to hold back from a subcontractor a preliminary section that perhaps limits for example working hours or places additional emphasis on control of dust or noise. If subcontractors have not seen this documentation, they cannot be deemed to include for it in bids.

Figure 3.6 includes a typical enquiry letter to subcontractors. They need to be aware at bid stage about what type of contract will be used if they are successful in their bid, and what payments terms apply. The length of a bidding window will depend upon the size and complexity of projects. Four weeks may typically be allowed for a medium-sized project of modest complexity. In traditional systems,

---

Dear Sirs

Re: Project: xxx

We write to invite you to submit your tender to carry out the plastering works on the above contract.

We attach the following documentation:

- Drawings – numbers 10 rev A, 11 rev B, 12 rev A and 13 rev D
- Specification page numbers 1–10 and 35–46 inclusive
- Bills of quantities items 58 A to C, 59 A to C and F to K, 60 A to M.

The following conditions will apply:

- ABC Construction Standard Subcontract Agreement (2016): amendments to standard form detailed within specification pages 1–10.
- Submission of invoices: monthly on the last working day of each month.
- Payment period: thirty-five (35) days after receipt of invoices.
- Retention percentage: 5%.
- Rectification period: six months.
- Programme details: to commence within seven days of written notice and to complete within six weeks. Circa June to August 2017.
- Damages: £5,000 per week or part thereof.
- Include for 2.5% main contractor's discount.
- Fluctuations: fixed price.

Price to remain open for acceptance until: 01.05.17.

You may inspect the full contract documentation by appointment at our office. Please ensure your bid is returned by 5 December 2016.

Yours faithfully

Figure 3.6 Sample enquiry letter to subcontract bidder.

contractors may be most busy in the first week as they prepare enquiries to subcontractors and suppliers, and also in the last week when they need to bring together and formulate their bids. The middle weeks can be a time to estimate for any work contractors propose to execute themselves and to complete tender programmes and estimates for preliminaries. If contractors are truly integrating supply chains, key staff should also be 'out and about' speaking to specialists about their elements of bids. Otherwise the middle weeks are sat back, waiting for subcontractors and suppliers to do their pricing. When bids are received back, there needs to be an appraisal of all to ensure they are compliant with requests; alternatively a non-compliant bid may be submitted, perhaps offering a lower price for a material different to that in the specification. A judgement would need to be made about the risk of accepting or not that lower non-compliant bid. Table 3.10 includes an example of a pro forma to be used to appraise subcontractors' bids.

Task 3.1: Two figures in Table 3.10 are substituted with asterisks thus: 1*, 2*, etc. What are those figures?

Table 3.10 Pro forma to assess competing subcontractor bids

Project name	Budget Hotel, Deane Rd
Retention	5%
Trade	Plastering
Payment terms offered	35 days
Contract type / Programme period	Six weeks; circa June to August 2017
Discount requested	2.50%
Damages	£5k per week
Fluctuations / Date open for acceptance	01.05.17
Fixed price	Fixed price
Method of pricing	Bills of quantities

Subcontractor name	Gross price £	Discount allowed %	Net price £	Daywork rate £	Percentage addition to daywork rate; labour	Ditto plant	Ditto material	Non-compliant parts of bid
A	48,236.15	0.00	48,236.15	12.00	75	20	15	Discount not included. Alternative plasterboard manufacturer specified. Assume design team will accept.
B	63,985.00	0.00	1*	32.00	0	30	30	JCT 2005 subcontract agreement
C	49,555.00	2.50	48,316.13	13.62	100	25	25	
D	51,259.00	2.50	2*	12.85	105	20	20	30 days
E								
F	53,256.00	2.50	51,924.60	14.00	10	20	20	Declined retention; damages £1k/week

Subcontractor taken forward to tender bid: A — Amount £ 49472.97

Comments: Subco A price increased to take account of discount. Closest prices A and C. 'A' choice supported by lower daywork rates. Non-return from subco 'E'; to review for future projects. Subco 'B' unusually high price.

Performance on previous projects: A excellent. B very good. C good. D satisfactory. E poor.

Complied by estimator: Tommy Smith

Approved by supply chain manager: Gerry Byrne by e-mail

3.12 Preliminaries and the tender programme

The Building Cost Information Service (BCIS, 2016) reports that the mean value of preliminaries, expressed as percentage of the remaining contract sum, was 13.1% in 2015, based on a sample of 114 projects. Thus £100k preliminaries based upon a £1m tender sum is 100k/900k = 11.11%. It is not however to estimate the cost of measured works and merely add a percentage figure for preliminaries. BCIS also report figures as low as 6.1% and as high as 21.7%. The estimate for preliminaries can make a significant impact upon whether bids are successful. For building projects, the New Rules of Measurement 2 (NRM2), and for civil engineering works the Civil Engineering Standard Method of Measurement 4 (CESMM4), detail all those items that can be considered as part of preliminary works. NRM2 (2012, p. 22) defines preliminaries as "the cost of administering a project and providing plant, site staff, facilities, site based services, and other items not included in the rates for measured works". Templates are provided in NRM2 (2012, pp. 269–271) to support contractors in their pricing of preliminaries. The most costly item is often site staff. This chapter, section 3.13, illustrates the cost of employing one professional team member as over £68k per year. Contractors make judgements about how many staff they need on projects. One contractor who estimates 10 staff will have a competitive advantage in this element of the bid compared to a contractor who estimates 11 staff. It is however not good that when work starts, the contractor with 10 staff finds that 11 or indeed more are required in order to support speedy, safe and high-quality work. In its costing systems, a loss will be identified. It is often the case that if projects start to fall behind programme, a key way to recover that time is to employ more staff. It can be very fraught for a contractor, trying to minimise the possibility of damages due to late completion and also suffering the 'double-whammy' of paying staff salaries to support acceleration of work with no budget available to cover these costs. NRM (2012, pp. 50–119) details an exhaustive list of preliminaries. If bills of quantities are used, a preliminary section is written by the employer's representative such that contractors may price as many items as they wish. NRM2 provides for two 'categories' each of which has two 'sections'. The two categories are first for main contractors, and second (to be used in construction management contracts) for works package contractors. In traditional contracts, subcontractors do not get the opportunity to price their preliminary costs, and therefore need to add items such as the cost of supervision to their unit rates. NRM2 is therefore giving works package contractors the chance to price their preliminaries separately. Within each of these two 'categories' the two sections are (i) information and requirements, and (ii) pricing schedule. Contractors have their own pro formas that they use to price preliminaries. They will then spread sensibly the money from their preliminarily pro formas to the bills of quantities.

Table 3.11 Build-up to a contractor's preliminary element of a bid

Project name	Hotel project	Duration	50 weeks	Fixed charge		Approx value (£)	Time related	3,200,000 %		Total amount
Component	Sub-component	Comments	Quantity	Rate	Amount	Duration in weeks	Rate	Amount		Total amount
Employer's requirements and accommodation	Site cabins	4 No inc toilets				200	70	14,000		14,000
	Mobilisation	Transport and fit out			1,500					1,500
	Demobilisation	Transport and strip out			600					600
Supervision	Construction manager and foreman	Manager 52 weeks				52	1,300	67,600		67,600
	Site engineer	Engineer 25 weeks, foreman 25 weeks				50	950	47,500		47,500
	Planner and QS	Planner 20% time = 10 weeks. QS 40% 20 weeks				30	950	28,500		28,500
General labour	Attendant labour	Including cleaning				52	560	29,120		29,120
	General attendance	For subcontractors			2,000					2,000
Site facilities	Administration	Security				50	250	12,500		12,500
	Services	Telephone, water, drainage, electricity			2,200	50	150	7,500		9,700
Temporary works	Mobilisation				2,500					2,500
	Maintenance	Temp road	300m²	20	1* 6,000					6,000

							Total
Mechanical plant	Mobilisation/removal Temporary work	Temp fencing	180m	15		2,700	2,700
		Protection of finishes				5,000	5,000
	Lifting	Craneage		10	2,500	25,000	25,000
	Transporting	Deliveries from yard		4	1,000	4,000	4,000
	Concreting	Included in rates					
Non-mechanical plant	Scaffold	Subcontract quote					15,326
	Instruments	GPS		15	72	1,080	1,080
	Miscellaneous						1,0000
	Small tools	Contingency			4*	5,000	5,000
Contract conditions	Insurances					3,200	8,600
	Bonds	3% of bond value	320,000		0.03	5* 9,600	9,600
Miscellaneous	Winter working	Included in rates					
	Quality assurance	Testing concrete				250	250
	Safety	PPE				750	750
	Setting out consumables					500	500
	Specialist clean on completion					3,000	3,000
Total						**Total**	**312,326**
Approx total tender	3,200,000					**Total tender minus prelims** 2,887,674	**Percentage** 312,326 / 2,887,674 * 100 **10.81**

An important tool to support estimating for preliminaries is the tender programme. This programme may not be as detailed as a contract master programme which is prepared by the winning contractor, but it does give durations of key activities. The programme is used to determine how long many key cost components will be required on-site. If for example, it is decided a setting-out engineer is required for substructure and superstructure work, what is the duration of those activities? How long is the engineer needed? If a tower crane is required for the construction of the superstructure frame, how long is that duration? Staff and cranes for example are significant cost items related to time, and accurate estimates are required to support the submission of competitive bids. Table 3.11 illustrates the build-up to a contractors' preliminary bid. Some items such as scaffolding will require estimates from specialist subcontractors. Lump sums may be used for many items, and these may not appear to be calculated so accurately as unit rates for measured works. For some items such as area of temporary roads or length of temporary fencing, it may be necessary to take off approximate quantities and build up prices using composite rates.

Task 3.2: Five figures in Table 3.11 are substituted with asterisks thus: 1*, 2*, etc. What are those figures?

3.13 Plant; hire, buy or subcontract

Contractors who undertake production work themselves with direct labour will often require some form of mechanical or non-mechanical plant. Work classified as 'preliminary items' also may require plant, for example in the construction of temporary roads and temporary fencing. Some plant may be classified as non-mechanical; that is with no motorised moving parts such as scaffolding, adjustable props, concrete skips *et al*. Mechanical plant can include hand-held tools such as heavy breakers, plant that is steered or operated such as vibrating plates, vibrating rollers; also mechanical plant that is driven, such as dumpers and excavators.

The principles of whether companies should hire or buy also apply to company cars that are provided as part of remuneration packages for professional staff.

The choice between hire and but may appear to be stark. Hire perhaps a hand drill for six months (26 weeks) at £25 per week; cost £750. Alternatively, buy the drill for a lump sum of £500. At first glance the choice to buy is clear, but there are many complicating factors such as: (i) if the drill breaks down, there could be potential delays on-site whilst it is replaced, (ii) if the drill is repairable, who will repair it?, (iii) who will service it and hold spare parts?, (iv) where will it be stored during periods when it is not required on-site?, (v) what happens if large stocks are purchased to suit buoyant markets, and there is a sudden downturn?, (vi) where will the £500, or many thousands of £s come from, for the lump sum payment?

Companies who choose to buy plant may need large support facilities, including in the context of larger plant, storage yards, maintenance sheds and garages, spare parts, tools, mechanics and fitters, administrative staff, security staff, supervisors and managers *et al.* All these items also have overhead costs, such as business rate charges, utility payments, IT equipment and software *et al.* Some companies may establish separate plant divisions or limited companies (under a parent company), and hire items to sites at commercial rates. It may be the case that companies instruct their sites that they must where possible, hire internally, even though hire rates might be higher than those available externally. Companies may set rates at high levels, since if plant hire limited companies make good profits, that money can be easily reinvested before payment of tax by purchasing new plant.

Whilst on the one hand, the hire or buy decision is based upon cost, also of primary importance is the cashflow issue. There may be many drills required at £500 each, or it may be several major items of plant that could cost over £100k. A company may just not have large sums of money available to invest, but can pay smaller sums over a long period. Using the small cost of the drill again, £25 per week even over 52 weeks may be a better option than the £500 lump sum. It may be possible to take loans to purchase large items of plant, or increase bank overdraft facilities, however some companies may be nervous about having high levels of debt. This is illustrated in the way company cars are often procured, where across all sectors of the economy, arguably most companies lease rather than buy.

Companies will make their own judgements to suit their own circumstances about hire or buy. There may be a different approach for different sites or it may use a combination of both. A company on a site some long distance from its head or regional office may find it more effective to hire. Some companies may take the view that their expertise is in the management of people and projects, and that provision of plant is for specialists. Other companies may operate on a 'management only basis', and prefer to subcontract all trades. The hire or buy decision is then one for subcontractors.

3.14 Discounts

Discounts have been used widely in construction; they have been pervasive, also pervasive in high street sales. In construction, their contractual status was established in nominated subcontract and supplier contracts that were popular in JCT 1998 and earlier forms of contract.

The discount amounts specified historically in JCT contracts was 2.5% for nominated subcontractors and 5% for nominated suppliers. These discounts were highly transparent or visible; design teams instruct main contractors to place orders with specialists at a specific sum, inclusive of main contractor's discount as noted in section 3.5.2.

Contractors also routinely ask all their subcontractors and suppliers to include for discounts in prices; very often that is 2.5%. When specialists prepare their bids, they need to add an amount on to the money they wish to be paid, to allow contractors to deduct the discount. The arithmetic does not work if 2.5% or 5%

is added, to take off the same 2.5% or 5%; the appropriate additions are $1/39$ th for 2.5% and $1/19$th for 5%. Examples:

A subcontractor wishes to receive a payment of £10,000.00 for its work.

The incorrect calculation is £10,000 plus 2.5% = £10,250; of the £10,250 the main contractor would pay the nominated subcontractor 97.5% and retain 2.5%. 97.5% of £10,250 is £9,993.75, and the discount £256.25.

The correct calculation is £10,000 plus $1/39$th = £10,256.41; of the £10,256.41 the main contractor would pay the nominated subcontractor 97.5% and retain 2.5%. 97.5% of £10,256.41 is £10,000, and the discount £256.41.

The $1/39$th factor arises from the fact that the total figure of £10,256.41 can be considered in two portions; 97.5% and 2.5%. The starting figure of £10,000 is the 97.5% portion. Note that 2.50 / 97.50 = 39; therefore to increase the 97.5% portion to 100% add $1/39$th.

The same principles apply if there is a discount of 5% thus:

A nominated supplier wishes to receive a payment of £10,000.00 for its work.

The incorrect calculation is £10,000 plus 5% = £10,500; of the £10,500.00 the main contractor would pay the nominated supplier 95% and retain 5%. 95% of £10,500.00 is £9,975.00, and the discount £525.00.

The correct calculation is £10,000.00 plus $1/19$th = £10,526.31; of the £10,526.31 the main contractor would pay the nominated supplier 95% and retain 5%. 95% of £10,526.31 is £10,000, and the discount £526.31.

The $1/19$th factor arises from the fact that the total figure of £10,526.31 can be considered in two portions; 95% and 5%. The starting figure of £10,000.00 is the 95% portion. Note that 5.00 / 95.00 = 19; therefore to increase the 95% portion to 100% add $1/19$th.

Exercise 3.4

A kitchen supplier in the high street wishes to attract sales by offering a 40% discount for a typical range of cabinets. It wishes to receive £10,000.00, so that it still makes a profit. What is the amount that the kitchen should be advertised at less 40% discount, so that the supplier receives its £10,000.00? Or put another way, what amount should be added to the £10,000.00 so that it can be taken off again as a 40% discount?

The original purpose of discounts was for payment on time. By convention, subcontractors and suppliers who submit invoices to main contractors at the end of a month are due payment at the end of the following month. Therefore an invoice submitted at the end of January is due for payment at the end of February. If main contractors do not pay on time, they waive their right to the discount. Thus an invoice of £100k minus 5% discount gives an amount payable at the end of

February of £95k; the value of the payment should be £100k if it is delayed until March or beyond. However, in practice this may often not work. Subcontractors or suppliers who receive late payment are so relieved to get that 95% payment (the alternative is that payment is never received if main contractors go into liquidation), it is just not worth the hassle of chasing the 5% and risk losing the next order with that contractor.

Whilst discounts lose their intended power of securing payment on time, they do become a convenient tool to be used by main contractors in securing lower prices from supply chains, or indeed supply chains offering lower prices to secure work. If as part of negotiations, lower prices are agreed, it may be cumbersome to alter all rates if a bills of quantities or schedule of rates is employed. The simple solution is to increase the initial discount offer, such that what may have been an initial bid inclusive of 2.5% discount becomes an agreed order with a 3%, 4% or even higher discount. There is a need to be mindful that some commentators refer to discounts as 'funny money'. That is always be mindful of the net amount to be paid, and better to work from the net amount required to a discount figure rather than the other way round. Do not be blind to the real money consequences of a change in discount rate; the change in value on a £100k order from 2.5% to 5% is clearly £2.5k. Is that the amount of the profit for the subcontractor?

A more recent trend is that contractors do not invite prices with a discount; they just prefer the net price. Whilst on the one hand, perhaps most subcontractors will not chase return of discounts if payments are made late, some will, particularly sophisticated subcontractors who have large balance sheets and funds to employ legal specialists. A simpler solution therefore, especially in the context of contractors who have a propensity to pay late, is to seek net prices.

3.14.1 Who owns the discount?

Since contractors' discounts are transparent through supply chains, they may be viewed as extra profit for contractors. However, that is not necessarily the case. Contractors may be keen to win projects, and to do so they need to submit competitive prices for their bids to be lowest. Therefore, they simply take a sum of money off their bids, equal to the value of discounts offered by all parties, such that employers get the benefit. In exercise 3.4, if this were a small builder bidding for supply and installation of the kitchen, including labour and other costs, it is likely in competitive situations that the builder will need to price the materials element at £10,000 not £16,667.

For large projects, contractors may propose that a substantial amount of work is executed by subcontract specialists. For example, on a proposed £10m project with £8m value of subcontract work, a 2.5% discount on £8m amounts to £200k. In competitive markets, a contractor may wish to bid at £9.8m to submit the lowest possible bid. In this situation, the employer benefits from, or 'owns' the discount. In buoyant markets, the contractor may bid at £10m.

3.15 Company head office overheads

All businesses, other than micro and small companies, incur head office overhead costs that need to be recovered in prices that are charged to customers; or in a construction context from employers. Larger companies will have both head office and regional office costs. In the UK, many consultants, contractors and specialists may have a head office in London, or other major city, and satellite offices in those regions of the country where their businesses may have lots of work. The largest companies also carry the cost of international headquarters, perhaps in the UK or elsewhere around the world. Organisations often re-structure themselves. As work grows, companies may be in the news as they open up new offices to ensure they give best service and get close to customers. On other occasions, as work may be diminishing or for other reasons, offices may be closed as part of cost savings initiatives. Some head and regional office costs are visible and transparent, such as the cost of buildings themselves and professional and administrative support staff that occupy them. Other costs are perhaps less visible, such as software licence fees and utility charges.

When consultants bid for projects, their prices are based on estimates of the amount of time that will be required for designers 'on the drawing board', and in management and administration work. There are lots of other costs too that can be directly associated with projects, such as travel, and these will be identified separately in bids. The amount of time that designers require is multiplied by an hourly rate to determine a bid amount. On some occasions, rather than bid for work on a lump sum basis, employers may engage consultants on an agreed hourly rate. This may be because there is some complexity in the work, and it is difficult to estimate how long tasks may take.

When members of the public seek professional advice on the high street, perhaps from accountants or lawyers, they may be charged an hourly fee in excess of £250 per hour, plus VAT. This does not mean that the salary of these professionals is £250 per hour x 40 hours per week x 46 working weeks per year = £460k per year. Each professional needs to charge an hourly rate, such that there is a contribution to company overheads. Construction consultants too, need to estimate based upon hourly rates that allow contribution to overheads. Consultants calculate the hourly cost of staff, to include for salaries and all associated employment costs such as holiday pay, pension contributions, employers national insurance contributions *et al.* Rates may be based upon level of seniority; senior partner, partner, director, senior chartered professional, chartered professional, assistant, trainee. A benchmark figure that consultants may add to the hourly rate to cover overheads is the payroll amount plus 150%, such that a person with an hourly payroll cost of £40 per hour would have an hourly rate of £100. Alternatively, some design practices may charge a 200% uplift, or three times the salary amount such that a £40 payroll amount is charged at £120.

Contractors need to include in their bids, such that sites make a contribution to overheads. A benchmark figure that contractors may add to forecast site costs in tenders to cover such overheads is 7.5%. This figure is substantially lower than the 150% used by consultants, since it is added to the work of all trades including subcontractors, as illustrated thus:

Forecast site costs including preliminaries; labour, plant and materials for contractor's own work; and subcontractors' work,	£1,000,000
+ 7.5% contractor's overheads	£75,000
Total bid cost, excluding profit	£1,075,000

Contractors take this £75k, plus other sums of money from other projects, to fund overhead costs. One of the most significant costs within head or regional office overheads is staff salaries. There is a multitude of people required. Consultants need to cover the costs of partners or directors. Contractors also have specialist full-time contracts managers, estimators, planning surveyors or health and safety officers. Both organisations may have teams for accountancy, wages and salaries, legal, marketing, human resource, information communications technology (ICT), reception and secretarial work, administration, security and cleaning. Each post at head or regional office has a budget value. Table 3.12 illustrates a budget value for a contractor's senior health and safety manager is £68,528.10. The salary of £49,500 per annum is based on the salary survey by Building (2015) on a post located in North West England. These budget values need to be under constant review, since the same survey reported average rises in contractor staff salaries of 7.4% in 2014 and 5% in 2015, and also the cost of items such as fuel for company cars can be volatile. Contractors have their own systems; some will budget for company vehicles and small ICT provision separately, whilst some would count these costs as being attributable to the post; in Table 3.12, they are attributed to the post. The weekly cost of this position is £1,249.09 say £1,250 and the overall cost of £68,528.10 is 38.44% higher than the salary of £49,500. Some companies may calculate in detail a weekly figure for each post; others may use a rule-of-thumb figure of say 40% that they add to market salaries. Those staff that do not have company cars and ICT equipment to support mobile working will have a lower overhead charge to their salaries, perhaps 25%.

Table 3.13 illustrates the budget for the year ahead for company overheads for a small/medium-sized contractor with one office. The contractor initially forecasts its costs at £998,368.00 and then decides that the salaries of two contracts managers will not be allocated to head office overheads, but instead to the preliminaries element of individual sites. Therefore, the company has projected overhead charges of £848,168.

Contractors need to work in two ways to ensure that costs for overheads do not exceed budgets. The first is to ensure that they generate sufficient work to support this overhead. If company turnover is less than forecast, they will not have

Table 3.12 Budget value for a contractor's senior health and safety manager

Item	Comments	Amount (£)
Salary including holidays		49,500.00
Employer company pension contribution	Defined contribution scheme industry average 6.1%	3,019.50
Personal learning fund	To support continual professional development; lump sum	750.00
Employer national insurance contribution 13.8%	Based on salary	6,831.00
Sick pay and redundancy allowance	Allow 2% of salary to support employment of freelance staff as temporary cover	990.00
Private health care	Blanket company policy; average monthly premium per employee £40.00 x 12 months	480.00
Company car; family saloon	Three years, permitting up to 30,000 miles per year including private mileage. £325 x 12 months	3,900.00
Fuel	3 days visiting sites @120 miles per day plus 2 days at office = 50 miles. Total 460 miles per week x 46 weeks per year = 21,160 miles per year. 50 mpg = 11.45 miles per litre. 21,160 / 11.45 = 1,848 litres @ £1.20 per litre.	2,217.60
Mobile telephone	12 months @£45.00/month	540.00
Lap top	Lump sum: £500.00 / 3-year life expectancy	166.67
iPad	Lump sum: £400.00 / 3-year life expectancy	133.33
Total per year		**68,528.10**
Cost per week £68,528.10 / 52		**1,317.84**
Percentage overheads on base salary: £68,528.10 – £49,500 / £49,500 =		**38.44% say 40%**

Table 3.13 Budget for the year ahead for company overheads for a small/medium-sized contractor with one office

Item	Sub-item	Amount
Office space; 10-year lease including service charge for security, insurance and communal cleaning. 300m² x 10.76 = 3,228 ft² x £33.50/ft²		£108,138
Uniform business rates (Council tax)	National multiplier £0.513 x year 2008 rental value say £80,000	£41,040

Item	Sub-item	Amount
Utility charges	Electricity 12 months x £200	£2,400
	Water	£3,000
	Broadband including telephone; 12 months x £75.00	£900
Bank charges and interest payments		£12,500
Staffing; 11 No	*Managing director	£105,000
Salaries x £1.40 for overheads if car and mobile ICT equipment provided marked thus* e.g. Managing director £75k x £1.32 = £99k. Those without cars salaries x £1.25.	*Quantity surveying director	£98,000
	*Accountancy director	£98,000
	*Contracts manager x 2	£151,200
	*Human resource manager including health and safety	£69,300
	*Estimator	£69,300
	*Assistant estimator	£44,800
	Director's secretary	£42,000
	Administrator/receptionist	£26,400
	*Graduate	£30,800
Consultant design fees to support tenders	For design and build bids; assume 20 days each year x £300/day	£6,000
Staff training, education and continual professional development; including industry conference fees and travel	Some monies included in 40% overhead addition; allow additionally lump sum	£5,000
Legal consultancy support	Assume 15 days each year x £1,000/day	£12,000
Audit fee for annual returns	Lump sum	£3,000
Furniture; all capital cost / 10 years' life	Work stations; desk, chair, filing cabinet, book cases. 11 No x £400 / 10	£440
	Shared document storage	£180
	Boardroom table and cabinets	£300
	Quiet work rooms	£100
	Reception area for visitors	£150
Work station information communication technology	Work stations; PC, printer, 2 x VDU. Allow 12 No x £1,200 / 3-year life including toner	£4,800
Shared information communication technology	Photocopier/scanner; lease 12 months x £75	£900
	Plotter; lease 12 months x £60	£720
	Data projector and screen	£2,000
	Server; purchase. £6k / 3 years	£2,000
Software licences	Microsoft, BCIS, BIM	£15,000

(*Continued*)

Table 3.13 (Continued)

Item	Sub-item	Amount
Postage		£1,000
Stationery including A0 paper		£1,500
Kitchen, water fountain, WC and cleaning consumables	Allow 50 weeks x £40 week	£2,000
Web site development and ICT support	Allow 20 days x £500/day	£10,000
Marketing activities; company hospitality		£10,000
Corporate social responsibility	Sponsorship of local schools and sports teams	£2,000
Sundries		£5,000
Contingency		£12,500
Sub-total		£999,368
Deduct contracts managers' salaries		£151,200
Total		**£848,168**

Note: Staff marked with an asterisk, thus* have 40% mark-up on salaries to include for company cars. Those without cars have 25% mark-up.

sufficient receipts from sites. Directors and estimators always have their minds focused on securing work. If sufficient turnover is not achieved, often savings in overhead charges will be sought. Since many costs can be classified as 'fixed', for example office rental charges, the area most frequently targeted for cost cutting is staff. In buoyant markets it works the other way too, and if turnover is more than anticipated, that will generate extra income which may be invested in more staff.

Exercise 3.5

A contractor adds 7.5% to its estimates to cover for head office overhead costs. Its forecast overhead costs for the next year are £848,168. What is the total amount value of work or turnover it will need to achieve to sustain its overhead costs?

The second way in which overhead costs must be monitored is to compare on a monthly basis if expenditure on individual items is exceeding budget; for example, is monthly expenditure for marketing higher than budget, and if yes, can costs be curtailed so that the overall annual budget is not exceeded?

3.16 Company profits

All companies are in business to make a profit. After taxation, profits are distributed to shareholders or re-invested in businesses to, it is hoped, make more profits next year. That investment may be in plant and machinery, land, people, research and development *et al.* It may also be used to pay off debts such that in subsequent years, costs incurred in interest charges are reduced. Directors and other paid employees may take a share of profits; indeed, to incentivise staff, remuneration schemes may include for profit sharing. Companies may also prefer to hold some profits as cash in the bank to bolster their cashflow position.

When companies bid for projects, they forecast the actual costs that they will incur if they are successful in securing the work. They will then add a sum of money for profit. If market conditions are buoyant, the amount of profit added may be higher than if there is not much work around. In tough conditions, companies may not add any profit to their bids or bid at less than cost; both known as 'suicide bidding'. The motive behind suicide bids is to achieve sufficient turnover to stay in business; that is generate some contribution to fixed overhead charges and to keep key people employed. If losses are incurred, the company has cash reserves to sustain its survival (or financial support from elsewhere), and then in years ahead, market conditions will improve. Companies may also try to mitigate the consequence of suicide bids by asking supply chains to reduce their prices. There are examples of many large companies in all industries that have reported losses, but they survive and go on to be profitable in later years. In construction, as house prices dropped in the UK, many private house builders were losing substantive amounts of money in the period 2008 to circa 2012; the same companies were reporting very healthy profits in subsequent years. There are also examples of construction companies who submitted bids that were arguably suicidal, and these bids led to business failure and liquidation.

Profit can be added to bids by a lump sum addition, or by a percentage to forecast costs. In the construction press, consultants report positively if their profits are in the range of 10% to 15% of turnover. Private housebuilders are similarly pleased with profit values in these ranges. However, percentage profit for contractors is often lower, with some words used to describe their profit performance thus:

- 1 to 2%: acceptable since the company survives; hope for better next year;
- 2.5%: acceptable since sector norm;
- 3%: good;
- 4%: very good;
- 5%: excellent;
- > 5%: outstanding.

The question that could be asked is that 'why are contractors happy with 2.5% profit, when this amount can be secured by merely sitting back and investing in banks; why take all the risks and expend blood, sweat and tears?' The answer is that it is not profit on turnover that matters; it is profit on the amount of money invested, or using accountancy terminology, return on capital employed (ROCE).

As a rule of thumb, contractors may be able to generate £6 worth of turnover for every £1 of capital employed, as illustrated thus:

Main contractor:	market capital	£500,000
	turnover	£3,000,000
	profit	£75,000
	profit on turnover	= 2.5%
	profit on capital employed	= 15%

Whether it be a group of investors, or an individual entrepreneur, an investment that appreciates from £500k to £575k in one year, giving a 15% return, is arguably worth the risk and the 'blood, sweat and tears'. For contractors therefore, a benchmark profit figure on turnover that generates an acceptable return on capital employed is 2.5%. Site teams and the construction press focus almost exclusively on profits on turnover; it is for head office accountants to calculate ROCE.

Similar to overhead percentage additions, contractors add profit to the work executed by specialists lower down the supply chain. For specialists, the rule of thumb '£6 worth of turnover for every £1 of capital employed' does not apply. The ratio, likely to be far less for most, will depend on the type of work. Clearly mechanical and electrical specialists will be different to some of the simpler trades.

Exercise 3.6

A subcontractor has a market value of £500,000, a turnover of £1,250,000 and a profit of £75,000. Calculate its profit on turnover, profit on capital employed and the ratio of market value to turnover.

3.17 Model answers to discussion points

Discussion point 3.1: Why do you think it is so important for employers to use contractor pre-qualification as part of the tender process?

The use of pre-qualification allows employers to check prospective of the financial standing of contractors, their suitability for projects, the quality of their past projects *et al.* Ultimately by using pre-qualification processes, employers can reduce the risk of projects going wrong once contracts have been signed and works have commenced on-site.

Discussion point 3.2: How would you determine whether to proceed with tendering for a project if the employer invited you to submit a bid?

When contractors receive invitations to express an interest in bidding for projects, this can often come with conditions attached. Some employers for example will advise that if the contractor agrees to tender and subsequently withdraws, this could affect their position on the employers' approved list of contractors.

It is therefore essential that contractors arrive at considered and balanced decisions, considering factors like:

- Project value and duration – implications for cashflow and resourcing;
- The nature of the project – is it in the contractor's areas of expertise?
- The employer – have they encountered problems with them in the past? Are they a serial employer? A new employer with a potentially large forward construction programme?
- Potential profitability – could a better return be realised elsewhere?
- Conditions of contract – are the terms agreeable, for instance has the employer amended the contract to remove variation clauses, putting change risk on the contractor?
- Payment terms – will the employer be a prompt payer or will the contractor have to wait for 60, 90 or even 120 days for payment?
- Supply chain complexity – will the project need complex supply chains and management agreements?

Discussion point 3.3: If you were invited to bid for a design and build project what bid documents would you expect to receive?

The bid documents should include:

- A letter of invitation to tender;
- The form of tender;
- Preliminaries – pre-construction information, BIM Protocol, conditions of contract and proposed amendments to standard forms;
- BIM Model – model enabling amendments;
- Tender pricing document – this will be a contract sum analysis for design and build projects;
- Employer's requirements document (specification);
- Contractor proposals documents;
- Drawings related to the project, e.g. outline planning drawings. This could also be provided as a BIM model.

Discussion point 3.4: An employer is concerned a contractor may have submitted a suicide bid that could lead to excessive claims post-contract. What actions should the employer take?

In the form of tender the employer should include an 'escape clause', typically as detailed thus:

We understand that you are not bound to accept the lowest or any tender received, nor assign a reason for the rejection of any tender.

In this way the PQS can advise the employer that the lowest tender is unlikely to be viable and should not be accepted due to the risks. Ultimately, however, the decision will rest with the employer.

Discussion point 3.5: A contractor submits the lowest tender for a project, but it exceeds the employer's budget. As part of the process to negotiate a reduction in the tender price, the contractor states that it has a strong 'balance sheet' and suggests that the employer omits the requirement for a bond. Advise the employer.

The employer may reasonably omit the bond, particularly if the contractor is long established and can provide additional and more up-to-date information beyond that available to the employer in the public domain that its current position is strong. The premium that the contractor has included in its bid can be deleted.

Discussion point 3.6: A contractor has three performance bonds lodged with its bank for different projects and also has a large overdraft facility. The contractor applies to the bank for a fourth bond, but there is concern that the bank may be taking on board too much risk. Advise the bank.

The bank may chose not to offer a bond for the fourth project. In negotiations, the bank may alternatively ask the contractor to reduce its reliance on the overdraft, or to wait until one of the other three projects is successfully completed, and thus a bond 'released'.

Discussion point 3.7: A national contractor undergoes substantive reorganisation. Each region in the UK is registered as a separate PLC. Advise an insurance company which is considering an application for a bond from the North West–based PLC.

Contractors may establish separate PLCs in each region, such that if one region was to incur substantive losses, there would not be a requirement for other profitable regions to subsidise those losses. The loss-making region may be allowed to fail and go into liquidation. The insurance company may seek a parent company guarantee, such that the parent company (assuming it is still in business) will pay the bond amount in the event of non-performance.

Discussion point 3.8: Why do you think insurance is so important to construction?

As with any production process, construction is a high-risk business. As a result, employers, their professional advisors and the construction supply chain need to protect themselves against as much of this risk as possible. One way of reducing their exposure to risk is to transfer that risk to other parties who are better equipped to cope with the financial consequences of the risk should it be realised. Risk transfer attracts a financial premium.

Discussion point 3.9: Why do you think insurances are covered by an express clause in standard forms of contract?

This follows on from discussion point 3.8. It was argued in 3.5 that construction is a high-risk activity and insurance is therefore needed to deal with its financial consequences. Construction is also a high-value (high-cost) process; should something go seriously wrong both parties to the contract could face business-ending levels of financial liability. To ensure the contract protects both employer and contractor from these catastrophic events, risk insurance clauses are added, mandating insurance policies are taken to cover these potential risk events.

Discussion point 3.10: You are drafting tender documents for the refurbishment of 20 private terraced properties as part of a housing regeneration project. Each property will be fully occupied during the work. Which insurance option in JCT would be most appropriate?

The nature of the project is inconsequential; equally the fact that the residents and/or their landlords will have insurance (contents, buildings or landlord cover) is also inconsequential to the letting of the work. Assuming the works are let using the JCT SBC 2011, the project should be insured using Insurance Option C '*Insurance for existing structures and insurance of the works in, or extensions to, existing structures*'.

Discussion point 3.11: What are the professional indemnity requirements for firms registered with the RICS (these are in the Code of Conduct for Firms)?

If you are preparing to sit your APC final assessment, or will be looking to move towards RICS Professional Membership (MRICS), you are required to evidence your understanding of the Code of Conduct. As part of the questions in this area, it is highly likely you will be asked about PI insurance. The requirements for firms or individual self-employed surveyors are stipulated in the RICS Rules of Conduct for Firms (regulation 9) and include:

• Insurance is taken out on a 'each and every' claim basis;
• The policy meets the RICS minimum wording requirements;
• The firm has the following minimum levels of cover:

  ○ Turnover < £100,000 needs to have a minimum of £250,000 of cover.
  ○ Turnover £100,001–£200,000 needs to have a minimum of £500,000 of cover.
  ○ Turnover > £500,000 needs to have a minimum of £1,000,000 of cover.

Discussion point 3.12: What other types of common risk do you think contractors should make allowance for in their tenders?

The list of potential risks captured in the contractors tender will be linked to both the nature of the project and its complexity; however risks to consider would include:

• Weather;
• Liquidated or delay damages;
• Supplier lead-in periods;
• Price fluctuations;
• Labour supply;
• Supply chain claims and potential insolvency;
• Defective workmanship;
• Loss and/or damage to works through theft, damage *et al.*
• Health and safety – accidents and the like;
• Damage to adjacent buildings and/or highway infrastructure.

Discussion point 3.13: Assuming you are applying for an extension of time for exceptionally inclement weather, how would you prove the weather was exceptionally inclement?

The JCT Standard Form of Contract is silent on what it terms 'exceptional'; as a result contractors will need to validate this in their application. Simply stating it has rained in November would not be exceptional. However, for example, in late April 2016, parts of Lancashire in North West England were covered with 100mm of snow. That would be exceptionally inclement weather. To prove this, contractors could purchase detailed weather information from the Met Office which would show the norms for the month and the actual weather observed. This data would support a contractor's application that snow at the end of April is an exceptional event and an extension of time should be forthcoming.

Discussion point 3.14: It is the policy of an employer not to include contingencies in projects. A design team knows that whilst there are some uncertainties in the ground, there are many risks that the employer should best retain in the design of superstructure elements. How may the design team ensure it has some monies somewhere in the tender documentation to pay for unforeseens?

Whilst arguably it should not be done, the employer's quantity surveyor may include in bills of quantities some provisional quantities for extra excavation, hardcore backfill or concrete. It is hoped substantive extra excavation will not be required, and there will be monies available that can be used for superstructure risks. The presentation of provisional quantities would be presented in a bills of quantities thus:

Foundation excavation, not exceeding 2m deep   1,000m^3
Extra over provisional quantity ditto (to be deducted if not required)   200m^3

Discussion point 3.15: A supermarket employer has an excellent long-standing supply and maintenance relationship with a specialist manufacturer of food refrigeration equipment. It has a new project which is put out to tender to six main contractors. How can the employer ensure its favoured specialist secures the work without using nomination provisions?

As an alternative, design teams may 'name' specialists in tender documentation, though to avoid the legal consequences that stem from nomination, it may be appropriate to give contractors a choice of two, or better three, so that ultimately it is contractors who choose and therefore take responsibility for the performance of the parties they contract with. Contractors may be steered to the favoured specialist by design teams specifying equipment models only manufactured by that specialist, although there may be a need alongside the specification to state the model type and then 'or similar approved'. There is clearly a risk that another supplier will win the order, though if indeed the existing relationship between the employer and the favoured specialist is strong, that should give efficiency gains that can be reflected in prices. Whichever specialist is selected, the main contractor will then enter into conventional main contractor/subcontractor or supplier contractual arrangements.

3.18 Model answers to exercises

Exercise 3.1

Your employer is proposing to construct a new hotel complex. The value of the project is £8 million and the project duration is 15 months. Explain what the insurance provisions are within JCT Standard Building Contract 2011 for this project and advise the employer on which provisions they should use, including details of any appropriate financial limitations.

SUGGESTED ANSWER

It is important to differentiate between indemnity and insurance. Indemnity means to provide security against loss. In JCT SBC, the contractor is required under clause 6 to indemnify the employer against various losses or incidents. For example clause 6.1 requires the contractor to indemnify the employer against personal injury or death of any person arising out of, caused by or in the course of carrying out the work. Insurance on the other hand, provides a guarantee of compensation for a specified loss, damage, illness or death in return for a payment. So it transfers the indemnified risk to a specialist third-party organisation. In JCT forms there are various clauses providing insurances the employer could demand; these include:

- Clauses 6.1–6.3 – Insurance for personal injury or death; insurance for property not incorporated into the work.
- Clause 6.5 – additional insurance for injury or damage to property not caused by negligence of the contractor e.g. damage to buildings during routine piling works as a result of unavoidable vibration. If the work is in a city centre this would be a useful additional protection, but potentially expensive.
- Clause 6.7 – Insurance for the works and site materials (options A, B and C) – in relation to this project it is expected option A would be selected. The policy would be in joint names.

Exercise 3.2

In 2011 the government proposed a new procurement route called 'Integrated Project Insurance' the procurement route is focused on the employer holding a competition to appoint an integrated project team to deliver the design and construction of the scheme. However, the novelty in this route is the use of a single insurance policy covering all the insurance policies that exist in other projects.

The policy would cover the top slice of commercial risks covering any cost over-runs in the project above and beyond the pain-share threshold set in the contract and apportioned between the employer and the contractor and its supply chain. To what extent do you think this proposal will work? Would it remove the blame culture and adversarial nature of construction and make insurance claims and liability easier to proportion and resolve?

SUGGESTED ANSWER

The fundamental aim of this new procurement route is to reinforce collaboration and trust, and therefore to drive the industry away from an embedded adversarial culture. With regards to insurance, if the risk of financial overrun is removed, it is hoped the insurance would prompt a closer working relationship and improve working practices. Ultimately, this will require more than an adjustment to insurance provisions, but if it removes the triggers for dispute it can only be a positive step for the industry.

Exercise 3.3

Working as the estimator for a major contractor you have been tasked with evaluating the risk associated with the tight programme the employer has insisted upon in the tender documents. Using the following data determine the risk allowance to be included in the bid:

- LDs are set at £3,450 per week or part thereof
- 20% chance of 1-week delay
- 24% chance of 2-week delay
- 18% chance of 1-month delay
- 22% chance of 6-week delay
- 16% chance of 2-month delay

Table 3.14 Expected monetary value (EMV) calculation for liquidated damages

Outcome (i)	Probability $P(X_i)$	LDs (£)	$X_iP(X_i)$
1	0.20	£3,450	£690
2	0.24	£6,900	£1,656
3	0.18	£13,800	£2,484
4	0.22	£20,700	£4,554
5	0.16	£27,600	£4,416
Financial E(X)			**£13,800**

SUGGESTED ANSWER

The impact of the risk event is computed using Expected Monetary Value (EMV) as follows:

As a result of the analysis is it suggested the contractor allows for £13,800 within their bid for the risk of damages.

Exercise 3.4

A kitchen supplier in the high street wishes to attract sales by offering a 40% discount for a typical range of cabinets. It wishes to receive £10,000.00, so that it still makes a profit. What is the amount that the kitchen should be advertised at less 40% discount, so that the supplier receives its £10,000? Or put another way, what amount should be added to the £10,000 so that it can be taken off again as a 40% discount?

The calculation is:

Net amount required from a sale £10k; this amount constitutes 60% of the money on an invoice
Therefore, £10k / 60 = £166.67; which constitutes 1% of the money
£166.67 x 100 = £16,667; which constitutes 100% of the money
Invoice value £16,667 – 40% discount (£16,667 x 0.40 = £6,667)
Net amount paid = £16,667 – £6,667 = £10,000
Amount added to the £10,000 = £6,667

Exercise 3.5

A contractor adds 7.5% to its estimates to cover for head office overhead costs. Its forecast overhead costs for the next year are £848,168. What is the total amount value of work or turnover it will need to achieve to sustain its overhead costs?

The calculation is:

Forecast overhead costs = £848,168
What is the net figure, such that when 7.5% is added, it will amount to £848,168?
£848,168 / 7.5 = £113,089
£113,089 x 100 = £11,308,907
Check calculation: £11,308,907 x 0.075 = £848,168
Total value of work required = £11,308,907 + £848,168 = £12,157,075

Exercise 3.6

A subcontractor has a market value of £500,000, a turnover of £1,250,000 and a profit of £75,000. Calculate its profit on turnover, profit on capital employed and the ratio of market value to turnover.

The calculation is:

Profit on turnover = £75,000 / £1,250,000 x 100 = 6%
Profit on capital employed = £75,000 / £500,000 = 15%
Ratio of market value to turnover = £500,000 / £1,250,000 = 1:2.5

3.19 Model answers to tasks

Task 3.1: Two figures in Table 3.10 are substituted with asterisks thus: 1*, 2*, etc. What are those figures?

Task 3.1 answers: 1* = £63,985.00; 2* = £49,977.53

Task 3.2: Five figures in Table 3.11 are substituted with asterisks thus: 1*, 2*, etc. What are those figures?

Task 3.2 answers: 1* = £14,000; 2* = £67,600; 3* = £6,000; 4* = £5,000; 5* = £9,600.

References

ABI (2014a) Public Liability Insurance. Association of British Insurers. Available at: www.abi.org.uk/Insurance-and-savings/Products/Business-insurance/Liability-insurance/Public-liability-insurance Accessed 03.05.16.

ABI (2014b) Professional Indemnity Insurance. Association of British Insurers. Available at: www.abi.org.uk/Insurance-and-savings/Products/Business-insurance/Liability-insurance/Professional-indemnity-insurance Accessed 03.05.16.

ABI (2016) Model Form of Guarantee Bond. Association of British Insurers. Available at: www.biba-credit-and-bonds.com/files/ABI%20BOND%20NEW.pdf Accessed 24.03.16.

Association of Consultant Architects (2013) ACA PPC2000 Standard Form of Contract for Project Partnering (amended 2013). Kent. Association of Consultant Architects.

BCIS (2016) Building Cost Information Service – Independent data for the built environment. Available by subscription at: service.bcis.co.uk/BCISOnline/Indices Accessed 26.03.16.

BIS (2013) Construction 2025: Industrial Strategy for Construction – Government and Industry in Partnership. Available at: www.gov.uk/government/publications/construction-2025-strategy Accessed 26.08.15.

Building (2015) Contractors' Salary Survey 2015: Rise and Shine. Available at: www.building.co.uk/professional/careers/salary-surveys/contractors-salary-survey-2015-rise-and-shine/5078898.article Accessed 27.03.16.

Cabinet Office (2011) *Government Construction Strategy*. London: Cabinet Office.

Cantor, M. (2007) Insurance for Building Design and Construction. Available at: www.designingbuildings.co.uk/wiki/Insurance_for_building_design_and_construction Accessed 25.08.16.

Capita (2016) Constructionline. Available at: www.constructionline.co.uk/static/about-us.html Accessed 03.05.16.

Chan, E.H.W. and Au, M.C.Y. (2007) Building contractors behavioural pattern in pricing weather risks. *International Journal of Project Management* 25 (6), pp. 615–626.

Dunning, A. (2008) *What Is Contractors' All Risk Insurance?* London: Nabarro.

Flanagan, R. and Jewell, C. (2016) *CIOB Code of Estimating Practice*. 8th Edition. Oxford: Wiley-Blackwell.

Gov.uk (2016) Rates and Thresholds for Employers 2015 to 2016. Available at: www.gov.uk/guidance/rates-and-thresholds-for-employers-2015-to-2016 Accessed 27.03.16.

JCT (2012) SBC/Q2011 Standard Form of Building Contract. The Joint Contracts Tribunal.

Klien, R. (2015) Cash retentions: Tell them where to stick it. *Building*. Available at: www.building.co.uk/professional/legal/cash-retentions-tell-them-where-to-stick-it/5073989.article Accessed 29.03.16.

Lock, D. (2013) *Project Management.* 10th edition. Aldershot: Gower Publishing.

Met Office (N.D. a) Location Based Averages. Available at: www.metoffice.gov.uk/media/pdf/4/j/Location_based_monthly_planning_averages.pdf Accessed 17.04.16.

Met Office (N.D. b) Data and Statistics. Available at: www.metoffice.gov.uk/media/pdf/r/m/Data_and_statistics.pdf Accessed 17.04.16.

NEC (2005) *Engineering and Construction Contract.* London: Thomas Telford.

NJCC (1995) Performance Bonds. National Joint Consultative Committee

North West Construction Hub (N.D.) North West Construction Hub. Available at: nwconstructionhub.org Accessed 03.05.16.

RICS (2016a) Price Adjustment Formulae Indices Online (PAFI). Available at: www.rics.org/uk/knowledge/bcis/online-products/price-adjustment-formulae-indices-online-pafi/ Accessed 27.03.16.

RICS (2016b) PAFI List of Contents. Available at: www.rics.org/uk/knowledge/bcis/online-products/price-adjustment-formulae-indices-online-pafi/ Accessed 27.03.16.

Swarb (2016a) Tai Hing Cotton Mill Ltd -v- Liu Chong Hing Bank Ltd; PC 1986. Available at: swarb.co.uk/tai-hing-cotton-mill-ltd-v-liu-chong-hing-bank-ltd-pc-1986/ Accessed 26.05.16.

Swarb (2016b) Donoghue (or McAlister) -v- Stevenson; HL 26 May 1932. Available at: swarb.co.uk/donoghue-or-mcalister-v-stevenson-hl-26-may-1932-3/ Accessed 26.05.16.

Swarb (2016c) Murphy -v- Brentwood District Council; HL 26 July 1990. Available at: swarb.co.uk/murphy-v-brentwood-district-council-hl-26-jul-1990/ Accessed 26.05.16.

4 Design and consultancy teams managing finance and risk for employers

4.1 Introduction

This chapter focuses on the wider services professional quantity surveyors (PQS) provide, including but not limited to value and risk management. The focus of government procurement legislation is on the value added benefit achieved by construction projects, and therefore management of both project risk and the value projects deliver are important and growing parts of QS practice. In addition, the chapter introduces the role of the PQS related to post-contract financial management and reporting. Following the signing of main building contracts, this latter element of service typically accounts for 40% of fees, so it is a major area that needs appropriate attention. As a result, chapter 4 aims to develop your understanding of these different elements of the cost consultant's pre-contract and post-contract financial management roles, and how these impact on projects and how they are managed. Finally the text introduces the important area of company-level financial management. Although this is the preserve of chapters 8 and 9, this chapter begins this process by looking at the PQS role in managing employer funds within practices and also the terms of engagement put in place between consultants and employers.

4.2 Cost prediction accuracy

Most construction projects are financed using either or a combination of:

- Equity – internal funds of employers such as retained profits or shareholder capital;
- Debt – sourced from a variety of sources ranging from conventional bank loans to company bond issues traded on the stock exchange.

As a result, it is important that any assessment of finance needs, including accompanying applications for debt funding, are supported by detailed and accurate financial information from design teams. Equally, the continued issue of detailed cost plans and supporting reports will either give employers the confidence to proceed with projects, or they could ultimately decide they have insufficient funds to proceed. It is therefore critical for quantity surveyors to develop accurate cost models.

Chapter 1 provided a very detailed review of the various methods and techniques available to allow quantity surveyors to develop pre-contract cost estimates for projects and to implement pre-contract cost control. Yet, the literature relating to pre-contract cost control is abound by references to the lack of accuracy in the cost predictions developed, with numerous reasons and explanations provided for this lack of accuracy.

As professionals, quantity surveyors argue that levels of accuracy achieved are dependent on the amount of information provided, and that accuracy will generally improve as designs evolve. However, for employers, who use estimates as the basis for financial decision making within their organisations, it is critical they play their part in providing information so that estimates are as accurate as possible.

The American Association of Cost Engineers (AACE) is regarded as one of the leading organisations focused on the application of scientific principles to the commercial management of construction projects in the US. It has developed detailed guidance relating to the development of cost plans, and importantly the accuracy levels quantity surveyors or cost consultants should be achieving at the various stages in the project evolution, as illustrated in Table 4.1. Although these make reference to the AACE's work stages, additional information is provided to show mapping to the corresponding stages in NRM1.

Discussion point 4.1: How useful do you think the AACE's work stage model would be to an employer who is trying to determine the level of accuracy their quantity surveyor may provide?

Cost planning is an art that requires continued practice and requires quantity surveyors and cost consultants to develop a clear understanding of the project in question. Pre-contract cost control is certainly not merely a slavish series of very detailed calculations produced on a very large spreadsheet using techniques and principles surveyors studied at university or gleaned from textbooks. There are many occasions when there is a need to exercise professional judgements. It is important to achieve the levels of accuracy required in Table 4.1. However quantity surveyors must also ensure they do not over-inflate estimates, nor produce estimates that reflect what design teams or employers think projects should cost, or the amount of money available. Quantity surveyors have an ethical obligation to provide a robust and expert service to employers. Only then can accusations of negligence be avoided, or if needed robustly defended.

Table 4.1 AACE estimating accuracy guidance

Estimate class	Design maturity level	End usage	Expected accuracy range	Equivalent RICS work stage
Class 5	0% to 2%	Concept screening	L: −20% to −50% H: +30% to +100%	
Class 4	1% to 15%	Study of feasibility	L: −15% to −30% H: +30% to +50%	Order of cost estimate
Class 3	10% to 40%	Budget authorisation or control	L: −10% to −20% H: +10% to +30%	Formal cost plan 1
Class 2	30% to 75%	Control or bid/tender	L: −5% to −15% H: +5% to +20%	Formal cost plan 2
Class 1	65% to 100%	Check estimate or bid/tender	L: −3% to −10% H: +3% to +15%	Formal cost plan 3

Source: Association of American Cost Engineers (2005).

Discussion point 4.2: To what extent do you agree with the view that cost planning is more art than science?

4.3 Value management and value engineering

Value management is not regarded as one of the main services offered by consultancy practices in the UK. However, despite this, the commissioning of value management services is beginning to increase as employers seek to embed both continuous improvement and best practice in the services they receive. Many employers are becoming ever more focused on the value adding benefits consultants can bring to projects, other than just lowest fees. The provision of a value management service can often be seen as a way of achieving the increasingly higher level of service employers expect.

Value management can be defined as a service that allows the team of consultants working for employers to:

- Improve the value for money projects deliver by helping employers to fully understand their strategic and business demands, and how buildings interface with business needs;
- Improve the function, performance and quality of buildings;
- Improve business procedures and project times.

4.3.1 Defining and explaining value management

Value management is often defined in a more technical way as:

> *The name given to a service in which the sponsor of a project, the employer, transmits a clear statement of the value requirements of that project to the project designers.*

(Eaton and Kotapski, 2008, p. 50)

Some would argue this working definition to be more reflective of simple contracts between employers and their design teams. Employers work with designers to develop detailed briefs for projects which designers interpret and progress. However, this would be to overlook core arguments and indeed core benefits of implementing value management. Traditionally, the briefs of architects are developed in situations where there is clear unequal distribution of knowledge. Also the requirements of employers can be misinterpreted and lost as teams develop, present and re-develop various iterations of schemes within limited timeframes. Design teams seek the approval of employers as they attempt to refine design briefs and better understand their employers.

Value management tries to improve this process by changing the approach to one that benefits employers. The strategic needs of both employers and other project stakeholders are drawn out to try to find a suitable balance between their wants (the non-essential added benefit) and needs (the functional requirements for successful projects) on the one hand, and the available budget on the other. Employers are asked to identify the required balance between cost, time, quality and other objectives they require projects to deliver. To achieve this Kelly *et al.* (2014) advise that these dimensions must be capable of both description and measurement. For this reason they advocate the measurement of quality to be achieved by considering several key variables including:

* Environment (sustainability);
* Exchange;
* Politics/community/popularity;
* Esteem;
* Flexibility;
* Comfort.

Using Kelly *et al.*'s view of the employer value system, value workshops need to consider the overall scope of projects by evaluating the importance of these facets, along with cost (capital and maintenance) and time. The results of this process provides employers, stakeholders and design teams with a base from which projects can be advanced.

This process is best illustrated through an example. For this we return to the hypothetical scenario introduced in chapter 1:

> *AZX Property Ltd has recently completed the purchase of an old 1960s office building with a view to developing a budget hotel. Structural analysis of*

the existing building discovered it was unsuitable for rehabilitation. Consequently the existing office building will be demolished, the site remediated and a 5-storey, 64-bedroom hotel constructed on the site.

We will start by assuming the hotel is to be developed using a traditional approach. In this situation, AZX Property Ltd appoint a design team including an architect, engineers and quantity surveyor based on their experience and fees. The architect engages with the employer to develop a brief and develop some initial designs for the hotel, holding progress meetings with the employer. As this is a simple budget hotel, and wishing to minimise the risk to the employer, the design team recommends a design and build contract is used, and tender the project at the conceptual design stage. They find a contractor who can build the hotel to the required standard (stated in the employer requirements), by a set date and under the conditions of the contract. After signing the contract the employer has limited further influence on the hotel design. The contractor takes control of the detailed design and construction of the building, seeking to ensure it achieves the return the employer requires and that the building is delivered within the cost forecasts developed when bidding for the project.

Clearly this approach does not provide much room for the employer to impose its needs on the project. If these are misunderstood or lost in confusion or a rush to develop bid documents (the employer's requirements), the building may not provide the value benefits the employer had hoped for.

Now re-run the scenario, but with the design team commissioned to manage the construction process also providing a complete VM study as part of their package.

The architects and design teams engage with the employer to develop a brief and again develop initial designs. However, at the same time the employer, designers, a representative from the hotel franchiser, cost consultants and other construction professionals undertake a VM workshop facilitated by an experienced VM consultant to establish a basic set of values for the project. A document called the *job plan* starts to be assembled. In the job plan, the employer specifies expectations of the project, the reason they commissioned the project and the amount of money they have available to deliver the scheme. Following this initial workshop, the full cycle of VM, explained later in this section, takes place and during it, the needs and wants of the various stakeholders emerge and are explored.

An example of this could be seen when the VM team looked at the strategic drivers for the hotel project, looking at the scheme's focus, scale and location. Using Kelly *et al.*'s (2014) framework, the VM team revaluate the hotel brief, trading off competing time, cost and quality objectives through the simple matrix model illustrated in Table 4.2. As a result of this process, the team identify the major parameters of the employer's value system, which for this project require a focus on:

1   The speed of delivery for the project. As this is a budget hotel, a short build time is required to facilitate early return on capital employed.
2   Ensuring the budget is tightly controlled. Again as a budget hotel, with low rental income from room occupancy, the return for the developer is critical. So reducing capital build costs will be seen as a priority.

Table 4.2 Client Value System Model

A - Capital cost							
A	B – Operating costs						
C	C	C - Time					
A	B	C	D - Environment				
E	E	C	D	E - Esteem			
A	B	C	D	E	F - Flexibility		
A	B	C	G	G	G	G - Comfort	
A	B	C	H	E	H	G	H – Politics /Community/Accessibility

A	B	C	D	E	F	G	H
5	4	7	2	3	0	4	2

Source: Kelly *et al.* (2014).

3   Running or operating costs. Given the tight margins in the highly competitive hotel market, it is important these are protected through low running costs for the tenant of the building, not just savings in the cost of construction.

4   Operating costs achieved the same score as comfort. Comfort here is seen as the internal feel and aesthetics of the building from the perspective of users. Again this is important to the success of the hotel scheme, which will be reliant on commercial employers and repeat business.

As a result of this process the design team will now devise the employer requirements and select a procurement approach that will maximise benefit against key features identified in the employer value system.

This approach is what value management is all about – systematically aiding the process of design evolution of projects to achieve best value for employers whilst preserving the overall function and cost of projects. The subsequent sections of this chapter provide more detailed explanations of value management, its history, key techniques and the process of implementing a value management study in full on a proposed project.

Discussion point 4.3: An employer is considering a major business relocation to a large, modern warehouse facility adjacent a major motorway network. What do you think, if anything, value management could add?

4.3.2 The value management framework – definition of key terms

The term 'value management' may cause significant confusion, with many often misunderstanding the fundamental differences that exist between *value management* and *value engineering*. There are numerous examples of these two concepts being incorrectly applied or demonstrated in the construction management and quantity surveying literature. To disentangle the confusion and differentiate between these two processes, it is important to go back to the very origins of the whole discipline.

Value and value thinking has been around for thousands of years, after being proposed in the philosophies of Aristotle and Plato, and evolved by economists and philosophers such as Adam Smith (exchange value), Jeremy Bentham, John Stuart Mill (welfare economics and utilitarianism), Bernoulli and Jevons and many others. Value engineering (not value management) is a relatively new discipline originating from the North American manufacturing industry in the 1940s. The concept of value engineering evolved in the 1940s through the work of Lawrence Miles, a purchasing engineer at General Electric Company during the Second World War. Faced with a perfect storm of an industry running at full capacity to service the demands of the military, and the shortage of essential components, alternatives to designed components that performed the same function were needed. It was soon discovered that many of the alternatives provided, equaled or out-performed the original at a reduced cost. In 1954 this led to a formal programme of design review by the US Department of Defense Bureau of Ships and was termed 'value engineering' (RICS, 2012b).

Later in the 20th century, value engineering started to spread across the globe. By the 1960s, after some slight modifications to adapt the concept to a European audience, value engineering was slowly embedded into the UK manufacturing sector. This was eventually "formalised by the establishment, in 1966, of the Value Engineering Association. This organisation changed its name in 1972 to the Institute of Value Management (IVM)" (RICS, 2012a). Consequently, the European community in its SPRINT (Strategic Programme for Innovation and Technology) adopted the term 'Value Management' as an official term. It described the same philosophical concept but in terms that are more acceptable to a European management style. As a result, 'value engineering' and 'value management' are commonly used terms internationally, although 'value engineering' is the predominant term used in the US (Eaton and Kotapski, 2008, p. 53). In the UK, 'value management' is also used as an umbrella term, encompassing all value techniques whether applied at the strategic or tactical level as illustrated in Figure 4.1 and defined below:

- **Value management** – the overarching term used to define all value processes.
- **Value planning** – carried out at the strategic stage of projects, value planning is used to identify and define employer value systems, which are usually linked to overall business strategies leading up to decisions to commission construction work.
- **Value engineering** – used during the detailed and technical design phases to eliminate unnecessary cost from schemes by evaluating alternative materials (discussed in more detail in section 4.3.2).

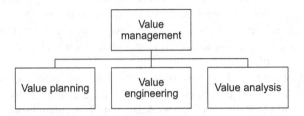

Figure 4.1 Value management hierarchy.

- **Value analysis** – very similar to value engineering, however, value analysis is used to carry out *ex-post* or retrospective value reviews of completed buildings.

Despite this framework of established definitions, in April 2000 the British Standards Institute issued BS EN 12973, which suggests construction professionals often discard *value planning* and *value analysis* labels. Instead, the term 'value management' is used in place of 'value planning', rather than the umbrella term suggested above. As a result the following definitions are considered more aligned to current industry practice:

- **Value management** – in construction generally considered to be applicable to the project schemes being considered by employers and defined as the process of making explicit the functional benefits of projects and appraising those benefits against value systems determined by employers.
- **Value engineering** – in construction industry, is technically focused, looking at space, elements and components of buildings. VE is commonly defined as an organised approach to providing the necessary functions of buildings, elements or components at lowest cost whilst maintaining the specified level of quality (discussed in more detail later).

4.3.3 The value management process

At the root of the value management process is the structured *job plan.* The job plan focuses the efforts of VM teams by providing them with a logical, sequential approach to the study of value. Although the job plan was initially developed for manufacturing industry, it does have uniform applications across other industries; as a result, the core principles of the job plan can be applied to the construction sector. Research conducted by Kelly *et al.* (2014), reported in their seminal text 'Value Management of Construction Projects', questioned the validity of the sequential structure of the job plan. They argued that operating value management in a sequential process can have a negative impact on projects, as it can hinder flexibility and innovation within value teams, whilst also impacting on team dynamics. As a result, it is recommended individual value management consultants, in negotiation with teams, decide whether they wish to strictly adhere to job plans or if they prefer to slightly amend the process and use their own bespoke versions.

There is some confusion about the structure of the traditional job plan with some commentators suggesting it contained as many as eight individual phases. The latest version of the job plan, mapped out in the latest version of the Society of American Value Engineers (SAVE, 2015, p. 4) international value standard identifies six principal phases, which can be summarised as follows:

- **Phase 1: Information stage** – the VM team reviews and defines the current conditions of the project and identifies the goals of the study.
- **Phase 2: Function analysis** – during this phase the VM team will define the various functions of the project. Functions are always defined using a two-word active verb/measurable noun context. The team reviews and analyses these functions to determine which need improvement, elimination or creation to meet the project's goals.
- **Phase 3: Creative phase** – the VM team employs creative techniques to identify other ways to perform the project's function(s).
- **Phase 4: Evaluation phase** – the VM team follows a structured evaluation process to select those ideas that offer the potential for value improvement whilst delivering the project's function(s) and considering performance requirements and the limits of the employer's financial and other resources.
- **Phase 5: Development phase** – the VM team will start to develop the selected ideas into alternatives or proposals. These will be supported by sufficient documentation to allow decision makers (this could be the design team, employer, key stakeholders or a combination) to determine if the suggested alternative should be implemented.
- **Phase 6: Presentation phase** – this is the final phase in the VM job plan. At this point the team leader (VM consultant) will write a detailed report (or give a presentation) for/to the employer that documents and conveys the adequacy of the alternative(s) developed by the VM team and the associated value improvement this is likely to deliver.

The implementation of these phases of activity is further demonstrated in SAVE's (2015, p. 4) international value standard as shown in Figure 4.2. This shows the role of the three phases of pre-workshop, workshop and post-workshop.

The use of workshops in value management is critically important, since its core values are to operate on the principle of using expert groups to evaluate problems, identify value-based opportunities that will enhance benefits, evaluate options and finally make decisions and provide recommendations for

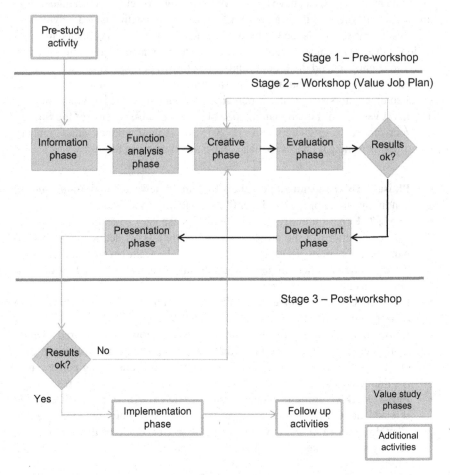

Figure 4.2 SAVE Value Work Plan Framework.
Source: SAVE (2015, p. 4).

employers. Value management is an approach that provides a framework for project stakeholders to reach a consensus decision on value aspects that will enhance projects. As Figure 4.3 illustrates it is important value management studies are undertaken early in the life cycle of projects to ensure potential benefits are maximised.

To identify the most suitable points for the implementation of value management and value engineering, Kelly *et al.* (2014) conducted a benchmarking study where they investigated the whole project life cycle to identify the points in projects that would be most suitable for value management intervention. They concluded that there are five points in the project life cycle where the implementation of value management would have a positive impact and result in benefits for

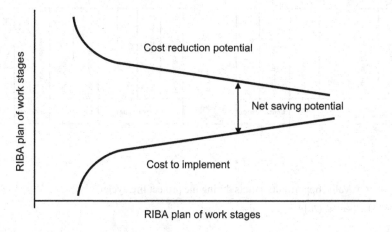

Figure 4.3 Value management – cost of implementation.
Source: Eaton and Kotapski (2008, p. 56).

various project stakeholders. As a result they recommended the following value workshops (studies) are conducted:

1 Strategic briefing study;
2 Project briefing study;
3 Concept design workshop;
4 Detailed design workshop;
5 Operations workshop.

To add further clarity to these workshops, they are mapped, as illustrated in Figure 4.4, against the RIBA plan of work; this mapping has been updated to reflect the issue of the revised RIBA plan of work in 2013.

VM1 – Strategic briefing study

This is the first value management study. It is conducted at the end of the strategic definition stage (RIBA stage 0) of projects. The primary aim of this workshop is to first establish whether or not projects should proceed before employers are committed to significant expenditure. The VM team might ascertain that the employer's objectives, for instance expanding manufacturing capacity or expanding into a new market, could be achieved without the need for new building. In this phase, all possibilities will be considered to see if the objectives could be achieved in another, less expensive way. The second objective of the workshop is to identify the value system of employers. This allows design teams to fully understand employer requirements. Only after careful evaluation is the decision 'to build' taken, and projects advance to the next step.

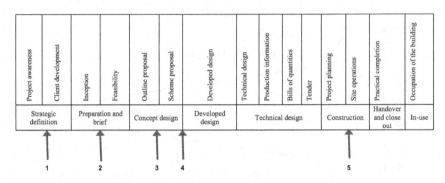

Figure 4.4 Value opportunity points during the project life cycle

Source: Kelly *et al.* (2014).

VM2 – Project briefing study

Assuming the outcome of VM1 is positive in terms of the decision to build, the technical phase of developments can be considered. VM2 is conducted during the Preparation and Briefing stage (RIBA Stage 1). Adopting a more technical view, project briefing studies seek to determine whether any changes need to be made to design objectives originally formulated. This in effect produces the response of construction teams to employer value systems. As a result, the following should now emerge:

- The aim of designs;
- The functions and activities of employers;
- The size and configuration of facilities;
- Project budgets.

VM3 – Concept design workshop

Once construction teams have agreed how they can meet the briefs of employers, this next stage starts to evaluate potential design solutions for projects. This workshop would normally take place when design teams have focused on a selection of priced (order of cost estimates) outline schemes. VM teams will be asked to evaluate options proposed against the value systems of employers. The workshop will also look to identify potential marginal value improvements that could be selected for designs. This process, Green (1994) suggested, would be best undertaken using simple decision trees to allocate weighting to criteria identified in VM study 1 and refined in VM study 2, as illustrated in Figure 4.5.

These criteria could then be transformed into a simple scoring matrix, similar to that shown in Figure 4.6, allowing design proposals to be evaluated against pre-determined criteria. The best design solution identified from this process of

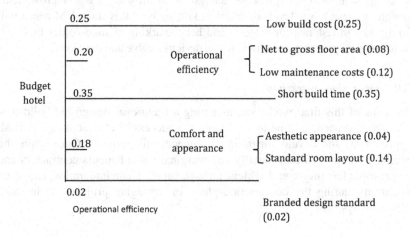

Figure 4.5 Simple value decision tree for the hotel project.

	Factors							
Weighting of importance	Low build cost	Net to gross floor area	Low maintenance	Short build time	Aesthetic appearance	Standard room layout	Branded design	Total
	0.25	0.08	0.12	0.35	0.04	0.14	0.02	1.0
Options								
Design A								
Design B								
Design C								

The factors' score is inserted in the top portion of the box. Below that, the score multiplied by the weighting is inserted

Scores for each option are totalled together and entered here.

Figure 4.6 Simple scoring matrix.

analysis will then move forwards to the more advanced stages of design (Detailed and Technical Design in the RIBA plan of work).

VM4 – Detail design workshop

Once VM teams have agreed conceptual designs for schemes, this workshop focuses on developing technical plans for delivery. Designs will be evaluated at elemental and component level using function analysis to identify any value improvements possible at these lower levels of detail. Additionally at this stage VM teams will identify key milestones for projects and benchmarking methodologies they will use to measure progress and efficiency as designs evolve and develop.

VM5 – Operations workshop

The focus of this final workshop, assuming a traditional design-bid-build procurement approach, will be to work with the successful contractor to establish supply chains and develop operational sequences for projects. Once again, the focus will be on trying to identify any marginal value benefits contractors and design teams feel they could achieve through supply chain integration, enhanced buildability during the construction phase or managing project risk in more effective ways.

4.3.4 Reconciling value and cost

The discussion above has largely focused on the management of value within projects. Value management is often seen as a wide evaluation of projects that uses structured functional analysis and other problem-solving tools and techniques in order to determine explicitly the needs and wants of employers. It is critically important that the results of value studies are also rationalised against the budget of employers and the cost planning processes discussed in chapter 1. The ultimate purpose of this process is to identify designs solutions that provide employers with best value for money. This is demonstrated by the equation function divided by cost. Schemes that deliver enhanced functionality at the same price will be seen to deliver enhanced value for money to employers. As a result, the costs of VM studies can be easily offset against these improvements in value for money achieved from schemes.

Discussion point 4.4: Identify the main types of construction employer. How would you define their value system?

Discussion point 4.5: 'Sustainability is becoming a critical aspect of many employers' corporate values, but this does not appear to be reflected in current value management practice'. Critically evaluate this statement.

4.3.5 Value engineering

Value engineering was briefly mentioned in preceding sections of this chapter. However, as the book focuses on financial management it seems appropriate to give value engineering its own section to explore in more detail how value engineering could be used alongside value management to enhance project outcomes. Value engineering is a technique used during both the detailed and technical design phases of projects to eliminate unnecessary costs by evaluating alternative materials which do not compromise the overall functionality of elements or components.

Returning to the hotel project, an example of value engineering can be seen when the VM team examines the specification for the external walls. The design team had included a traditional brick and blockwork specification, which would be wrapped around the steel frame of the building. After evaluating the function of the wall, the VM team proposed replacing the internal leaf of blockwork and the cavity insulation with structurally insulated panels (SIPS). This would not detract from the employer's requirements for the building's appearance, the size of rooms nor the performance of the wall. It was suggested significant time savings (33% – 50% is suggested by the manufacturer's literature) could be achieved given the building would be wind and weather tight quicker, shortening the overall construction phase by several weeks. Although, as Table 4.3 demonstrates, the use of structurally insulated panels would significantly increase the cost of the external wall cladding. In spite of this the walling solution would remain within the agreed cost limit of £213,030 (see Table 1.12 in chapter 1). As a result the outcome of the value engineering exercise is to recommend a SIPS solution, as this would

Table 4.3 Comparative analysis of external wall specifications

Specification	Price	Price per m² (EUR)	Total estimate	Anticipated construction duration
Conventional cavity wall consisting of facing brickwork, insulation and internal blockwork	Blockwork internal leaf £16.76/m² Cavity insulation board £10.69/ m² Brickwork outer skin £45.75/m² Allowance for sundries £10.00/m²	£83.20/m²	£83.20 x 1,578m² = £131,290	Brickwork facing 1.71hrs/m² Blockwork internal leaf 0.90hrs/m² Insulation board 0.11hrs/m² Total time 2.72hrs/m²
SIPS panels with external facing brickwork cladding	SIPS panel £70/ m² Brickwork outer skin £45.75/m² Allowance for sundries £10.00/m²	£125.75m²	£125.75 x 1,578m² = £198,434	SIPS Panel 2+1 Gang 0.08hrs/ m² Brickwork facing 1.71hrs/m² Total time 1.79hrs/m²

Source: Spons Architects and Builders Price Book 2015 (AECOM, 2014).

Table 4.4 Comparative cost analysis for internal partition systems

Specification	Price/m²	Total estimate
Stud partition of 75 x 50 SW studs at 400 centres, covered both sides with 12.5 plasterboard, scrim and 3mm skim plastered finish painted with two coats emulsion paint	£52.00/m²	£52.00 x 1,160m² = £60,320
70mm steel studs and channels covered both sides with 12.5 plasterboard, scrim and 3 skim plastered finish painted with two coats emulsion paint	£39.00/m²	£39.00 x 1,160m² = £45,240

Source: Spons Architects and Builders Price Book 2015 (AECOM, 2014).

accelerate the construction process, thus achieving the employer's objective without significantly increasing the budget.

Value engineering can also be applied 'in reverse' as a tool to lower costs due to overspend in elemental cost plans. Take for example internal walls from the hotel development and assuming the initial design solution may be to use timber stud partitions throughout the hotel. However, this significantly exceeds the cost limit set for this element of the building. As illustrated in Table 4.4 timber would cost £52.00/m² or £60,320 overall, against a cost limit for internal walls of £47,560 (Table 1.12 in chapter 1). Whilst some of these funds could be recovered from underspend elsewhere, this is a risky solution as design development is not yet complete. The second option would be to use value engineering to evaluate the function of walls and identify alternative specifications. As a result of the value engineering process, it has been suggested that the timber partition could be replaced with a proprietary metal stud system, which would receive the same plasterboard and plaster finish. Clearly this option would not compromise the performance of the wall. Table 4.4 illustrates this amendment to the specification would reduce the cost of internal walls to £45,240, and bring the sub-element back in line with the overall project budget.

Discussion point 4.6: How do value management and value engineering align with the UK Treasury's call for improved value for money in public sector construction and civil engineering projects?

4.3.6 Summary

When consultants are commissioned to provide a value management service, it would be normal practice to apply both value management and value engineering processes.

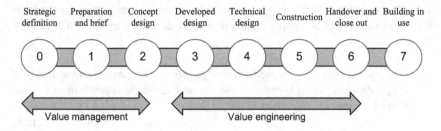

Figure 4.7 Value processes mapped against RIBA plan of work.

As illustrated in Figure 4.7, consultants will initially apply value management to projects in early stages to guide employers in the development of their business cases, and to draw out key criteria that are likely to inform option evaluations or project appraisals. For example, value management could be used to better understand the needs of employers for buildings and the location of buildings to allow potential sites to be evaluated. Or indeed value management could be used to better understand the need for new buildings, as opposed to more economical refurbishment of existing premises. As projects evolve and designs develop, value engineering will then be applied. Ideally this will be proactive and targeted; however, value engineering is also often seen as a tool for the retrospective elimination of cost when the cost limit has been exceeded.

4.4 Risk management

Risk management is now a mainstay of the construction curriculum and a key topic for quantity surveyors and construction managers when studying financial management. Writing in 1996, for the Construction Industry Research and Information Association (CIRIA) Godfrey (1996, p. 1) observed:

> *For a long time the construction industry, its employers and the public have suffered the painful consequences of the failure to manage risk, exemplified in many major projects by long disputes and out of control schedules and cost budgets. . . . Sir Michael Latham concludes that real savings of up to 30 per cent of construction costs are possible with a will to change.*

Twenty years on, much as changed, and risk management as a discipline has evolved significantly. Construction professionals have gained an enhanced understanding of risk management and now have access to highly skilled consultants. The successful implementation of robust risk management processes on large high-profile UK projects has increased awareness of its capabilities. The New Engineering and Construction Contract (NEC3) mandates the use of risk management processes through contractual obligations, and thus reinforces the benefits of proactive approaches to risk. Increasing use of NEC3 (accounting for up to 40%

of UK construction output) supports a shift in views of risk away from the 'old school' perception that risk is something to be considered as part of the procurement and transferred to supply chains to a view described by the RICS (2015), thus:

> *The success of construction projects arguably can be gauged on the ability of the professional team to mitigate threats and maximise opportunities in relation to the overall objectives of the project.*
>
> (RICS, 2015, p. 8)

As a result, risk is now proactively managed throughout the construction life cycle, informing both construction budgets and triggering more strategically focused approaches to project procurement and contract strategy development.

In consequence this section of the book provides an overview of risk management throughout the project life cycle. Although it is not possible to provide the depth of coverage provided in many specialist risk management texts, this book adopts a practical focus on risk management.

Discussion point 4.7: Differentiate between risk and uncertainty.

4.4.1 Risk management and the role of construction professionals

Throughout our lives everything we do has some level of risk associated with it; even crossing the road presents us with the need to evaluate and manage risk. Risk is common place in other parts of our lives too, and events that might prevent the achievement of stated objectives if they occur can be categorised as risk events. Unfortunately for us, we have decided to forge our careers in an industry that is incredibly risky. Sir Michael Latham (1994, p. 14) observed:

> *No construction project is risk free. Risk can be managed, minimised, shared, transferred or accepted. It cannot be ignored.*

Projects by definition try to introduce some form of change – new buildings, or modifications to existing buildings, new forms of production or something similar. Change involves uncertainty, which in turn means projects are more likely to be blown off course by potential future events. In other words, projects are inherently risky undertakings.

Risk management provides project teams with a means of dealing with uncertainty created within projects. This can be by identifying sources of uncertainty and the associated risks, and then managing those risks such that negative outcomes are minimised (or avoided) and positive outcomes are capitalised upon. The process is to appraise project life cycles and identify potential risk events, so that measures can be put in place to address them. Project teams who pro-actively manage risks will increase the chances of successfully delivering projects. Pro-actively managing risk should give lower costs, quicker time completions and buildings that are more likely to meet employer requirements.

Effective risk management is team orientated, and quantity surveyors or construction managers alone cannot implement risk management. A shift in project culture is required if effective risk management is to be successful. All stakeholders must be 'risk aware', that is understand the benefits of risk management and buy into approaches used to identify and control project risks. This is essential, as the two key features required for the deployment of risk management are:

1   To plan and take management action to achieve the aims of removing or reducing the likelihood and effects of risks before they occur and dealing with actual problems when they do; and
2   To continuously monitor potential impacts of risks, review associated action plans, and provide and manage adequate financial and schedule contingencies for risks should they occur.

In this context, risks can be defined as potential events, either internal or external to projects that, if they occur, may cause projects to fail to meet one or more of their objectives. A risk therefore has two aspects:

•   The expected likelihood (or probability) of that event occurring;
•   The expected impact of the event on the project if it does occur.

Effective risk management seeks to identify these events as early as possible so their possible effects can be fully understood. Preventative (pro-active) steps are then taken to minimise either the likelihood of events, their impact on projects or ideally both. This process of risk management follows a standard process as illustrated in Figure 4.8. The process starts with project teams identifying potential risks. This is followed by the assessment and evaluation stage that looks to qualify and quantify risks against a set of criteria using various risk analysis techniques to gauge likelihood and impact. From this a series of actions is proposed and implemented before the cycle closes with a final stage of monitoring and reflection of how well risk management solutions have performed. Thus allowing what is termed 'experiential learning' to take place within professional teams that can be transferred to subsequent projects.

However, as with value management, the extent of risk management is a question of resource balance. Not all risks can be controlled, and not all risks should be controlled. Some risks are so minor in terms of probability and impact that it would be inefficient to attempt to manage them. Other risks could have catastrophic impacts on the project, but they are so remote (termed 'never risk events' in the NHS) that controlling for them would be both highly complex and potentially unattainable. It is therefore essential for project teams to identify the optimum balance for risk control. This is illustrated in Figure 4.9.

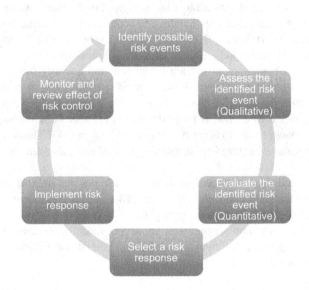

Figure 4.8 Risk management process.

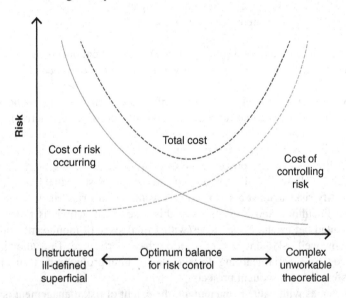

Figure 4.9 Balancing risk and control.

Source: Burtonshaw-Gunn (2009).

Discussion point 4.8: Explain the 'optimum balance of risk management' in the context of a major construction project.

4.4.2 Taxonomy of risk

To help project teams think about risk in a structured and pro-active way, most of the seminal guidance on the use of risk management (Godfrey, 1996; Lewis *et al.*, 2014; RICS, 2015) advocate the use of a risk taxonomy. That is putting risk into a classification system to aid collective understanding of it, so we can visualise the full range of potential risk events that are likely to impact on projects.

Project risks – these are risk events within project boundaries that could have an impact if something was to go wrong during execution. These are essentially risks that could materially affect successful delivery. Project risks are generally considered as events that impact on either time, cost, quality or other aspects of employer project objectives, or a combination of these elements.

Programme risks – risks that impact on the programme as a whole, rather than individual projects. These risks concern decisions that transform strategies into action. Typical risk areas could include funding, organisational and cultural issues, quality, business continuity and so on. Also when project risks exceed set criteria, and affect programme objectives, then they would be impacting at programme level (RICS, 2015).

Business risks – these are project, programme, consequential or benefit risks that break into the public domain and subsequently have an adverse effect on businesses as on-going concerns (RICS, 2015). This could also include risks that affect the business of employers during projects, or risks originating from projects that affect business outcomes once buildings have been completed.

Political risks – strategic level risk associated with changes in fiscal and other policies at local, regional or national government level. For example a change of government could result in certain policy and spending commitments changing that have a material effect on the overall direction of projects. For example in 2010 the UK coalition government implemented a strategic spending review, and suspended £7bn of government projects whilst it was being conducted.

Environmental risks – risks associated with natural phenomenon for example ground conditions, earthquake *et al.* that cannot be controlled by human hand. Frequently these are termed 'force majeure events' in standard forms of building contract. In JCT Standard Form of Contract 2011, force majeure is included as a relevant event for an adjustment to the contract completion date (an extension of time).

Benefit risks – the failure of projects to deliver expected performance, leading to a long-term business case being undermined. For example compliance with planning requirements may limit the size of schemes and hence revenues through reduced net-lettable space. Project teams do not have the power to eliminate this risk; however, employers should safeguard against it through risk assessments during business case development (RICS, 2015).

Consequential risks – risk that may occur as a result of other risks, which leads to knock-on effects. Consequential or secondary risks may occur within projects. These risks are secondary, because they arise from other risks and can also have an effect on projects overall (RICS, 2015).

In addition to these general classifications, NRM1 (RICS, 2012c) identifies and categorises risks into four main classifications:

- *Design development risk* – an allowance for use during the design process to provide for the risks associated with design development, changes in estimating data, third-party risks (e.g. planning requirements, legal agreements *et al.*).
- *Construction risks* – an allowance for use during the construction process to provide for the risks associated with site conditions (e.g. access restrictions, existing buildings *et al.*).
- *Employer change risk* – an allowance for use during both the design process and the construction process to provide for the risks of employer-driven changes (changes in scope of works or brief *et al.*).
- *Employer other risks* – an allowance for other employer risks (e.g. early handover, postponement, acceleration *et al*).

4.4.3 Risk identification

Risk identification involves determining which risks might affect projects and documenting the characteristics of each risk event encountered. This activity is normally undertaken by project teams and is often used in facilitated risk management workshops. It is vitally important that the correct people are invited to attend to ensure the full spectrum of risk events is considered. As a result workshop participants need to be selected based on their ability to identify risks in a given technical or management area. As a result Burtonshaw-Gunn (2009, p. 38) recommends the following people as members of a workshop team:

- The project team;
- The risk management consultancy team;
- Subject experts (e.g. specialist engineers *et al*);
- The employer;
- End users if they are identified and differ from the employer;
- Project managers who have experience of the type of project but may not be involved in the one under review;
- A selection of key stakeholders (user and/or employer will determine);
- Outside experts (e.g. local authority representatives *et al*).

As risk management is an iterative process, it would be impossible, and potentially ill advised to invite all the above to one workshop. For the London Olympics in 2012, the Olympic Delivery Authority (ODA, 2010) implemented a hierarchy of risk identification workshops ranging from strategic down to tactical. Each workshop focused on different levels of risk, related to professional skills, knowledge and perspective of participants. However, as this is iterative, each workshop would not be siloed, since risks were fed from the project team (tactical level) up into the larger and more strategically focused workshops.

There are a number of different tools risk identification workshops can utilise. The main techniques used by workshop facilitators include:

- **Brainstorming sessions** – probably the most frequently used technique; the goal is to obtain a comprehensive set of risks that be addressed later during analysis and evaluation. The lack of structure here allows workshop teams to 'think outside the box' or to explore issues which seem at first to be irrelevant.
- **Diagramming techniques** – used extensively in engineering, some organisations find the use of cause-and-effect illustrations such as the Ishikawa Fish Bone diagram helpful to identify cause of risks and seek out the root cause from the effect.
- **Checklists** – although these can be seen as constraining for project teams, the use of checklists developed from previous projects can provide a vehicle for transferring experiential learning and knowledge from one project to the next.
- **Risk breakdown structures** – providing a semi-structured workshop, a risk breakdown structure acts as a mid-way point between the very structured checklist approach and the completely unstructured brainstorming workshop. The risk breakdown structure lists the typical risk environments encountered on most construction projects; a number of these exist, and this text uses the 'SCLEEPT' mnemonic developed by Professor David Eaton at the University of Salford, although a similar list can be found in the RICS Risk Management Guidance Note (RICS, 2015, p. 8). SCLEEPT stands for:

  ○ Societal
  ○ Cultural
  ○ Legal
  ○ Economic
  ○ Environmental (natural)
  ○ Political
  ○ Technological

Risk events are then identified in a hierarchical format, under these headline risk environments. Following the identification of the fundamental risk events associated with the project, the risk management process moves to the next phase: analysis and evaluation.

Discussion point 4.9: Comment on the purpose of risk breakdown structures, explaining how their adoption would enhance project team ability to manage risks.

Discussion point 4.10: Explain the purpose of a risk management workshop. How would you manage the workshop environment? Would facilitation or leadership be required?

Exercise 4.1

Your employer, XZA Group, is considering development of a range of basic 2*
hotels throughout the UK, following success in the Irish Republic. The model
is based on the construction of 40–50 bedroom hotels adjacent to major infrastructure transport links. Despite objections from local residents, XZA Group has
recently entered into negotiations to purchase a disused asbestos factory adjacent
to a major UK motorway. Using the SCLEEPT mnemonic identify two possible
risks related to the project that would fall under each category.

4.4.4 Qualitative risk analysis

Risk analysis is essentially an extension of the earlier risk identification stage. At its
simplest level, risk analysis will be limited to project teams identifying and describ-
ing risks. However, in other situations, a more in-depth evaluation of identified risk
events will be required. Risk analysis is qualitative, in that it involves non-mathe-
matical evaluation of risks, and the focus instead aims to understand and prioritise
risk events identified in terms of their likelihood or probability of occurrence and,
should they occur, their impact projects in overall terms. Through this process, risk
events are prioritised allowing project teams to identify potential effects, decide
appropriate risk responses, whilst also trying to maintain an important balance
between ineffective, overly bureaucratic and therefore expensive risk control.

As a result, risk analysis requires the probability and consequences of identified risks to be evaluated. This is undertaken using established qualitative analysis
methods and tools including:

- Risk probability and impact analysis;
- Probability impact matrix;
- Ishikawa cause-and-effect diagram (Fish Bone diagram);
- Fault and decision trees;
- Failure mode and effect analysis (FMEA).

Other risk analysis tools

The first two are the most common approaches to risk analysis used by construction professionals in the UK. The latter three techniques originate from the engineering sector where they have been used as quality enhancement tools. They
can equally be applied to construction risk management, although because they
require a more detailed understanding of the risk event, they are not used frequently. A description of each follows.

Risk probability and impact analysis

Analysis of risk probability and impact is one of the most widely used techniques
when undertaking qualitative risk analysis. The process looks to analyse and

classify individual risk events based on (i) the likelihood of the event occurring and (ii) the impact of the event should it actually happen. These two elements are then multiplied together and given a severity rating. Likelihood and impact are normally measured using scales such as:

- Very high (VH)
- High (H)
- Medium (M)
- Low (L)
- Very low (VL)

Letters are often substituted for numbers to overcome the confusion of trying to multiply the letters together to establish a severity ranking. The use of this approach first requires project teams to define what constitutes risk at each of these levels. Tables 4.5 and 4.6 provide examples of possible scales that could be used to evaluate the likelihood and impact of a series of project risks.

Table 4.5 Likelihood scale

Description	Scenario	Probability (%)
Very high (5)	Almost certain to occur	75–99
High (4)	More likely to occur than not	50–75
Medium (3)	Fairly likely to happen	25–50
Low (2)	Low but not impossible	5–25
Very low (1)	Extremely unlikely to happen	0–5

Source: RICS (2015, p. 11).

Table 4.6 Impact scale

Description	Scenario	Cost (%) of overall project value	Time
Very high (5)	Critical impact on the delivery of the project and overall performance. Huge impact on project costs.	2.00	4–6 month delay
High (4)	Major impact on costs and project objectives. Serious impact on the completion time for the project. Detrimental impact on project functionality.	1.50	2–4 month delay
Medium (3)	Reduced project viability, impact on project costs and time for completion.	1.00	1–2 months delay
Low (2)	Minor delay, and some slight increase to the overall project cost.	0.50	3–4 week delay
Very low (1)	Minimal impact on both project cost and the time to completion.	0.25	1–5 day delay

Source: RICS (2015, p. 11).

Referring back to the hotel for example, one of the risks identified for the project relates to a section 278 agreement (a planning requirement making certain highway improvements compulsory). As a result the team has concluded this risk is likely to have 'high likelihood' and a 'very high impact' on time, as the highway works are located on the critical path for the project.

Calculated using the example with letters, it is difficult to gauge an outcome:

- Risk Consequence = Likelihood x Impact, so H x VH = VH.

However, replace letters with numbers, and this equation becomes:

- Risk Consequence = Likelihood x Impact, so 4 (high) x 5 (very high) = 20 out of a maximum score of 5 (very high) x 5 (very high) = 25. Scores of 20 out of a possible 25 advise project teams that these risks need robust risk management.

Probability impact matrix

Once the results have been individually quantified, the risks can be displayed in table format; this table is called the probability impact matrix. It is often combined with a 'traffic light' coding system, or heat map to allow users to easily identify safe, problem or danger zones of risk. This is illustrated in Table 4.7, which combines the data from Tables 4.5 and 4.6 into a risk impact matrix.

Once the risk matrix is produced, it is also possible for project teams to set colour systems to suit the risk appetite of employers. This is the amount of risk employers are willing to accept or absorb into projects before they require teams to implement some form of risk mitigation, which will almost always come at a cost to employers.

Discussion point 4.11: Comment on the limitations of the risk matrix model.

The *'Fish Bone' cause-and-effect diagram* developed by Karou Ishikawa in the 1960s, illustrated in Figure 4.10, explores all the potential or actual inputs

Table 4.7 Risk impact matrix

		Probability (likelihood) of risk event				
		Very high (5)	High (4)	Medium (3)	Low (2)	Very low (1)
Impact	Very high (5)	25 red	20 red	15 red	10 amber	5 green
	High (4)	20 red	16 red	12 red	8 amber	4 green
	Medium (3)	15 red	12 red	9 amber	6 amber	3 green
	Low (2)	10 amber	8 amber	6 amber	4 green	2 green
	Very low (1)	5 green	4 green	3 green	2 green	1 green

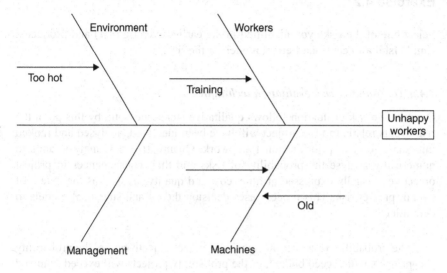

Figure 4.10 Risk impact matrix.

(causes) resulting in a single output (effect). In terms of risk management, Ishikawa diagrams help quantity surveyors or construction managers recognise likely triggers of specific risk events.

Fault tree analysis (FTA) is a method of analysis commonly used by reliability and safety engineers to analyse fault scenarios in design and construction. Most often associated with engineering systems, FTA seeks to understand the causes of machine or operation failure by starting with the effect (a part failure, or accident) and working backwards to identify possible causes. For example, if the light in a house failed, the FTA could be used to try to develop an understanding of the potential causes of failure. This would likely include power, lamp (blown or faulty) or switch failure. These would be then analysed hierarchically to identify the root cause. For example power failure could have resulted from either power outage, battery failure due to lack of charge, disconnection for not paying bills, blown fuse, contact failure and so on. Clearly this technique has compatibility problems with risk management in construction, as analysis is conducted from the opposite viewpoint. The process is to first identify all possible risks (causes) and then assess their probable effects on the project.

Failure mode and effect analysis (FMEA) is very similar to fault tree analysis but looks at the problem for a bottom-up rather than top-down perspective. This time the method starts by considering risk events and then proceeds to predict their possible effects on projects.

Discussion point 4.12: Explain the principles of the Ishikawa cause-and-effect model and comment on its applicability to construction risk analysis at project level.

Exercise 4.2

Select one of the risks you identified in the earlier exercise (4.1) and produce a simple Ishikawa cause-and-effect model for the risk.

4.4.5 Quantitative risk evaluation techniques

Quantitative risk evaluation follows qualitative techniques, and by this point the risk events relating to the project will have been identified, analysed and ranked into some form of prioritisation framework. Quantitative risk analysis aims to numerically analyse the probability of risks and their consequences to project objectives (usually expressed at time, cost and quality) as well as the extent of overall project risk. This process uses decision theory and statistical models to determine:

- The probability of achieving specified project objectives; i.e. the probability projects will exceed budget, or the probability projects will exceed time and incur the payment of damages to employers for example;
- Quantify the risk exposure for projects and determines the amount of contingency needed to deliver projects;
- Identify risks that need attention by quantifying their relative contributions to project risk;
- Identify realistic and achievable cost, time or other targets.

A number of quantitative risk analysis techniques emerging from both decision theory and finance have moved over into construction risk management. Although explaining all these techniques and their theoretical foundations is well beyond the scope of the text, the most common techniques used in the UK construction sector are evaluated, thus:

- Monte Carlo simulation
- Sensitivity analysis
- Probabilistic or Expected Monetary Value (EMV) analysis

Probability simulations

Probability simulations such as Monte Carlo and Sensitivity Analysis are mathematical models that allow quantity surveyors or construction managers to experiment with problems by looking at how they might impact on projects (for example 40 days of rain, 80 days of rain, 120 days of rain). Sometimes, with the use of computer algorithms, multiple risk events can be added to experiments to understand the effect of each collectively on projects, and the consequential risks arising or transferring from the one risk event to another. For example, the impact of rain on the exposure to delay damage payments.

These models are defined as either:

1 Deterministic models – the variables added into the model are known for certain, so the impacts can be modelled with absolute certainty. Unfortunately in construction, we do not know for certain rain will start on the 2nd December and end on the 8th February. So, generally speaking deterministic models are avoided for risk analysis.
2 Stochastic (probability) models – the variables added to this model are not known with any certainty. As a result the variables added to the model are input in ranges rather than single figures. So we would add an expectation for rain on the hotel project of between 30 and 50 days.

Both Monte Carlo simulation and sensitivity analysis are therefore classified as stochastic models, as they use extensive calculations to simulate their effects on projects, allowing quantification of the impact of a series of risk events.

1 Monte Carlo simulation

One approach to evaluating a risk event is to run a detailed stochastic simulation for that risk. Let's return to the hotel and select the risk of poor ground conditions leading to a variation and additional costs. The quantity surveyor could use a Monte Carlo simulation to evaluate the probability of this risk occurring, and to develop an idea of the financial effects of the risk should it occur on the project. To run the simulation the QS would need to identify and input likely financial impacts of the risk. So that would be the *minima* or the minimum possible financial impact of the risk, assume say £10,000. The most likely financial impact estimate, assume say £40,000. Finally the *maxima*, the maximum financial impact of the risk is identified, so assume say £70,000. So we have a clear range for the financial impact of poor ground (£10,000–£70,000)

The software would then use this data to generate random values and will run up to 1,000 individual simulations of the event; from this data it will be able to understand the risk profile. To test the likelihood of a particular result, we simply count how many times the model returned that result in the simulation. A simplified version of the simulation is illustrated in Table 4.8.

Table 4.8 Monte Carlo simulation output

Financial impact	Number of times (out of 1,000)	Percentage of total (rounded up)
10,000	4	0%
15,000	103	1%
40,000	345	35%
45,000	647	65%
50,000	874	87%
55,000	998	100%
70,000	1,000	100%

The original estimate for the most likely financial impact of poor ground conditions on the hotel project was £40,000. From the Monte Carlo simulation, however, it can be seen that out of 1,000 trials (simulations) using random values, the financial impact was £40,000 or less in only 35% of the cases. On the other hand, the simulation identified that there was a 65% chance of a financial impact of £45,000 or less, and an 87% chance of the financial impact being £50,000 or less.

2 Sensitivity analysis

Sensitivity analysis, also known as 'what if' analysis, is a simpler form of simulation, and one surveyors or construction managers can comfortably perform using a spreadsheet. Sensitivity analysis is a practical method of analysing risks on projects by varying the key factors and measuring outcomes. It does not give mathematical outcomes to the depth of Monte Carlo analysis, but it does allow teams to evaluate the impact of variables on overall project outcomes.

For example the cost of finance on the development could be adjusted by several per cent to see how the change affected the overall project. This is illustrated in Table 4.9. Once risks have been modelled in this way, or as banks term it, 'stress tested', the viability of developments can be reviewed and decisions made.

Probabilistic or expected monetary value (EMV) analysis

A different way of looking at probability is to argue that it reflects the long run performance of variables (in this case risk events). It provides an indication of how, on average, variables will perform. The method is technically and mathematically very restrictive, as we cannot say for certain how any risk will perform on given projects. The method nevertheless provides a relatively simple way for surveyors to evaluate expected value based on the probability of occurrence (likelihood) and associated impact. EMV analysis develops the simple risk impact assessment discussed earlier, as:

Expected Monetary Value = Likelihood x Impact

Table 4.9 Sensitivity testing outputs for a development appraisal

	Residual valuation	Rent decreases by 10%	Landlords yield increases by 10%	Construction costs increase by 10%	Finance costs rise by 10%	Construction time increased by 25%
Profit on costs	24.27%	12.20%	13.17%	15.87%	23.47%	22.75%
Profit on GDV	19.53%	10.87%	11.64%	13.70%	19.01%	18.53%
Developers yield	12.51%	11.30%	12.54%	11.67%	12.43%	12.36%

However, EMV analysis adds a valuation to the outcome of the process to develop project contingency funds. So for a risk event, such as the poor ground conditions, the cost plan in Table 1.12, chapter 1, includes an allowance of £182,450 for substructure. This is the base estimate and does not include an allowance for potential risk. However, the quantity surveyor along with the project team have determined there are four possible outcomes:

- Outcome 1: 10% likelihood that poor ground will increase substructure costs by 3% (£5,500);
- Outcome 2: 35% likelihood that poor ground will increase substructure costs by 7% (£12,780);
- Outcome 3: 35% likelihood that poor ground will increase substructure costs by 9% (£16,420);
- Outcome 4: 20% likelihood that poor ground will increase substructure costs by 12% (£21,894).

Using this information, we can now use expected value to determine the size of the contingency allowance the quantity surveyor should include for this risk event as illustrated in Table 4.10.

The above example provides a reasonably simplistic overview of expected monetary value. The technique is also discussed in the context of risk estimating during bid development in chapter 3.

Discussion point 4.13: Now you have studied both qualitative and quantitative risk management techniques, can you explain the difference between the two?

4.4.6 Risk response and risk mitigation

Once the risks exposed on projects have been identified, analysed and evaluated, project teams need to understand the *risk appetite* or the amount of risk employers are willing to accept to ensure their objectives are achieved. The final stage in the risk management cycle is then to develop suitable risk response and/or mitigation

Table 4.10 Expected monetary value analysis for ground risk

Outcome (i)	Probability {P(X_i)}	X_i	X_iP(X_i)
1	0.10	£5,500	£550
2	0.35	£12,780	£4473
3	0.35	£16,420	£5747
4	0.20	£21,894	£4379
		E(X)	**£15,149**

Table 4.11 Example risk mitigation approach

Description	Action required
Unacceptable; red	Comprehensive action required immediately. Must eliminate or transfer the risk.
Highly undesirable; amber	Attempt to manage, avoid or transfer risk. Some immediate action is required along with the development of a comprehensive action plan.
Manageable; yellow	The team is uneasy about carrying this risk. Retain and manage the risk.
Negligible; grey	The team will carry the risk but it needs to be actively managed.

Source: RICS (2015, p. 12).

strategies. The RICS Guidance Note on Risk Management (RICS, 2015) asserts that project teams should agree a series of actions for the various levels of risk events, before mitigation or responses are planned and implemented. As with the risk assessment matrix, this usually adopts a colour coding scheme, allowing project teams to track and prioritise their risk response and mitigation actions. An example of this, based on the RICS Guidance Note, is shown in Table 4.11.

Once the risk attitude of employers has been determined, project teams can start to consider risk response and mitigation measures available to them. As risk is an expensive commodity, teams will first try and reduce risks within projects, either in terms of their anticipated likelihood of occurrence or their impact on projects, to bring them to levels employers feel are acceptable (within their risk appetite). If that is not possible, then teams will seek to mitigate risks in other ways. Risk response and mitigation will normally include:

1   *Risk treatment* – seek to reduce risks to acceptable levels. Risk mitigation can take one of several forms including:

   a   Implementing new processes;
   b   Changing specifications e.g. including more off-site manufacture or innovative materials;
   c   Undertaking more facilitating works before main contracts;
   d   Using named suppliers or supply chain collaboration;
   e   Changing contractual conditions;
   f   Modifying procurement routes proposed;
   g   Adding further resources to projects.

2   *Risk transfer* – risk transfer seeks to move the consequences of risks to third-party organisations. Whilst this does not remove the risk from projects, it does protect employers of effects should they occur. Risk transfer can include:

   a   Transferring the risk to contractors through a combination of techniques:

      i   Amendments to contract clauses;
      ii   Procurement route selection e.g. design and build;
      iii   Adoption of performance specifications;

      iv   Use of contractor design.

   b   Transferring risk further down the supply chain, to tier 2 or 3 contractors or specialist suppliers.

   c   Insuring risks with specialist insurers, in return for payment of suitable premiums.

3   *Tolerate risks* – project teams may decide that risks either fall below the risk tolerance of employers, so they can be accommodated within their risk appetite, or the financial effects of transferring or insuring the risk make the strategy too prohibitive.

4   *Risk termination or avoidance* – risks are so significant and potential impacts so damaging that project teams decide the only viable solution is to change project plans to avoid risks completely. That may involve re-design of parts of buildings, or change in positioning or orientation, or even change proposed sites of buildings. Such levels of pre-emptive action are really a last resort for project teams and will only be implemented when other possibilities have been exhausted.

5   *Risk sharing* – encouraged under project management–based contracts such as NEC3, where risk sharing is a type of risk transfer, but it is more collaborative. Risk sharing will look at the skills of competencies of employers and contractors and allocate risks to parties most able to deal with them, and then making them contractual obligations.

4.4.7 Risk management strategy – a project management approach

The final part of this section seeks to bring all these phases of activity together and looks towards the implementation of an overall risk management strategy. Many texts written on risk management suggest the implementation of risk management strategies form an essential part of the successful management of projects. When using NEC3 (clause 11.2(14) and clause 16) the development of contract risk registers and risk review meetings are obligatory. As the RICS Guidance Note on Risk Management (RICS, 2015) and the two CIRIA research publications looking at risk management in construction (Godfrey, 1996; Lewis *et al.*, 2014) make clear, it is best practice when managing risk to implement an effective risk management process from pre-contract through to completion of the construction phase.

For risk management processes to be successful, RICS (2015, p. 15) asserts that project teams should produce, enact, prepare or instigate the following project deliverables:

- Risk management plans;
- Risk registers;
- Risk ranking and critical risk identification;
- Quantitative cost risk analysis results;
- Quantitative schedule risk analysis results;

- Risk response plans;
- Risk-response project reviews;
- Risk management reports;
- Risk maturity assessments;
- Procurement options reviews;
- Tender return risk reviews.

Some of these issues have already been discussed in the preceding sections, and others such as procurement risks are discussed in chapter 2. As a result, the discussion in this section will be focused towards what the authors consider to be the major outputs most project teams should be developing, thus:

- Risk management plans;
- Risk review meetings;
- Risk registers.

Risk management plans

Risk management plans, also called risk management strategies, are essential parts of project governance systems. They sit alongside other key project documents such as strategic procurement plans, health and safety plans, and contract strategies as key implementation and control documents. Risk management plans are normally drafted by project managers and either construction managers or central risk managers for contractors. Documents should provide:

- Details of employer risk appetite and risk tolerance;
- The identification of the member within project teams with responsibility for risk management;
- A detailed description of how risk will be identified, analysed, evaluated, managed and reviewed;
- Frequency of risk review meetings (note NEC3 requires these monthly post-contract);
- Risk management software, tools and techniques project teams will use as part of the overall risk management strategy;
- Reporting – the forms and structure of risk reviews and employer risk reports.

Risk review meetings

As with all meetings, the frequency of them needs to be decided and agreed by project teams, to ensure they remain valuable opportunities to discuss possible risk events and identify risk mitigation and risk reduction opportunities. Having too many risk review meetings can cause them to be seen as a massive inconvenience by most, an opportunity to apportion blame or just an opportunity to network. It is important these meetings are held, but not too frequently.

Chaired by the person nominated as risk manager, risk review meetings are designed to give project teams opportunities to reflect on the overall risk situation on projects, identify and discuss critical red-flagged risks and to evaluate any new risks identified as projects move forwards. Atom Risk (Hillson and Simon, 2012) recommends these meetings take place every quarter and last approximately half a day. It is worth also noting that as part of the early warning system (an element of the risk management framework) in NEC3, project managers can mandate that these workshops take place and also mandate attendance.

Exercise 4.3

You are responsible for the risk management of XZA Group's new hotel development; prepare an agenda for the first risk review meeting.

Risk registers

Risk management texts often discuss the importance of risk registers, although few provide visual examples. This book will do so, and provide an actual example based for the hypothetical hotel project.

Thus far the role of risk registers has been the point of some contention. Some risk experts suggest risk registers are merely reporting tools, and their purpose is to merely document risk management processes, thus giving them a marginal role. Conversely, others have elevated risk registers to a slightly higher position, suggesting they are useful recording tools in risk management processes, as they track risk events and provide a framework around which risk management can take place. Finally, the third view treats risk registers as a principal document, arguing it sits at the very core of the risk management process. For example Barry (1995) defines a risk register as a comprehensive risk assessment system, used as a formal method of identifying, quantifying and categorising risks and providing the means of developing a cost-effective method of controlling them. This view has gathered pace following the publication of collaborative contracts such as NEC3 and PPC2000, which both put risk registers at the core of risk management. It is clear that risk registers are now seen as essential project documents.

It is important at this point to distinguish between project risk registers and contract risk registers. The project risk register is a document, created at the inception of projects, which monitors risk through the entire project life cycle. This prompts project teams to make effective risk management decisions at the earliest possible stages of projects, when the costs of change are minimised. It must also be remembered that the risk register is not a static document, it is a live and ever-evolving record of risk events. As risk review or reduction meetings are held, risk registers should be reviewed and updated to reflect the latest range of risk information. Contract risk registers are developed as a direct requirement of the express provisions of contract strategies selected by employers.

In terms of structure and content, risk registers are often described as tools used to record and document information generated through the use of project risk management processes, which enables users to consciously evaluate and manage risks as part of decision-making processes. Godfrey (1996, p. 42) suggests risk registers are likely to include the following information:

- Details of each risk event anticipated for projects; these could be sub-divided into more detailed clusters of risk, with some form of hierarchy;
- Records of any possible risk reduction or mitigation actions proposed by project management teams;
- A measure of risk likelihood (probability) and consequences;
- Identification of risk owners;
- Importance, cost and acceptability of risks;
- Practicality of mitigation actions;
- Cost of actions;
- Change in the likelihood or consequences of risks following mitigation;
- Some measure of costs against benefits.

Table 4.12 illustrates a risk register for the hotel project initially introduced in chapter 1. It includes the majority of the features identified above, and a variety of risks from both the design and construction phase, in addition to employer and externally generated risks. For live projects, this document would include hundreds of different risks, which are likely to have an impact; however, to keep things simple, the risk register for the hotel project only identifies three design and three construction risks to give an illustration of how risk registers are developed.

Exercise 4.4

Returning to the XZA Group hotel scenario, produce a risk register for the project using the risks you identified when completing Exercise 4.1. You should use the example risk register provided in Table 4.12 to assist you with this task.

4.4.8 Summary

Risk management systems are important tools to be used by all parties in construction supply chains. It is important that quantity surveyors, project managers and construction managers engage with these systems since they form the basis for many strategic and operational decisions.

Risk management is essentially a means to an end which should be encouraged as a systematic tool for decision making, rather than just something that is implemented post-contract at the bequest of contracts such as NEC3 or PPC2000. Undertaken proficiently, risk management adds real value to projects.

Table 4.12 Example risk register for hotel development project

Risk Identification and Mitigation					Residual Risk Assessment				Action Plan for Residual Risk			
Ref	Cat.	Risk	Potential impact	Completed mitigation action (to date)	Prob.	Impact	Risk score	Financial impact	Action plan	Action by which party	Next action target date	Achieved
1	Client	Changes to brief and/or scope of works	Agreed strategies and plans require review and amendment. Cost and time implications.	Change Control procedure enforced to provide auditable process.	1	3	3 amber	£175,000	Project Reviews implemented to predict and mitigate potential changes.	Design team		
9	External	Neighbours – dust, noise and the proximity of the local college; Students – lack of barrier around site.	Stoppages to construction due to complaints/ delay costs.	Management issue of main contractor's site logistics.	1	2	2 green	£30,000	Main Contractor to provide regular progress reports on the management of the site.	Contractor, PM	On-going	
12	Design	Building regulations approvals.	Delayed approved by Building Control results in late design changes.	Independent review before submission for approval.	1	2	2 green	£4,000	Continuous communication required to ensure that user expectations are managed during the project.	Design team/ client/ employer/ users	On-going	
23	Design	Floor finishes for circulation are being reviewed with Client FM team.	Changes to the current design may delay works.		1	2	2 green	£26,000	Review underway. Monitor to not affect Wallis programme.	Architect	On-going	

(Continued)

Table 4.12 (Continued)

	Risk Identification and Mitigation				Residual Risk Assessment				Action Plan for Residual Risk			
Ref	Cat.	Risk	Potential impact	Completed mitigation action (to date)	Prob.	Impact	Risk score	Financial impact	Action plan	Action by which party	Next action target date	Achieved
46	Construction	Decants of existing office building in progress – if these cannot be completed by the time (and within the duration) currently assumed the construction schedule may be affected.	Delay to the project.	Decant Manager to assist and liaise with client over the decants.	2	3	6 red	£10,000	Monitor situation and work with client to ensure legal speedy decant prior to demolition.	Decant Manager Client	On-going	
49	Construction	Extreme weather.	Significant project delays.	Design and construction methodology will contemplate this and reasonable measures put in place. Time risk allowance incorporated into programme in general.	3	2	6 red	£36,000	Monitor	Main Contractor	August	

Please note the risk register does not include the scheduling impact of the risk events identified; if this information is included an additional column would be added to the residual risk assessment to include these details.

4.5 Establishing project cashflow – the perspective of employers

Cashflow is often called *the lifeblood of the construction industry* (RICS, 2011). Cashflow is a very different concept to profit; most people seem to evaluate the success of a firm not by its ability to simply survive in a very competitive and volatile market such as construction, but by the on-paper profits the company presents to its shareholders at the end of the year, or the company reports to Companies House. As a result, businesses can have really excellent returns on capital, shown by high profits and excellent returns to shareholders, but they can also have terrible cashflows, as a result of not implementing effective financial management systems or simply not getting paid on time or at all.

As many championship football clubs and indeed public limited companies demonstrate, assuming shareholders will bail the company out from time to time by injecting more capital, profitability is not the key to survival, as a business can survive without profit. On the other hand having the ability to pay the company's bills, therefore having cash in the business, is critical to survival. As with your own bank account, a business's ability to pay the bills is dependent on cashflow.

More construction projects and indeed construction companies fail due to a lack of cashflow than for any other reason in the UK (Mutti and Hughes, 2002). Yet in construction management and construction finance texts, cashflow is typically explored from the perspective of main contractors and their supply chains. This is usually the case as most consider detailed project level cashflow analysis to be the preserve of the commercial manager, and a key part of the contractor's post-contract cost management system, and accompanying internal accounting procedures. This undoubtedly includes a 'project dashboard' of key financial indicators including cashflow, cost to value reconciliation, the financial management of the supply chain (package tendering amongst other things), analysis of return on capital employed (ROCE) from the project and so on.

As would be expected this theme is continued to an extent in this book, with chapter 6 providing comprehensive coverage of the key issues associated with cashflow, including the theory and practice of cashflow management. As such for a comprehensive review of this the reader should first read this section of chapter 6.

However, when we say *cashflow is the lifeblood of construction* that concept does not begin at tier 1 in the supply chain with the main contractor. It is equally important to explore cashflow from another perspective, that of employers. Cashflow will be critical to ensuring they can both afford the project and more important have sufficient liquidity to make the required payments to the main contractor. For those avid viewers of Channel Four's *Grand Designs* or the BBC Two's *The House That £100,000 Built*, you will have noticed that one common trait exhibited by the majority of the people the programmes follow is some level of financial difficulty. This is usually related to cashflow and the availability of funds. For the majority this is followed by a scramble for funds to keep paying builders.

What you are seeing is an employer who decided not to employ the expertise of quantity surveyors, project managers or construction managers to provide post-contract financial management and advise on cashflow demands and finance. This text attempts to 'break the mould' and looks at post-contract cashflow from multiple perspectives, including that of employers. As a result, this part of chapter 4 is devoted to the management of the employer's cashflow.

4.5.1 Employer's need for post-contract financial management

The importance of post-contract financial management from the perspective of employers has significantly increased in recent years. Part of this is due to the stringent financial monitoring companies are required to implement as part of international accountancy standards and partly due to the fact that employers have accelerated and changed the way they commission, procure and manage projects. As a result, levels of pre-contract design and financial management have diminished, making it even more critical to monitor projects post-contract. Most contracts require contractors to provide employers with cashflow forecasts, normally as part of pre-commencement documentation, so employers know the amount and timing of monetary commitments (RICS, 2011). It is critical that quantity surveyors implement rigorous cashflow models that are capable of monitoring and providing real-time financial information related to projects, similar to the one for the hotel development illustrated in Figure 4.11. The cashflow forecast provides:

- 'Snap shot' of the financial state of the work, usually after each payment;
- The progress of the contractor from month to month as compared with the pre-commencement cashflow and programme;
- Anticipated final account for the project;
- Predicted impact of variations issued by the architect (if JCT Standard Form of Contract);
- Impact of possible loss and expense claims arising from relevant matters (if JCT Standard Form of Contract);
- Financial impact of compensation events (if NEC3 Standard Form of Contract);
- Details of other extra payments made or likely to be made to the contractor;
- Impact of fluctuations, if the clause is in use.

4.5.2 Finance and bank drawdown facilities

For the vast majority of employers, the majority of funding for construction projects is likely to be debt (loans) rather than equity (employers' own money), and as a result, they will have sourced some form of finance from banks or other financial organisations. As part of the due diligence process, banks will engage the services of quantity surveyors to evaluate proposed costs of projects, along with property experts to advise on out-turn valuations. Assuming loan to value (LTV) ratios are acceptable, and the level of risk presented to financial institutions is within their risk tolerances, they will approve finance for schemes.

However, unlike unsecured personal loans members of the public source from banks, these loans will not be released in full to borrowers (the construction

Client/Employer Project Cashflow 2017	Jan	Feb	Mar	Apr	May	June	July	Aug	Sept	Oct	Nov	Dec	Total
Land costs													
Land purchase													
Site purchase		125,000	125,000	125,000	125,000	125,000	125,000	125,000	125,000	125,000	125,000	125,000	125,000
Interest (payable on 75% of site purchase costs)			1,875			1,875			1,875			1,875	7,500
Repayments	0.00	0.00	0.00	0.00	0.00	0.00	0.00	0.00	0.00	0.00	0.00	0.00	0
Total land	125,000	125,000	125,000	125,000	125,000	125,000	125,000	125,000	125,000	125,000	125,000	125,000	125,000
Construction costs													
Substructure	27,368	109,470	45,612										182,450
Superstructure			112,727	178,978	234,912	201,678	135,736				56,459		920,490
Internal finishes							126,452	91,568					218,020
Fittings, fixtures and equipment		8,700									78,720	174,980	262,400
Services					21,000	130,404	276,200	256,180	105,936		46,440	140,000	976,160
External works	198,700							49,131				23,619	271,450
Main contractor prelims	110,400	23,418	23,418	23,418	23,418	23,418	23,418	23,418	23,418	23,418	23,446	23,418	368,026
Variations													0
Claims													0
Total construction expenditure	336,468	141,588	181,757	202,396	279,330	355,500	561,806	420,297	129,354	23,418	205,065	362,017	3,198,996
Debt finance for construction													
Debt finance (80% of construction costs)	269,174	113,270	145,406	161,917	223,464	284,400	449,445	336,238	103,483	18,734	164,052	289,614	2,559,197
Interest payment			2,639			3,349			4,446			2,362	12,796
Revolving advances	269,174	113,270	145,406	161,917	223,464	284,400	449,445	336,238	103,483	18,734	164,052	289,614	2,571,993

Figure 4.11a Example employer cashflow for proposed hotel development.

	Jan	Feb	Mar	Apr	May	June	July	Aug	Sept	Oct	Nov	Dec	Total
Client Project Cashflow	2017												*Total*
Professional and other fees	343,566	347,002	352,347	396,349	456,179	537,065	654,796	745,404	785,049	797,583	846,572	846,572	846,572
Land													
Balance (25% of purchase cost)	195,000												195,000
Construction													
Balance (20% of construction costs)				40,479	55,866	71,100	112,361	84,059	25,871	4,684	41,013	72,403	507,837
Finance													
Commitment fee	5,000												5,000
Land finance			1,875			1,875			1,875			1,875	7,500
Construction finance						3,349			4,446			2,362	10,157
Client's own finance		3,436	3,470	3,523	3,963	4,562	5,371	6,548	7,454	7,850	7,976	8,466	62,619
Professional fees													
Design team fees	143,566											95,710	239,276
Cumulative expenditure	343,566	347,002	352,347	396,349	456,179	537,065	654,796	745,404	785,049	797,583	846,572	1,027,389	1,873,961

Figure 4.11b Example employer cashflow for proposed hotel development.

employer). Banks implement a process called 'drawdown', often linked to project milestones, which reduces bank risk by relating the value of works completed to funds they release. As a result, banks require detailed cashflow forecasts, showing anticipated monthly valuations and the amount of capital likely to be drawn down, as illustrated in Figure 4.11. Employers also need to advise banks of contractual payment schedules to ensure funds are available in sufficient time to allow payments to be made.

As part of bank post-contract cost control, they will often engage the services of bank monitoring specialists (often quantity surveyors) to ensure that employers drawdown requests are in line with original cashflow forecasts and to provide explanations if they differ. Banks will reconcile any differences with original cost plans and valuations received for completed buildings to ensure risks are not significantly amended by any changes. For employers, it is critical that cashflow forecasts are accurate and drawdowns do not change significantly from those forecast. Any deviation from original agreements could be seen as increased risk and banks may impose higher interest rates, penalty charges or additional agreement fees for accelerated drawdown (RICS, 2011, p. 3).

> **Discussion point 4.14:** Why do you think cashflow management is as critical to employers as it is to contractors?

4.5.3 Financial reports

A core function of the PQS post-contract is to keep employers updated on the financial performance of projects. Using the cashflow information quantity surveyors develop monthly financial reports using a format similar to that illustrated in Figure 4.12. As part of internal project governance, it would be normal for these reports to be initially reviewed by project managers or contract administrators prior to their final issue to employers.

Monthly reports provide employers with a 'snap shot' of the financial position of projects and the short- to long-term forecast of potential adjustments and financial risks. Thus employers may anticipate and control current and future financial commitments, allowing them to prepare for any anticipated overspends by seeking additional bank loans or releasing further equity from assets or other sources to fund any differences.

For employers managing a programme of works using a pre-determined annual capital budget, as is common in both public sector and private organisations with large property portfolios, property directors may need to revisit annual asset management plans to identify areas of potential savings or look to delay later projects to free up additional funding.

Financial Statement

No: _____   As: _____
For: _____
Construction period   _____   Extension (weeks)   _____
  (weeks)
Date for completion: _____

Authorised Commitment
1   Contact Sum
2   Client revised requirements
3   Fluctuations

Variations and adjustments
1   Omission of provisional sums for risk and daywork
2   Adjustment of approximate quantities and provisional sums
3   Other architect's or contract administrator's instructions
4   Instructions confirmed by contractor but for which architect's or contract
      administrator's instructions have not been issued
5   Anticipated variations in cost for which no formal instructions have been issued
6   Amount of direct loss and/or expense resulting from disturbance to regular
      progress of the works

                ESTIMATED FINAL COST   £
                Forecasted Net Under or Overspend   £
Liquidated Damages
Following the issue by the architect of a certificate under clause 2.32.2.1 and/
  or 2.32.2.2 of the Conditions of Contract the employer may elect to claim
  liquidated and ascertained damages amounting to £_____unless further
  amendments to the original completion date (extensions of time) are granted by
  the architect or contract administrator, assuming completion by _____
Claims
Claims lodged by not included   £
  above

Figure 4.12 Monthly financial statement to be issued to the client.

4.6 Managing consultancy income and contracts with employers

When looking at the role of professional quantity surveyors or cost consultants, it is important to also examine the way consultancies are managed in-house. Most practices employ professional staff including accountants to manage the practice. As most readers will at some point be seeking professional membership and therefore chartership with either or both the CIOB and RICS, it is important that you understand the principles of corporate accountancy and the management of businesses.

4.6.1 Managing consultancy income

Fee income generated by consultancy activities must be managed and controlled. Practice managers provide day-to-day management of invoicing, chasing unpaid bills, paying creditors and other financial issues such as processing monthly salary payments. It is important consultants maintain positive cashflows, as discussed in chapter 6. However, managing consultancy income can also present some unique challenges; the biggest is *client/employer money*. For those seeking chartered membership of the RICS, you will be expected to understand these unique problems and be able to provide a reasoned answer as part of the ethics assessments in interviews related to the *RICS Rules of Conduct for Firms*.

4.6.2 *Appointment of quantity surveyors or cost consultants*

It is important that the appointment of quantity surveyors is set out in writing as soon as is practically possible, thereby forming a written contract between employers and the quantity surveying practice commissioned to provide financial management. Contracts (terms of appointment or consultant agreements) are important for both parties as they define the scope of services that are to be provided and fees that will be charged for those services. If quantity surveyors are to perform additional roles such as value managers or risk managers this must also be clearly agreed and included in terms of engagement.

A number of standard forms exist for the appointment of quantity surveyors including:

- RICS Standard Form of Consultant Appointment;
- NEC Professional Services Contract;
- JCT Consultancy Agreement 2011;
- CIC Consultants Contract.

In the event employers or consultants decide not to use standard forms of appointment in favour of developing their own bespoke terms, documents should include provision for:

- The date of the agreement;
- Scope of appointment;
- Details of the project and other members of the professional team;
- Details of services to be provided;
- Details of the employers' budget and programme requirements;
- Duty of care obligations;
- The basis of remuneration (payment) and detailed arrangements;
- Details of Professional Indemnity Insurance requirements;
- Details of any collateral warranty requirements;
- Copyright, intellectual property and document ownership;
- Assignment and sub-letting arrangements;
- Suspension of the agreement;
- Termination and dispute resolution;
- Confidentiality arrangements;
- Other statutory obligations.

As part of agreements, parties need to define the scope of the services quantity surveyors will provide. As most consultancy practices will be RICS regulated, it is normal practice for the majority to adopt the *RICS Standard Form of Consultant Appointment*. In support of this document, the RICS also publishes an append-age document entitled *RICS Quantity Surveyor Services*; the nine-page support-ing document provides a comprehensive breakdown of the services a quantity surveyor could be expected to provide, broken down into:

- Core services e.g. cost planning;
- Supplementary services – e.g. life cycle costing;
- Project-specific services (these are non-standard specialist services).

Using the document as a base to identify the services provided, employers and consultants can work through the extensive lists of potential services and identify what is to be included or excluded. This is illustrated in the excerpt from the *RICS Quantity Surveyor Services* document completed for the hotel project, shown in Figure 4.13; this subsequently forms part of contractual agreements.

In 1997 the Monopolies and Mergers Commission undertook a detailed review of the use of fee scales within the construction industry. Its investigation con-cluded the control of professional fees by professional bodies such as the RIBA, RICS, CIOB and others using pre-defined fee scales was anti-competitive and operated against the public interest. As a result, fee-scales for quantity surveyors and other RICS professionals were 'scrapped' in February 2000. It is now the norm for fees charged to be privately negotiated with employers who commission a service, or for consultants in bid for work in competition with others. The *Spons 2015 Architect and Builders Price Book* (AECOM, 2014) provides benchmark fee guidance which suggests for a new-build contract, priced at £3.5m (exclusive of VAT) the fee would range from 1.5 to 2.5%, with a mean of 2%. Similarly Designing Building Wiki (2016) suggests quantity surveying fees for commercial office buildings in London would range from 1–2% depending on the forecasted construction cost, as illustrated in Table 4.13.

Design (RIBA outline plan of work 2007)

- 1.3.1   Prepare, maintain and develop a cost plan and cashflow forecast.
- 1.3.2   Advise on the cost of the Professional Team's proposals, including effects of site usage, shape of buildings, alternative forms of design, procurement and construction, etc. Advise on any cost variances to the allowances contained in the cost plan.
- 1.3.3. Measure gross floor area.
- 1.3.4   Measure net lettable/saleable floor areas.
- 1.3.5   Confirm the scope of Building Contract to the Client and advise on additional works required by third parties.

Figure 4.13 Completed excerpt from RICS Quantity Surveyor Services for Hotel Devel-opment Project

Source: RICS (2008).

Table 4.13 Professional fees for cost consultants based on new build London commercial office space

Under £1.5m	£1.5m–£3.0m	£3.0–£10.0m	£10m–£25m	£25m–£50m	£50m +
2%	2%	1.5%	1.5%	1%	1%

Source: Designing Buildings Wiki (2016).

4.7 Chapter summary

Chapter 4 has provided a review of additional services provided by quantity surveyors, including value management, value engineering and risk management. The chapter also explains the role of professional quantity surveyors in post-contract financial reporting and management. In addition the management of employer funds by consultants and terms of engagement (contracts) between employers and professional quantity surveyors have been reviewed, along with the need for the quantity surveyors to ensure pre-contract cost forecasts are as accurate as reasonably possible.

It is hoped the chapter has demonstrated the importance of post-contract financial management from the perspective of employers, whilst also introducing readers to the important and growing area of additional services such as value and risk management. In an industry where consultants are operating in highly competitive markets for their commissions, demonstrating added value through the provision of services like value and risk management can make a real difference to their ability to win work.

4.8 Model answers to discussion points

Discussion point 4.1: How useful do you think the AACE's work stage model would be to an employer who is trying to determine the level of accuracy their quantity surveyor may provide?

There has been a lot of argument offered against the AACE's framework, with critics arguing it is not realistic and the design development suggested is rarely achieved. However, the framework provides a useful guide, against which the performance of the quantity surveyor can be reviewed.

Discussion point 4.2: To what extent do you agree with the view that cost planning is more art than science?

Forecasting is very much a skill held and honed by quantity surveyors over their careers. In the UK, unlike other countries such as Australia, where a scientific approach based on rate build-ups is employed, quantity surveyors are tasked with using historic cost data to which appropriate professional adjustments are applied to predict the cost of projects.

Discussion point 4.3: An employer is considering a major business relocation to a large, modern warehouse facility adjacent a major motorway network. What do you think, if anything, value management could add?

Implementing value management on this project will allow the employer to ensure the final scheme fully meets its requirements. By allowing a value management consultant to review the existing facility and gather data from its users, the project team will be able to ensure any problems and inefficiencies are designed out of the new development as much as possible. As a result the employer will receive a more functional space, which better meets its requirements.

Discussion point 4.4: Identify the main types of construction employer. How would you define their value system?

Construction employers can be sub-divided as follows:

- Public sector – value systems defined by UK and EU policy frameworks, based on collaboration, best practice, quality rather than cost and delivery of value for money.
- Private sector – price-driven, value system based on lowest possible cost and fast delivery. Usually seek to transfer risk to constructors.

Employers can also be divided into experienced and inexperienced. Experienced employers will have a far more comprehensive understanding of the construction process, and may have more complex demands that project teams need to deliver. Inexperienced employers will generally seek to appoint an employer's agent or will let architects lead them through the process. Inexperience may mean value systems are very simplistic, and they may not receiving the buildings they had hoped for.

Discussion point 4.5: 'Sustainability is becoming a critical aspect of many employers' corporate values, but this does not appear to be reflected in current value management practice'. Critically evaluate this statement.

This question could be answered as an extended essay. It is anticipated your answer will develop an argument based around the authenticity, validity and legitimacy of the statement.

- Define sustainability.
- Define value and discuss various constructs of value relating to sustainability such as ethical dimensions.
- Examine arguments that sustainability is becoming a strategic goal of employers, discussing issues such as corporate social responsibility and socially focused organisations such as co-operatives.
- Examine the links between sustainability and the definition of value.
- Discuss the central argument in the work of Abidin and Pasquire (2005); to what extent is the model/framework developed in Abidin and Pasquire's work and presented in this text an answer to the problem?
- Define and discuss the way ahead for value management practitioners in terms of the sustainability agenda and embedding value-based principles into practice.

Discussion point 4.6: How do value management and value engineering align with the UK Treasury's call for improved value for money in public sector construction and civil engineering projects?

The UK Treasury sees the attainment of value for money as a ratio between life cycle cost and function. If employers implement value management this could be seen as a key tool project teams can use to ensure delivery against this requirement. Value management allows teams to ensure maximum functionality is delivered against employer objectives. Value engineering allows project teams to ensure the most efficient use of resources is achieved in terms of both capital and life cycle expenditure by mapping specifications against employer requirements for buildings.

Discussion point 4.7: Differentiate between risk and uncertainty.

The two terms are often used interchangeably; however, uncertainty is something that is impossible to predict or measure. Risk, on the other hand, can be measured, quantified or forecast, allowing people faced with risks to make decisions as to whether they are prepared to accept them or not.

Discussion point 4.8: Explain the 'optimum balance of risk management' in the context of a major construction project.

Risk is an expensive commodity; if the employer tried to comprehensively manage all the risk associated with the project, the project would never be constructed or the cost would be so prohibitive it could not be afforded. As a result the employer needs to determine their risk tolerance or risk appetite, how much of this are they willing to accept and use this to identify the risk they will retain and the risk they will transfer.

Discussion point 4.9: Comment on the purpose of risk breakdown structures, explaining how their adoption would enhance project team ability to manage risks.

Risk breakdown structures provide a framework around which risk managers can run risk workshops. However, some argue risk breakdown structures can prevent risk events being identified if they fall outside the classifications provided.

Discussion point 4.10: Explain the purpose of a risk management workshop. How would you manage the workshop environment? Would facilitation or leadership be required?

The risk management workshop occurs at the risk identification stage of a project. It is designed to allow the project team to identify as many risk events as possible before considering their probability of occurrence and potential impact for the project. As such it is better if the workshop is facilitated and not led, as leadership is likely to curtail or prevent open discussion.

Discussion point 4.11: Comment on the limitations of the risk matrix model.

Risk matrix models are very popular, but they are based on absolute judgements and provide little space for borderline scenarios. The risk (probability x impact) is allocated to a box and colour coded based on the risk score. So a risk scoring 9.91 could be coloured amber, whilst a risk scoring 10.04 could be coloured red, with very different risk management systems applied, yet in reality, both risks should be treated in a similar way.

Discussion point 4.12: Explain the principles of the Ishikawa cause-and-effect model and comment on its applicability to construction risk analysis at project level.

The Ishikawa model is based on engineering and manufacturing processes, which dictate that events (effects) must have root causes. The model allows users to identify and map many possible causes of the risk, to allow them to identify the most likely one.

Discussion point 4.13: Now you have studied both qualitative and quantitative risk management techniques, can you explain the difference between the two?

Qualitative risk is more focused on the identification and prioritisation of risk events; most risk management in construction is qualitative. Quantitative risk management is highly statistical in nature and seeks to understand the impact or possible impact of risk events through extensive calculation. This is used extensively in real estate and financial management but less so in construction.

Discussion point 4.14: Why do you think cashflow management is as critical to employers as it is to contractors?

Most employers are highly reliant on debt funding to deliver projects. As a result they are likely to be using a financial institution to support them with money needed to build. Institutions will not provide the money without a clear understanding of how and when it will be spent. As a result clear cashflow predictions are needed to ensure the funds can be effectively drawn down and managed.

4.9 Model answers to exercises

Exercise 4.1

Your employer, XZA Group, is considering development of a range of basic 2* hotels throughout the UK, following success in the Irish Republic. The model is based on the construction of 40–50 bedroom hotels adjacent to major infrastructure transport links. Despite objections from local residents, XZA Group has recently entered into negotiations to purchase a disused asbestos factory adjacent to a major UK motorway. Using the SCLEEPT mnemonic identify two possible risks related to the project that would fall under each category.

Social – Lack of community support evident from resident backlash; failure to implement Considerate Constructors Scheme.
Cultural – Lack of understanding of UK regulatory system; project specification does not meet needs of target demographic.
Legal – Extensive planning conditions; difficulties with building control due to site contamination.

Economic – Ground remediation costs; risk of project overspend due to regulatory conditions.

Environmental – Pollution caused by asbestos contamination; non-attainment of BREEAM standards required by planning.

Political – Local intervention due to residents' campaign against development; difficulties securing planning permission due to political interference at planning committee stage.

Technological – Problems with extensive use of off-site manufacture; failures in proposed design.

Exercise 4.2

Select one of the risks you identified in the earlier exercise (4.1) and produce a simple Ishikawa cause-and-effect model for the risk.

Your solutions to this task will be individual to you; however, your Ishikawa diagram should identify the effect or outcome of the risk – say overspend due to planning conditions. The causes of the risk should then be mapped on the diagram; for example, you could argue the risk of overspend due to planning is caused by:

- Lack of understanding of planning system;
- Lack of financial management;
- Insufficient risk allowances made in risk register;
- Lack of a robust risk management system;
- Not employing planning consultant.

Exercise 4.3

You are responsible for the risk management of XZA Group's new hotel development; prepare an agenda for the first risk review meeting.

The meeting should be structured as follows:

a   Introductions;
b   Review red risks;
c   Review draft risks;
d   Consider new risks;
e   Review amber risks (if time allows);
f   Update risk register;
g   Close meeting.

Exercise 4.4

Returning to the XZA Group hotel scenario, produce a risk register for the project using the risks you identified when completing Exercise 4.1. You should use the example risk register provided in Table 4.12 to assist you with this task.

Your solutions to this task will be individual to you; however, you may have decided to use the example risk register illustrated in Table 4.12 to guide you in developing your own risk register.

References

AECOM (2014) *Spon's Architects' and Builders' Price Book 2015*. London: CRC Press.

American Association of Cost Engineers (1997) TMC Framework: 7.3 – Cost Estimating Framework. Available at: purchasing.borough.kenai.ak.us/docs/AACE_CLASSIFI CATION_SYSTEM.pdf Accessed 01.04.16.

Barry, L.J. (1995) Assessing Risk Systematically. *Risk Management*, 42 (1), pp. 12–17.

BSI (2000) *BS EN 12973:2000 Value Management*. London. British Standards Institute.

Burtonshaw-Gunn, S.A. (2009) *Risk and Financial Management in Construction*. Farnham: Gower Publishing Ltd.

Designing Buildings Wiki (2016) Building Design and Construction Fee. Available at: www.designingbuildings.co.uk/wiki/Building_design_and_construction_fees Accessed 03.04.16.

Eaton, D. and Kotapski, R. (2008) *Business Management in Construction Enterprise*. Warsaw: University of Technology.

Godfrey, P.S. (1996) *Special Publication 125: Control of Risk – A Guide to the Systematic Management of Risk for Construction*. London: Construction Industry Research and Information Association.

Green, S.D. (1994) Beyond value engineering: SMART value management for building projects. *International Journal of Project Management*, 12 (1), pp. 49–56.

Hillson, D. and Simon, P. (2012) *Practical Project Risk Management: The ATOM Methodology*. 2nd Edition. Tysons Corner, VA: Management Concepts Press.

JCT (2012) SBC/Q2011 Standard Form of Building Contract. The Joint Contracts Tribunal.

Kelly, J., Male, S. and Drummond, G. (2014) *Value Management of Construction Projects*. 2nd Edition. Oxon: Wiley-Blackwell.

Latham, M. (1994) *Constructing the Team*. London: HMSO.

Lewis, H., Allan, N., Ellinas, C. and Godfrey, P. (2014) *C747 – Engaging with Risk*. London: Construction Industry Research and Information Association.

Mutti, C.D.N. and Hughes, W. (2002) Cash flow management in construction firms. In: Greenwood, D. (Ed). *Proceedings 18th Annual ARCOM Conference, 2–4 September 2002*. Northumbria, UK: Association of Researchers in Construction Management. Vol. 1, 23–32.

NEC (2005) *Engineering and Construction Contract*. London: Thomas Telford.

Olympic Delivery Agency (2010) Managing Risk across the Olympic Programme. Available at: learninglegacy.independent.gov.uk/documents/pdfs/programme-organisation-and-project-management/112-managing-risk-popm.pdf Accessed 29.03.16.

Ramus, J., Birchall, S. and Griffiths, P. (2011) *Contract Practice for Surveyors*. 4th Edition. Oxon: Spon Press.

RIBA (2013) RIBA Plan of Work. Royal Institute of British Architects. Available at: www.ribaplanofwork.com/Download.aspx Accessed 02.05.16.

RICS (2007a) *Rules of Conduct for Firms*. Coventry: Royal Institution of Chartered Surveyors.

RICS (2007b) *Quantity Surveyor Services*. Coventry: Royal Institution of Chartered Surveyors.

RICS (2011) *RICS Practice Standards, UK: Cash Flow Forecasting*. Coventry: Royal Institution of Chartered Surveyors.

RICS (2012a) *New Rules of Measurement 2: NRM2.* Coventry: Royal Institution of Chartered Surveyors.

RICS (2012b) RICS Draft Guidance Note – Value Engineering and Value Management. Available at: consultations.rics.org/consult.ti/value_engineering/viewCompoundDoc? docid=863924&partid=864724&sessionid=&voteid=&employeruid Accessed 26.03.2015.

RICS (2012c) *NRM 1 – Order of Cost Estimating and Cost Planning for Capital Building Works1.* Coventry: Royal Institution of Chartered Surveyors.

RICS (2015) *Professional Guidance Note UK: Management of Risk.* Coventry: Royal Institution of Chartered Surveyors.

SAVE (2015) Value Methodology Standard. Available at: www.value-eng.org/pdf_docs/ monographs/vmstd.pdf. Accessed 27.03.2016.

5 Valuations and interim payments

5.1 Introduction to interim payments

A key feature of construction projects lasting over 45 days is the issue of interim payments throughout supply chains. It is normal practice, and a legal requirement under the *Housing Grants, Regeneration and Construction Act 1996* later amended as part of the *Local Democracy, Economic Development and Construction Act 2009* for contractors and subcontractors to receive regular payments at pre-specified intervals.

These intervals may be represented by discrete phases of activity, for example a payment once ground works are completed, another payment when the steel frame is erected and so on. This approach to payment is called an 'activity schedule'. The most common approach to payment is, however, to use regular timeframes for the issuing of funds; this could be weekly, fortnightly or monthly, depending on what parties to contracts agree.

The amount due for payment is stated in interim certificates and is usually, but not always, supported by a valuation of works completed. Interim valuations are normally undertaken by contractors' quantity surveyors, who initially submit applications for payment (valuations). Once received, these submissions are checked and adjusted (if required) by employers' quantity surveyors. The checking process may include meetings and a tour of sites to observe and record progress.

Valuations determine the value of works completed in monetary terms, so as to ascertain the amount due to contractors and subcontractors. Employers' quantity surveyors, through architects or project managers, make recommendations to employers about the amount of money due, and employers retain responsibility to ensure payments are made in a timely manner. It is not necessary to attain absolute accuracy in determining interim valuation amounts, since the aim is 'just' to give supply chains some money on account to support their cashflow needs. On large projects it is usually no matter if contractors are paid £x000 more or less than they ought in interim payments, perhaps due to some minor misjudgement in determining how much work has been completed. Absolute accuracy is, however, required in final payments and final accounts, which are subject to internal and external audits. Quantity surveyors will be a little more cautious in valuing work towards the end of projects, or valuing work for individual specialists as they complete their elements of work, since it is important to avoid paying too much and later asking for refunds. In order to ascertain payment amounts, quantity surveyors take account of:

- The value of measured works completed;
- The value of extra works completed by contractors that have been authorised by contract administrators;
- The work of nominated subcontractors (if provision is made within contracts);
- The overheads of sites (preliminaries);
- The value of any materials delivered to the site;
- The value of any materials in secure storage away from the site, where these are agreed before contracts are signed.

Discussion point 5.1: Why do you think it is important that contractors, subcontractors and other specialist contractors receive monthly interim payments?

5.1.1 Requirements of the Construction Act

Cashflow is the lifeblood of the construction industry, and given high levels of competition and associated poor profit margins, many construction companies rely on positive cashflow outcomes to support their businesses. Contractors often undertake speculative developments to generate the profitable returns demanded by shareholders. The record of the UK construction industry on the issue of payment has been somewhat 'patchy', with many organisations deliberately holding money and delaying its flow through supply chains. Tactics used to delay payment include:

- *Pay when paid* – payment is made to subcontractors when it is received by main contractors; this was prohibited by section 113 of the 1996 *Housing, Grants, Regeneration and Construction Act*;
- *Pay when certified* – main contractors pay subcontractors when their payment is certified by employers;
- *Pay when notified* – a fairly uncommon situation that arises when employers and main contractors negotiate contract clauses that makes payment dates flexible; in other words payment are made at the whim of employers.

In an attempt to deal with unfair payment processes used in UK construction, successive governments have introduced legislative instruments to tackle these practices and to introduce fairer approaches to payment. The *Housing, Grants, Regeneration and Construction Act 1996* sections 109–113 provided provision to close loopholes and introduce more stringent payment practices. Additional rules were incorporated in legislation to further tighten these provisions, through the *Local Democracy, Economic Development and Construction Act 2009*. This Act introduced additional provisions related to notices and payment periods, including two routes to payment. The default route sets out payee- (contractors or subcontractors) led processes, whereby contractors will start by making applications, usually about one week before the agreed due date. The full process is as follows, and is further illustrated in Figure 5.1:

- *Contractor's application for payment* – usually the main or subcontractor's valuation; under JCT this would be submitted seven days (one week) prior to the agreed valuation date (due date);
- *Due date* – the payment date defined in the contract and agreed by the parties;
- *Payment notice by the payer* – this is usually the Interim Certificate, outlining what will be paid and how the sum has been determined;
- *Payee notice* – issued in the event of a default, in this situation the failure to issue a payer notice. This is allowable under section 110B(2) of the *Local Democracy, Economic Development and Construction Act* 2009. It could simply be the contractor's initial valuation. Assuming it is correctly formatted it will become the default notice. This process is illustrated in Figure 5.2. This approach would be preferable for contractors, as waiting to issue a payee notice would delay final payments;
- *Pay less notice (withholding notice)* – issued by employers using section 111(1) of the *Local Democracy, Economic Development and Construction Act*. In the event of a default notice from contractors, the pay less notice allows employers to reduce sums claimed; this must be undertaken at least five days before the final date for payment;
- *Final date for payment* – this is a set period of time from the due date; the exact period is stipulated in contracts and will determine when employers must make payment.

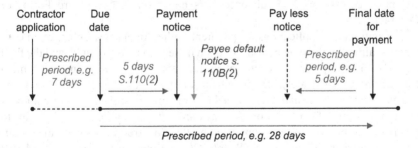

Figure 5.1 Default payment process under Construction Act 2009.

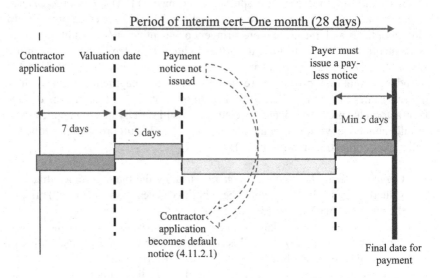

Figure 5.2 Contractor's application becomes default notice based on JCT Standard Building Contract 2011.

The second process facilitated for within the legislation is the *alternative procedure*. Unlike the default procedure, the *alternative procedure* is payer-led, so the requirement for a contractor application is removed from the process. Therefore valuations to determine the amount of payments are undertaken by employers' quantity surveyors in consultation with their counterparts in contractors. The process this time is simpler, but requires similar notifications to be issued at critical points. The alternative approach is illustrated in Figure 5.3.

The analysis above has thus far considered the effects of the legislation for individuals, with the procedural explanation focused on the interplay between employers and main contractors, who reside as the first tier in supply chains. Whilst the procedures are the same for subcontractors in lower tiers, how the various levels interact is also a critical issue, and one quantity surveyors must understand. Figure 5.4 therefore illustrates the full interplay between employers, main contractors and subcontractors based on JCT Standard Forms of Contract.

Subcontractors' applications for payment (valuations) come before main contractors, although the time period for this will need to be negotiated, and the final payment to supply chains occurs seven days after employer payments to main contractors. Unfortunately, despite changes in the *Local Democracy, Economic Development and Construction Act*, subcontractors are still struggling with payment, and terms of up to 90 or 120 days are not uncommon. In an attempt to resolve this, the government supported by the Construction Leadership Council introduced the *Construction Supply Chain Payment Charter* in 2014 that

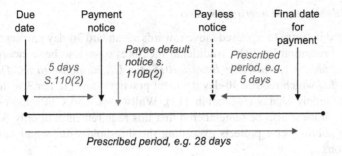

Figure 5.3 Alternative payment process under Construction Act 2009.

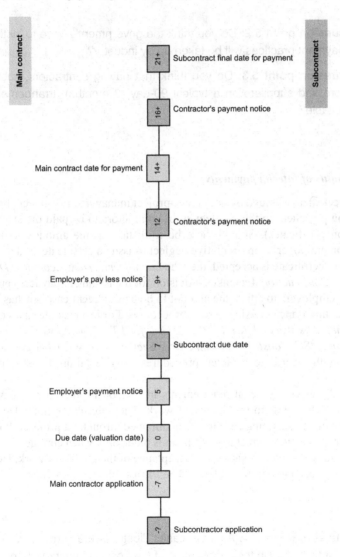

Figure 5.4 Relationship between main and subcontract valuation and interim payment based on JCT Standard Building Contract with Quantities 2011.

committed industry to a phased move towards a standard 30-day payment period with zero retention by 2025. Additional measures have also been incorporated into the *Public Contract Regulations 2015, the UK Response to EU Directive 2014/24/EU* which require 30-day payment practices down to tier 3 within construction supply chains (regulation 113). Whilst this marks an important step forwards, it must also be remembered that this requirement will only be applicable to public sector projects exceeding the threshold value stipulated in the legislation.

Discussion point 5.2: Do you think the government's drive to instigate fair payment practices will be taken up by industry?

Discussion point 5.3: Do you think that paying contractors, subcontractors and suppliers on a typical 90-day (3 months) arrangement is acceptable?

5.1.2 Status of interim payments

Interim certificates issued by architects, project managers or engineers provide a condition precedent to the entitlement of contractors to be paid the stated sums (sums on certificates). Should there be a mistake in the amount certified, or should an employer's representative neglect to issue a certificate, and a contractor default certificate is accepted, then the *Local Democracy, Economic Development and Construction Act* makes provision for the issuing a 'pay less certificate' to allow employers to adjust the amount to be paid. Recent case law has emphasised the importance of using correct procedures. For instance the cases of *Leeds City Council vs Wacco UK Ltd [2015], Galliford Try Limited vs Estura Limited [2015] and ISG Construction Limited vs Seevic College [2014]* are all worth reading with regards to payment procedures and the failure to issue pay less notices.

Second, despite the legal requirements in relation to interim payments, they are not conclusive as to the quality of work. They are merely intended to help supply chains by ensuring cashflow is maintained throughout projects. It must be remembered that in contract law, payment is made only when contracts have been completed, with payment signifying acceptance of the quality of work. However, such an approach in construction could be 'disastrous' for all parties.

Discussion point 5.4: As legal cases illustrate, the courts have been extremely precise in their application of the Local Democracy, Economic

Development and Construction Act 2009. Are courts being overly precise and not realising the complexities of construction?

5.2 Elements of the valuation (employer to main contractor)

Clause 4.10.2 of the JCT SBC/Q 2011 requires, where deemed necessary, architects to instruct quantity surveyors to determine the value of works completed
in order for interim payments to be made to contractors. Similarly, core clause
50.1 from NEC3 empowers project managers with the responsibility to determine
sums to be paid to main contractors. Normally project managers would instruct
quantity surveyors to 'value' works completed on their behalf and recommend
sums for payment.

The duty of quantity surveyors under terms of contract or engagement is to
value work carried out by contractors so that architects or project managers
can issue payment notices required under the terms of contracts and the *Local
Democracy, Economic Development and Construction Act 2009*. Valuations
are normally relatively straightforward, although there will be areas where
quantity surveyors will need to use their professional judgement to determine
sums due for payment. It is important quantity surveyors do not over-value
completed works, thereby putting employers in a precarious position should
contractors becomes insolvent. Or there is a need to terminate contracts; this
would be a breach of duty of care owed to employers. At the same time, contractors are contractually entitled for full payment in respect of works they
have undertaken.

The contract terms applicable to projects will ultimately determine the elements
quantity surveyors include in gross valuations of work. As most construction projects are awarded using either JCT or NEC3 standard forms, the discussion that
follows examines both. Under the JCT Standard Building Contract with Quantities 2011, the gross valuation is deemed to include:

- The total value of work properly executed by the contractor including:

  ○ Measured works;
  ○ Variations where the value of work has been agreed using the valuation
    rules;
  ○ Variations where a quotation has been accepted;

- The total value of materials and goods delivered to the site;
- The total value of any listed items (materials off-site);
- Any amounts due in respect of adjustments to the contract sum (e.g. statutory
  fees and charges; insurance payments *et al.*);
- Any amount ascertained for loss and expense;
- Any amount allowable under fluctuation clause A or B.

By contrast the gross valuation under the NEC3 is much less concise. The interim payment is explained in clause 50.1, which states that the amount due is:

- The price for work done to date;
- Plus other amounts to be paid to the contractor;
- Less amounts to be paid by or retained from the contractor.

The amount stated as due in interim certificates under both JCT and NEC is exclusive of VAT. However, VAT is payable on contract work, so all payments need to include an amount at the appropriate rate. It is usual practice that once valuations have been completed and payment certificates issued, for contractors to invoice employers for the certified sum plus VAT, which is then paid directly by employers. As a result, employers will have a VAT invoice for accounting purposes. Some construction work such a new build private housing is zero rated; in such cases housebuilders will not receive a VAT payment from members of the public when houses are sold, and claim back from HMRC monies they have to pay to builders merchants and the like.

In summary, contracts provide that valuations contain the following elements:

1   The works completed by the contractor (listed in the tender documents);
2   Variations (extras) completed;
3   Preliminaries (site overhead costs);
4   Unfixed materials on the site;
5   Any special pre-agreed materials stored off-site;
6   Claims for direct loss and/or expense from the contractor;
7   Any statutory fees or charges;
8   Any other changes in price levels assuming the employer agreed to cover the cost of these in the contract;

Items 1 to 5, from the above list along with the retention deduction will now be considered in detail. Finally an example is provided in section 5.4 to illustrate how the principles and procedures discussed are brought together by quantity surveyors to determine the final amount due in interim payments.

5.2.1 Completed works

There are several methodologies available to quantity surveyors for valuing construction works, these include:

- **Inspection** – assessment of percentage of works completed by quantity surveyors based on visits to sites;
- **Measurement** – physical site measurement, measurement from contract drawings (or revised drawings related to variations) or a combination of both. This is a very time-consuming process, so may be only adopted to determine a contractor's claim for re-measurement or for the agreement of final accounts;

- **Ogive curve** – identified as a method of valuation by Ross and Williams (2013, p. 362), the Ogive curve appraises the value of work completed with reference to a model of planned expenditure based on contract sums and an appropriate formula which predicts expected cumulative value to be achieved each month;
- **Gantt chart** – with the adoption of more complex software for management of contracts it is possible to link internal financial accounting processes and resources used by contractors to value of completed works based on programmes.

The latest *National Construction Contracts and Law Survey* produced and published by NBS (2015) suggests procurement practices are dominated by fixed price, lump sum design and build, traditional and design-bid-build. Despite the array of valuation methods available to quantity surveyors, the majority of practitioners have continued to value works completed on-site using inspection methods.

The inspection method of valuation is a subjective judgement-based process, which requires quantity surveyors (consultants, contractors or subcontractors) to visit projects and inspect work in progress. During inspections, quantity surveyors make assessments against each bill item of work, bill section or if design and build, sub-element or component of work and determine the percentage completeness of that work. This process is illustrated in Table 5.1, which depicts part of a valuation for a design and build project.

Whilst this is a fairly accurate method of valuation, since it is subjective, no two quantity surveyors will allocate the same percentage to the same work at the same time; it is, however, expected that professional judgements would be reasonably similar, say within 5–10% of each other. Contractors or subcontractors may instinctively try to value the work on the higher side to maximise potential cashflow into their business without being excessive, whereas quantity surveyors

Table 5.1 Example substructure partial valuation

Element	Percentage complete	Contract value (£)	Interim payment (£)
Substructure 270,000	75%	202,500	Breaking out existing floor slabs, piles and other existing substructure
540,000	100%	540,000	Piling Platform and mini piles and other works to boundary
320,000	20%	64,000	Foundations: bored piles with under-ream; ground beams; pile caps
737,200	5%	36,860	RC concrete basement slab
329,800	0%	Nil	Value of work completed
	843,360		

Source: AECOM (2015).

representing payer parties (employers or main contractors) may seek to value works on the lower side to ensure they do not over-value the works. Ultimately the valuation process operates based on judgement and negotiation.

For example say the steel frame is 45% erected at the date of valuation. When undertaking the valuation, the contractor's quantity surveyor will seek to maximise the valuation to enhance cashflow, so 55% is claimed. The employer's quantity surveyor, who does not want to over-value work completed, may suggest the steel work is only 40% complete. As most valuations are ultimately a negotiation, the two parties would usually settle somewhere between these different views.

A common practice between contractors and subcontractors is for subcontractors to issue invoices at the end of each month, and then for contractors to pay less than the amount claimed; this is referred to as 'paying short' and should be accompanied by a pay less notice. Reasons for paying short can be tenuous, and motivated merely as a way to keep cash within payers' businesses.

Discussion point 5.5: You have been tasked with valuing the completed works on a major office rehabilitation project. How appropriate do you think the use of the 'S curve' would be as a foundation for the valuation?

5.2.2 Valuation of variations

The JCT SBC.Q 2011 requires quantity surveyors to include the value of work completed as part of variations in monthly interim payments. Specifically the contract stipulates in clause 4.16.1.1 that gross valuations shall include the value of works completed as part of variations, provided that their value has been established under the contract. The principles for valuing change are well established under standard forms of contract. JCT SBC/Q 2011, for example, provides six approaches for the valuation of variations including:

- For similar items of work carried out under similar conditions with no significant difference in quantity, the work should be valued using bill rates (rates from the contractor's tender).
- For similar items of work which is completed under different conditions and/ or with different quantities, the work should be valued using contractors' tender prices as a basis. Tender rates will be positively or negatively adjusted based on differences encountered. These rates are also sometimes referred to as *Star Rates*.
- Work which is different to that included in tender documents and completed under different conditions or with significant differences in the quantity (100m^2 versus 2,000m^2 of brickwork for example) should be valued using *fair rates*. Fair rates are normally accepted as the rates available through commercially available price books.

- Where an approximate quantity provides a reasonable presentation of the amount of work completed, then bill rates (contractors' tender prices) are used for the variation.
- Where an approximate quantity does not provide a reasonably accurate forecast of the work required, the bill rate (contractors' tender prices) are used as a basis for valuation (*Star Rate*).
- Where the work required in the variation cannot be measured or quantified, the works are valued using daywork (daywork is discussed further in chapters 4 and 6).

An alternative approach to the valuation of variations, especially those requiring significant work, may be dealt with as a Schedule 2 quotation (JCT SBC/Q 2011). In this situation, contractors will be asked to provide quotations in advance for additional works before instructions are issued and before work commences. Interim valuations for variations are based on percentage judgements of work completed.

However, despite the clear rules for the valuation of variations in standard forms of contract, it is common in practice for two parties not to have agreed values for variations by the time work is partially of even fully complete. Indeed in many situations, contractors may still be awaiting for contract administrators to issue formal variation instructions even though works have commenced or are complete. As a result, money can become locked up in contracts, which ultimately has a detrimental impact on cashflow (Ross and Williams, 2013) unless employers are willing to release funds by payment 'on account'.

The authors recall one situation in practice where a main contractor threatened to terminate a contract and argue the case in court. The value of works completed for variations amounted to over £100k at 2006 prices, but they had not been formally authorised in instructions by the contract administrator. The employer's QS refused to include monies in the interim payment. The sum involved was crippling to cashflow and had the potential to lead to contractor insolvency. In this situation, the only barrier to payment had been the contract administrator's poor administration skills.

5.2.3 Preliminaries

Preliminaries represent contractors' project level overheads, or the costs of running the projects and supervising works for the duration of projects. During valuations quantity surveyors can only take account of the items of overhead priced in either the preliminaries bill (if a bill of quantities was used) or identified as such in other contract documents; for example on design and build projects, preliminaries could be included in documents dealing with the terms of contract or identified in employer requirements and priced as part of elemental breakdowns. This may only represent part of the contractor's full preliminary costs, as the others may be included in the measured works (e.g. scaffolding could be included in the external walls element).

It is normal practice for contractors to distribute the preliminary costs between:

- Works related to fixed charges, that are claimed at either the beginning or end of the contract period;
- Cost-based charges that are recovered as a representation of the overall contract value (unusual to use, due to the difficulties to determine the value of the claim when variations and other payment become due);
- Time-related charges, which are distributed over contract durations and claimed for on a weekly basis.

For contractors, the distribution of monies in preliminary sections of bills of quantities at tender stage needs careful consideration. The valuation of preliminaries is very difficult to control from the perspective of employers. On some occasions, contractors use fixed charges in preliminaries to front or back load funds, and thus generate better cashflow. For employer quantity surveyors it is important they do not overvalue preliminaries. It is difficult to challenge the distribution of fixed charges, and the payment of time-related preliminaries needs to be carefully managed, especially if contractors have accelerated works or more important are behind programme.

To illustrate this point, consider the example of the hotel project first introduced in chapter 1. Table 5.2 illustrates a section of the preliminaries breakdown provided by the contractor. It is assumed the contract is 48 weeks in duration, with the valuation is completed at week 16.

Table 5.2 Section of preliminary costs value based on programme completion

Preliminary item	Tender allowance (£)	Distribution of preliminaries			Progress to date	Value of prelims to date (£)
		Time-related (£)	Fixed charge (£) (commencement)	Fixed charge (£) (completion)		
Management and staff	140,000	140,000			$16/48$	80,000
Site establishment	20,000		15,000	5,000	75%	15,000
Temporary works	15,000		12,500	2,500	84%	12,500
Health and safety	5,000	5,000			$16/48$	4,167
Contract conditions	5,000	5,000			$16/48$	4,167
Insurances, bonds, etc.	12,000		12,000		100%	12,000
Temporary services	10,000		10,000		100%	10,000
Fees and charges	2,000	2,000			$16/48$	667
Control and protection	8,700	8,700			$16/48$	2,900
Security	18,000	18,000			$16/48$	6,261
Cleaning	2,400	2,400			$16/48$	800

The valuation of preliminaries is dependent on the progress that the contractor has made at the date of the valuation. The contractor will likely look as positively as possible at this, and will claim that the valuation is based on works completed at week 16, and therefore is entitled to:

- 100% of the fixed charges outlined for commencement;
- 16 weeks of the time-related charges, so each preliminary item is paid at $^{16}/_{48}$ of the value.

However, the employer's quantity surveyor may wish to safeguard against over-valuation by considering the contractor's progress against the programme. So for example, if the contractor is three weeks behind programme, the quantity surveyor will seek to reduce the amount of preliminaries payable, to the following:

- 100% of the fixed charges outlined for commencement;
- 13 weeks of the time-related charges, so each preliminary item is paid at $^{13}/_{48}$ of the value, to represent the actual works completed.

Consequently the value of works completed would be determined by the actual progress of the work rather than the anticipated rate for completion of the works identified on the contractor's master programme.

5.2.4 Materials on-site

A high proportion of the value work of construction work is for materials purchased, perhaps well in excess of 50% of project costs. For those projects that rely extensively on imports, proportionally material costs can be much higher. Some areas of the manufacturing sector have rationalised their supply chains to allow them to operate lean production principles and just-in-time material supply networks. The construction sector cannot usually operate the same factory-based mass production processes typically utilised by production and manufacturing organisations. Difficulties in construction are compounded by lack of component standardisation, and the bespoke nature of building designs. As a result, the efficient operation of construction processes requires site teams to ensure materials are available on-site to keep their workforce operating.

Most standard construction contracts make express provision for employer quantity surveyors to include the value of materials on-site within interim payments. This is an important area for quantity surveyors, as they are employed to safeguard the interests of employers and protect them from potential overpayments.

Valuing unfixed materials and goods can be seen as a problematic area, especially when it comes to legal ownership of materials included in valuations. This becomes critical should main contractors enter insolvency subsequent to a large material on-site valuation and subsequent payment. This is emphasised in the case of *Dawber Williamson Roofing vs Humberside County Council [1979]*. The contractor was paid by the employer for the subcontractor's materials on-site, and the contractor went into liquidation before monies were passed to the subcontractor. The court determined Dawber Williamson Roofing still held the legal title for the goods. As a result, the employer was forced to pay for the materials twice. It is therefore important that quantity surveyors ascertain who owns the legal title for materials and goods on-site. In the *Sale of Goods Act 1979*, title rests with suppliers until payment is received from main or subcontractors; so a vesting certificate maybe required to formally transfer ownership to employers, rather than simply relying on standard forms of contract.

Quantity surveyors need to be certain materials are only delivered in a timely manner and that they are intended for the given project. It has been known for contractors seeking to resolve a negative cashflow by 'flooding' sites with materials, in the hope that quantity surveyors will include their full value in interim payments. In extreme situations, it is also possible that contractors move materials around projects to boost cashflow and prevent potential business failure.

As a result, the inclusion of materials on-site in the interim valuation requires quantity surveyors to have knowledge of the law in relation to the legal ownership of unfixed materials and goods on-site and the ability to exercise judgement about whether or not materials should be included in valuations. Quantity surveyors need to consider:

- What materials are present on-site?
- Have they been delivered prematurely, and are they likely to be used before the next valuation? For example, you would not expect high-value windows to be delivered during the first month of contracts.
- Are the materials of the correct specification for the project?
- Are the materials adequately stored (according to manufacturer's instructions)?
- Are they protected from potential damage or theft?
- What materials were on-site last month?
- Has the contractor provided evidence of their legal ownership?

How materials on-site are included in valuations is decided by each individual quantity surveyor; a number of potential approaches exist, thus:

- Some will accept contractor's itemised statements of materials on-site and the correct valuation once they have satisfied themselves materials are present;
- Others will take a tour of the site documenting the materials they observe, making judgements as to the quantities of each; for example a surveyor may identify:
  - 20 packs of bricks; given a pack of facing bricks typically includes 400 bricks, there are 8,000 bricks on-site; and
  - 3 pallets of plasterboard; there are normally 80 sheets of plasterboard (12.5mm thick) on a pallet, so there are 240 sheets of plasterboard on-site.

They may then use their own knowledge of material prices, or ask contractors for basic material costs to determine amount due.

- The final way is to ask the contractor for copies of material delivery sheets and then to cross-check these with the loose materials and goods currently on-site. A valuation then can be determined.

Discussion point 5.6: How would you protect the interests of your employer if you included unfixed materials and goods in a valuation?

5.2.5 Listed materials (materials off-site)

In addition to materials on-site, it is also possible for contractors or subcontractors to receive payment for value of materials contractors have purchased that are yet to be delivered to sites, assuming the works are being undertaken under the JCT Standard Form, or an express clause has been added into contracts.

Prior to certifying the value of off-site materials quantity surveyors must ensure that main or subcontractors have satisfied the condition precedents listed in the clause 4.27 (JCT SBC/Q 2011). These include:

- The materials are pre-agreed and listed in the contract;
- The contractor has legal ownership of the materials and can provide proof or a vesting certificate upon payment to transfer legal ownership;
- The listed materials are insured against loss and/or damage;
- The materials are stored ready for dispatch and include:

  ○ Identification of the employer;
  ○ The address of the site;

- Potential provision of a bond to the employer to cover the value of the materials.

5.3 Retention

Retention is an amount of money held back by employers from contractors. It permeates down the supply chain so that it is also held by contractors from subcontractors and by subcontractors from sub-subcontractors. However, it is not normally held by contractors or subcontractors from suppliers.

The amount is specified in contracts. For example, in the JCT 2011 Standard Form of Building Contract, the recommended amount is 3%. However, employers are able to specify an alternative amount, and 5% may not be unusual. The retention percentage set by employers must be stated in tender documentation. If the amount is unduly large, contractors can take account of the risks or costs that that will impose and include extra sums in their bid if they feel it appropriate. Similarly for those lower down supply chains.

The calculation method of the amount to be paid and the amount to be retained (retention) is a move from the 'gross' to the 'net'; that is, start with gross valuation amounts, and then work to net values thus:

Gross valuation amount £100k; retention 5%
Net amount to be paid £100k x 0.95 = £95k
Retention amount £100k x 0.05 = £5k

The first half the retention would normally be released on completion; using NEC terminology that is 'completion of the whole works', and JCT upon the issue of a 'practical completion certificate'. To release the second half, NEC requires a 'defects certificate' and JCT a 'certificate of making good' of defects. Defects or rectification periods are often 6 or 12 months, and some contracts may specify both periods for different elements of the works. Traditional trades may be six months, but specialist installations such as mechanical and heating systems may need all four seasons of a year to ensure they work at all temperature extremities; also landscaping work may need four seasons to be sure that shrubs and plants have 'taken hold'. Table 5.3 illustrates a typical payment schedule for a six-month project, valued at £1.2m, with 5% retention, and six months' rectification period.

Note that 5% of the total project value is never retained; the maximum is 5% of valuation No 5. At valuation No 6, half retention is released, therefore the amount received at valuation 6 exceeds the value of work executed between valuations 5 and 6.

Retention is the subject of much debate and dispute. The ideological reason for retention is that it protects the party highest up the supply chain (employers in contracts with contractors, and contractors in contracts with subcontractors).

Table 5.3 Payments and retentions on a six-month project with six-month rectification period

	A	B	C	D	E
Valuation no	Assumed amount of valuation; cumulative (£)	Amount paid to contractor; cumulative (£)	Amount retained (retention) by client (£)	Gross value of work executed in the month (£)	Payment received by contractor each month (£)
		95% of 'A' = 'A' x 0.95	5% of 'A' = 'A' x 0.05		
1	90,000	85,500	4,500	90,000	85,500
2	250,000	237,500	12,500	160,000	152,000
3	500,000	475,000	25,000	250,000	237,500
4	750,000	712,500	37,500	250,000	237,500
5	910,000	864,500	45,500	160,000	152,000
6	1,000,000	975,000	25,000	90,000	110,500
12	1,000,000	25,000	0	0	25,000
Total		**1,000,000**			**1,000,000**

Protection may be needed if defective work is discovered after payment for that work has been made; some defects may only become apparent at a later stage. Money held as retention acts as a lever to 'encourage' or 'force' lower tier parties to rectify defects. If the lower-tier party does not rectify defects, providing the correct contractual procedures are followed, the higher party may employ someone else to do the work and deduct the cost from retention monies held.

However, there is a view that retention is not necessary. Contracting parties often work with each other on a repeat basis. Repeat work can be formally established in partnering or framework arrangements, but it also occurs in conventional standard forms of contract. For example, a subcontractor, through competitive tendering arrangements, may be able to secure a large number of contracts with the same main contractor. In such cases, the higher party has the lever of future work as a vehicle to ensure any defective work is rectified. Lower parties may argue they always want to ensure customer satisfaction and will always return to correct defective work, irrespective of retention. Further if a problem does arise, there are mechanisms in the contract to 'force' defect rectification, albeit those mechanisms can be costly. The reality though is that cash counts; from the highest parties' perspective, to argue about a defect with the money in your bank puts you in a stronger position.

Some argue that retention is not held as a lever to force the rectification of defects; more the case, it is about cashflow. Employers like it because they have the money in their bank and are able to either (i) if they are cash rich, earn interest on it, or (ii) if they are cash poor, pay less interest to their banks for overdrafts or loans, and use it to finance other parts of their businesses. Contractors like retention because they are able to hold from subcontractors amounts that are close to the amounts being held from them by employers; but more particularly, when contractors have retention released to them by employers, they may not promptly (or ever) release it to subcontractors. The party in the supply chain that suffers most from retention is subcontractors, especially those who complete their work at an early stage on projects e.g. specialist piling contractors. Normally subcontractors are only entitled to release of retention on completion of the main contract works, not on completion of their own part. Therefore specialist piling contractors and the like who complete their work in months one or two on large projects may have to wait possibly several years to get retention released.

Contractors may also use release of retention as a lever in final account negotiations to force subcontractors to accept a final account value that is lower than would otherwise be the case. For example, a subcontractor completes its work and a contractor makes a payment of £100k less 5% retention = £95k. Note the main contract work is not complete; the subcontractor gets release of the first half of its retention only when the entire project is complete. There are some extra works and variations, and the subcontractor submits its final account to the main contractor in the value of £105k, and asks for a total payment of £10k (£105k – £95k) including retention release. The main contractor disputes the £5k extra work and argues it should be more like £2k. A discussion ensues. The contractor makes a final offer of £102k, and therefore a £7k payment now including retention release or alternatively the subcontractor may wait several months or even longer for £10k. The subcontractor takes £7k now.

In some markets the supply chain may refuse to submit bids if retention is applied. Those markets may be where work is plentiful and there is little competition. The supply chain may submit 'conditional' bids; that is a bid in accordance with tender documentation, except for a stipulation that the bidder offers say a retention bond in lieu of retention. This may be the case for specialist subcontractors, again like piling, steelwork or cladding; that is trades who complete high-value work well before practical completion. Alternatively, employers may agree to the early release of retention (without a bond) to specialists who are involved at the early stages of projects.

As well as impacting on the cashflow of lower-tier organisations, there is the possibility that if the higher-tier organisation goes into liquidation, retention is never paid at all. Klein (2015) explains the case of a civil engineering company with a £4.5m turnover that lost £240k retention over five years due to upstream insolvencies. The amount held as retention may be equivalent to the profit, thus the supply chain has worked for nothing.

Section 3.2 explains the concept of retention bonds as an alternative to holding retention. Whether employers apply retention to projects is a market judgement for them. Lower parties in the supply chain argue they incur costs as a consequence of retention; for example (i) the administrative time used by chartered surveyors and other staff in processing release of retention, (ii) the bank and other finance charges that arise by having lower cash balances. The supply chain argues it must add these costs to bids submitted to contractors and employers. If the supply chain was not subject to retention, it argues it could submit lower bids. There are many enlightened employers who specify zero retention, recognising that supply chains may increase prices if retention is specified, and that in extreme cases, withholding retention may force companies into liquidation. The Government Construction Strategy (BIS, 2013) states that the vision for year 2025 is to have zero retention across all of construction.

5.4 Example interim payment valuation

Table 5.4 shows a summarised valuation statement for the first month of a project. For simplicity, group elements from the NRM1 are adopted, and these are not broken down to element, sub-element or component level. It has been assumed the project was let using the JCT SBC/Q 2011, with retention at 3%.

Exercise 5.1

The due date for the first valuation on a new school project is 7 March 2017, one month after the contractor took possession of the site. The contractor has not submitted an application for payment on this occasion. As the cost consultant for the project you are required to prepare the interim valuation. Consequently, you

Table 5.4 Summarised contractor's application for payment

	Contract Sum	% complete	Value (£)
Preliminaries	300,375	20%	60,075
Substructure	153,535	15%	23,031
Superstructure	1,092,475		
Internal finishes	153,525		
Fixtures, fittings and equipment	53,400		
Services	716,450		
External works	271,450	5%	13,573
Variations	0		
Sub-total			96,679
Materials on-site			10,000
• 30 nr 5.00m driven piles			
• 10 tonne MOT type 1			
Sub-total			106,679
Retention @ 3%			(3,200)
Previous payments			0
Payment due			**103,479**

have arranged to meet the contractor's quantity surveyor on-site to agree the first valuation.

From an initial inspection of the works you have found that the following works have been completed:

- Site set-up works including hoarding, temporary road and carpark, site accommodation, service connections *et al.* (100%)
- Site clearance and demolition of existing structures (100%)
- Cut and fill excavation works along with disposal of materials (75%) complete
- Hardcore base for piling rig established (20%)

You have also noticed the contractor has some materials on-site including:

- 40 tonnes of recycled aggregate (for piling rig base)

From the contract sum analysis you have noted the following rates:

- Prelims (10%) of £4,500,000–450,000 (no further breakdown)
- Substructure:

  ○ Site clearance and demolition   £150,000
  ○ Excavations   £380,000
  ○ Piling including sub-base   £520,000
  ○ RC concrete ground beams and floor   £145,000

You have also established that recycled aggregates cost approximately £12.00 per tonne delivered to site. Prepare the first interim valuation for this project in readiness for your meeting with the main contractor's quantity surveyor.

Exercise 5.2

The 7th interim payment is due on a new bus station the employer is developing in Manchester. The project has been let using a traditional (design-bid-build) procurement route using the JCT Standard Form of Building Contract with Quantities (2011) with several amended clauses. The contractor has recently submitted and agreed the terms of the valuation with you, and the contract administrator has asked you to provide the interim valuation breakdown so the payment notice can be prepared.

Using the data in Table 5.5, calculate the net sum due for an interim payment under the terms of the contract.

Exercise 5.3

Despite providing the financial data for the payment notice, the contract administrator has failed to process the document in time to meet the contract deadlines. Unsure about the specifics of JCT, the administrator has asked you to provide advice about what they should do next. You should fully explain how the Local Democracy, Economic Development and Construction Act 2009 enhanced payment provisions under the JCT and the position of the courts should the administrator decide to ignore the oversight and proceed as normal (issuing the final certificate four days late).

Exercise 5.4

There are just two agreed stage payments for construction of a detached house. The first is to achieve watertight for a building, and the second is completion. There is 5% retention to be applied to the first payment. On completion there will be half retention release and overall 2.5% retention. Final release of retention will be after six months. The first gross payment is £82,386 and the final payment £142,000. Calculate the three net payments the contractor will receive and the overall amount.

Table 5.5 Agreed interim valuation

Measured work (practical completion not achieved)	£6,750,000
Variations	£423,000
Local authority fees	£6,580
Materials on-site	£18,500
Loss and expense	£19,000
Fluctuations	£3,100
Deductions (work not in accordance with the specification)	£350
Previously certified	£5,470,000
Retention stated in the contract at 3%	

5.5 Model answers to discussion points

> **Discussion point 5.1:** Why do you think it is important that contractors, subcontractors and other specialist contractors receive monthly interim payments?

It is often remarked that *cashflow is the lifeblood of the construction industry*; this question is really asking you to reflect on this. Ultimately all companies need cash to survive. The problem with contract law is that people are not often paid until contracts are complete. Whilst this works fine for small short-term duration projects, the same cannot be said for long duration multi-million pound projects. To ensure supply chains can continue to trade whilst they complete projects, interim payments are made.

> **Discussion point 5.2:** Do you think the government's drive to instigate fair payment practices will be taken up by industry?

In mid-2016, the construction press reported that the government's fair payment charter is in trouble, with a number of the organisations who originally backed its creation identifying the unfairness of others not following their lead; therefore support slowly started to back away. Without legislation in the forthcoming years it would appear this initiative is in serious difficulty, despite the government's assertion that it will be mandated for public sector projects.

> **Discussion point 5.3:** Do you think that paying contractors, subcontractors and suppliers on a typical 90-day (3 months) arrangements is acceptable?

Just because something is common practice does not necessarily make it acceptable. The arguments for a strict 30-day payment term relate to the damage to a firm's cashflow such long waits for payment can have. If payment cycles are improved, there will likely be improved business survival rates and reduce the costs in construction. Risks and costs associated with long waits for payment are reflected in tender prices.

> **Discussion point 5.4:** As legal cases illustrate, the courts have been extremely precise in their application of the Local Democracy, Economic

Development and Construction Act 2009. Are courts being overly precise and not realising the complexities of construction?

Views may vary depending on from where in supply chains a perspective is taken. Tier 3 subcontractors would clearly argue precision is positive, whilst those advising clients may argue it is more restrictive. Overall, it can perhaps be argued to be positive and it is for construction professionals to ensure clients are not exposed to legal actions. PQS have a professional duty to ensure certificates are issued in a timely manner.

Discussion point 5.5: You have been tasked with valuing the completed works on a major office rehabilitation project. How appropriate do you think the use of the 'S curve' would be as a foundation for the valuation?

This would very much depend on how the project is procured and how the contract price mechanism works. However, it must be remembered that the S curve and others are theoretical models of the flow of money through a project. Whether you would want to link this to real-time programme data is a question for you. However, normal practice would be to use an inspection-based valuation technique to ensure over- or under-payment situations do not arise.

Discussion point 5.6: How would you protect the interests of your employer if you included unfixed materials and goods in a valuation?

This question is looking to test your understandings of the Sale of Goods Act 1979. JCT makes express provision at clause 2.24 for the transfer of materials on-site to employers upon payment. The same applies to the subcontractors at clause 3.9.2.1. These are vesting clauses, which overrule the provisions in the Sale of Goods Act 1979 and protect employers. However, some would go further and ask contractors to provide vesting certificates prior to payment to legally transfer the goods to employers.

5.6 Model answers to exercises

Exercise 5.1

The due date for the first valuation on a new school project is 7 March 2017, one month after the contractor took possession of the site. The contractor has not submitted an application for payment on this occasion. As the cost consultant for the project you are required to prepare the interim valuation. Consequently, you have arranged to meet the contractor's quantity surveyor on-site to agree the first valuation.

From an initial inspection of the works you have found that the following works have been completed:

- Site set-up works including hoarding, temporary road and carpark, site accommodation, service connections et al. (100%)
- Site clearance and demolition of existing structures (100%)
- Cut and fill excavation works along with disposal of materials (75%) complete
- Hardcore base for piling rig established (20%)

You have also noticed the contractor has some materials on-site including:

- 40 tonne of recycled aggregate (for piling rig base)

From the contract sum analysis you have noted the following rates:

- Prelims (10%) of £4,500,000–450,000 (no further breakdown)
- Substructure:

  - Site clearance and demolition   £150,000
  - Excavations   £380,000
  - Piling including sub-base   £520,000
  - RC concrete ground beams and floor   £145,000

You have also established that recycled aggregates cost approximately £12.00 per tonne delivered to site. Prepare the first interim valuation for this project in readiness for your meeting with the main contractor's quantity surveyor.

SUGGESTED SOLUTION

Table 5.6 illustrates how the main works and materials on-site can be valued; however, as the contractor has not yet provided a preliminaries breakdown, the value of those will need to be determined at the meeting. You will also notice the inspection revealed that the pile base as 20% complete. However, in the valuation only 5% has been certified against the contractor's contract sum analysis. This is because the element including the sub-base also includes provision for the full piling works. Clearly 20% would be an overvaluation in the case, so the quantity surveyor needs to make an appropriate judgement call on this item of work.

Exercise 5.2

The 7th interim payment is due on a new bus station the employer is developing in Manchester. The project has been let using a traditional (design-bid-build) procurement route using the JCT Standard Form of Building Contract with Quantities (2011) with several amended clauses. The contractor has recently submitted and agreed the terms of the valuation with you, and the contract administrator has asked you to provide the interim valuation breakdown so the payment notice can be prepared.

Using the data in Table 5.5, calculate the net sum due for an interim payment under the terms of the contract.

Table 5.5 Agreed interim valuation

Measured work (practical completion not achieved)	£6,750,000
Variations	£423,000
Local authority fees	£6,580
Materials on-site	£18,500
Loss and expense	£19,000
Fluctuations	£3,100
Deductions (work not in accordance with the specification)	£350
Previously certified	5,470,000
Retention stated in the contract at 3%	

Table 5.6 Solution for Exercise 5.1

Element	Contract value	Percentage complete	Interim payment (£)
Measured Works			
Substructure			
Site clearance and demolition	£150,000	100%	150,000
Excavation: cut and fill to level site and reduce levels to identified datum.	380,000	75%	285,000
Foundations: bored piling including piling platform and compacted aggregate base.	520,000	5%	26,000
RC concrete pile caps, ring beams and ground floor	145,000	0%	Nil
Preliminaries			
Commencement (fixed price)	TBC	TBC	TBC
Time related	TBC	4 weeks	TBC
Materials on-site			
300 tonnes recycled aggregate	12.00	300 tonnes	3,600
Value of work completed			464,600

Table 5.7 Solution for Exercise 5.2

Work subject to retention	Payment due (£)
Measurement work	6,750,000
Variations (correctly certified)	423,000
Work not in accordance with the specification	(350)
Sub-total	*7.172,965*
Works not subject to retention	
Local authority fees	6,580
Loss and expense claim	19,000
Fluctuations payment	3,100
Sub-total	*28,680*
Materials on-site	18,500
Gross valuation	**7,220,145**
Retention @ 3% on £7,172,965	(215,188.95)
Previous payments	(5,470,000)
Total	**1,534,956.10**

SOLUTION

Exercise 5.3

Despite providing the financial data for the payment notice, the contract administrator has failed to process the document in time to meet the contract deadlines. Unsure about the specifics of JCT, the administrator has asked you to provide advice about what they should do next. You should fully explain how the Local Democracy, Economic Development and Construction Act 2009 enhanced payment provisions under the JCT and the position of the courts should the administrator decide to ignore the oversight and proceed as normal (issuing the final certificate four days late).

SUGGESTED SOLUTION

From the scenario it is clear that the following has happened:

- The contractor has made an application for payment and submitted a valuation of work completed to date;
- The quantity surveyor has inspected the works and agreed a valuation sum with the contractor (not sure if this is the same as the application);
- The quantity surveyor has recommended a payment of £1,534,956.10 in respect of the interim payment number 7;
- The contract administrator has failed to act on this.

From the above, we can now provide advice. First the administrator cannot simply ignore the fact they failed to issue the payment notice within the defined period (five days from the due date S.110 of the Construction Act 2009). As four days have since passed and the contractor has not served a default notice the contractor's original application must now be accepted as the default notice. Issuing the payment notice late is not an option as the courts have indicated they will take the provision in the Construction Act 2009 as absolute obligations. So the contract administrator must immediately (before close of business today) issue the contractor with a pay less notice, adjusting the contractor's application to the sum approved by the cost consultant. If the administrator also fails to issue the pay less notice, the employer must pay the sum originally stated in the contractor's application and although this is an overpayment (assumed), the sum will be adjusted in interim payment 2.

References

AECOM (2014) *Spons Architects and Builders Price Book.* Oxon: Taylor and Francis.

BIS (2013) Construction 2025. Industrial Strategy: Government and Industry in Partnership. Available at: www.gov.uk/government/uploads/system/uploads/attachment_data/file/210099/bis-13–955-construction-2025-industrial-strategy.pdf Accessed 01.05.16.

JCT (2011) SBC/Q2011 Standard Form of Building Contract. The Joint Contracts Tribunal.

NBS (2015) National Construction Contracts and Survey. National Building Specification. RIBA Enterprises. Available at: www.thenbs.com/knowledge/nbs-national-construction-contracts-and-law-survey-2015-finds-disputes-continue-to-blight-construction-industry Accessed 22.03.16.

NEC (2005) *Engineering and Construction Contract.* London: Thomas Telford.

Ross, A. and Williams, P. (2013) *Financial Management in Construction Contracting.* Oxon: Wiley-Blackwell.

6 Post-contract

6.1 Contractors' cashflow introduction

Ensuring that all members of the supply chain have sufficient cash to fund their businesses is one of the most challenging aspects of construction. Most payment mechanisms are designed so that parties should be paid with 30 days of the date of invoice, but in practice all too often that does not happen. The Governments' Construction Strategy 2025 (BIS, 2013) reported its vision that 'Construction in 2025 is no longer characterised, as it once was, by late delivery, cost overruns, commercial friction, late payment, accidents, unfavourable workplaces, a workforce unrepresentative of society or as an industry slow to embrace change'. Late payment has been the subject of lots of initiatives over many years; some are successful, but problems still remain. Figure 6.1 illustrates how money may typically come into, and go out of, a contractor's account on traditional projects.

Arguably, most employers have a reputation for paying contractors on time. Problems start to occur when tier 1 contractors pay tier 2 subcontractors and suppliers; also when tier 2 pay tier 3 sub-subcontractors *et al*. It is often reported that construction is fundamentally under-capitalised; that is, businesses do not have sufficient cash reserves to withstand unexpected losses. That was also the case for many banks across the world in the period 2008–12, as some governments stepped in to help. In the UK subsequently, given the dependence of the whole economy on our banks, financial regulators insisted that banks recapitalise. One important way of doing that is to make sure they make profits and retain those profits within their businesses. Unfortunately, there are no regulators to make such demands of construction.

Profitability in construction, particularly for contractors, has had a reputation for many years as being too low. Retaining their own cash in their businesses can therefore be difficult. Some contractors do have strong balance sheets and cash deposits and are able to pay their supply chains promptly. This can give competitive advantage, since many suppliers will want to work with higher tiers who pay promptly, and suppliers may even offer lower bid prices or higher discounts for payment on time.

Other contractors may not be so well placed. To the extent that tier 1 contractors withhold payment, that may be because they simply do not have funds to pay (even though they have been paid by employers). Alternatively they may see the need

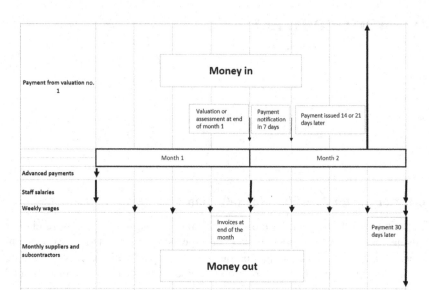

Figure 6.1 Money flow into and out of a contracting company.

to hold on to as much cash as possible. Strong cash balances can: (i) give respect in the market place, including with banks, and (ii) provide easily accessible cash in the event of unforeseen costs or to invest in profit-making ventures elsewhere.

Consider an individual who has £1,000 available in the bank, and debts of £900. Is it better to delay payment of the debt and remain with £1,000, or is it better to pay the debt and have cash of only £100. Many individuals may prefer the latter, but businesses, with many noughts added to the £1,000, may prefer the former.

There is also the position of say £500 available in the bank, and debts of £900. Is it better to remain in that position, or to have a £400 overdraft? Contractors may be very nervous about having overdraft facilities with banks, even overdrafts that are formally agreed. Overdrafts do attract facility fees and interest charges, and there is also the possibility that banks may withdraw them at very short notice.

Whilst traditionally, contracting has a reputation for low or poor profits, it has been perceived as being excellent for cashflow. It has only been excellent for cashflow since some contractors pay 'late' or 'later'. Paying 'late' may be when there are perhaps terms agreed for 30 days, or some other period, and payments are simply not made on time. Paying 'later' is where contractors or other party receives payments in perhaps 30 days, but agree contracts with lower-tier suppliers for payment in perhaps 45, 60, 90 or even 120 days. Some contractors may use a combination of the two, for example receiving payments in 30 days, contracting to pay supply chains in 45 and then actually paying in 60. These contractors may then use this cash to support speculative housing developments, which has a reverse reputation; that is very poor for cashflow, since payment is only received when properties are fully complete, but very good for profitability.

Site professional teams may be hampered in their attempts to push for good progress on sites if supply chains are not being paid promptly. Suppliers may refuse to deliver materials, and subcontractors may refuse to increase the amount of labour, or indeed proactively decrease it. Whilst contracts permit subcontractors to use adjudication proceedings to secure payments, subcontractors may be reluctant instigate proceedings, since they are mindful that such action may damage relationships with contractors and they may not get the next job. One fear that lower-tier suppliers may have if they are not being paid on time is the possibility that higher tiers may go into liquidation and payments are never received.

6.2 Assuring payments on time

Chapter 2 details the construction management method of procurement, which facilitates payments going directly from employers to works package contractors (subcontractors in traditional methods of procurement). In this circumstance, there is no main contractor through whom monies pass. There is, however, still the problem of prompt payment from works package contractors to their subcontractors and suppliers.

The government promotes the use of project bank accounts, as detailed in the Cabinet Office (2012) publication 'A Guide to the Implementation of Project Bank Accounts (PBAs) in Construction for Government Clients'. PBAs are used most often on public projects and are suitable for use where project values are as low as £1m. Employers pay money into project bank accounts; subcontractors submit invoices, and upon approval by employers and contractors, the money is paid promptly to subcontractors. At no time is money held by contractors, and there is no motive on behalf of contractors to delay payment; indeed contractors are keen to approve invoices quickly since promptly paid supply chains are motivated to progress works quickly on sites.

Another initiative is the Construction Supply Chain Charter, published by the Construction Industry Council (CIC, 2014). This is a voluntary code that has been signed up to by many contactors, with a commitment *inter alia*, to pay supply chains within 30 days by January 2018. Linked in to the supply chain charter is a government proposal that payment performance of companies should be published in annual reports. Fitzpatrick (2016) details proposals by Build UK to collate and publish publicly available information on payment by individual companies. When public bodies invite bids from contractors for projects, those who cannot demonstrate good payment performance may not be permitted to bid.

A growing initiative in recent years is a system colloquially known as 'reverse factoring' or 'supply chain finance systems'. Tier 1 contractors receive payments from employers in circa 30 days, but then make payments to low tier 2 and 3 subcontractors or suppliers in 120 days. However, these lower tiers may call for payment earlier than 120 days by offering the tier 1 a 'discount'. The discount may be at an extremely competitive rate, say 4%, and the payment can be called directly from tier 1 bank accounts as soon as invoices are approved electronically as being correct.

Exercise 6.1

A tier 2 subcontractor submits an invoice for £100,000.00 due for payment in 120 days. The tier 1 contractor approves the invoice quickly, and the subcontractor decides to take the payment after 30 days; that is 90 days early. What is the 'discount or 'charge' for taking early payment? Assume an interest rate of 4%.

Discussion point 6.1: What may the financial advantages be of reverse factoring?

Discussion point 6.2: If contractors are forced to pay supply chains in 30 days, what consequences may this have for those who have business models that depend upon holding the cash of others to fund their business? Is it viable that contractors can bid higher prices, and thus make employers pay instead?

6.3 Predicting the cashflow of projects

Employers and contractors need to know future cashflow projections for proposed projects. Employers need budgets for amounts and timing of payments to contractors, so they can plan for having money available. Contractors need budgets for both money coming in and money going out to their supply chains.

Payment terms from employers to contractors may vary. Some are modelled on stage payments e.g. payment 1 completion of substructure, payment 2 roof watertight, payment 3 completion. Such systems incentivise supply chains to complete work quickly. Larger projects would likely have many stages, so that money flows on a fairly regular basis. Most often payments are made monthly and are based on the amount of work actually completed; if good progress is made, payments will be higher than if progress is slow. NEC terminology is 'assessing the amount due', whilst JCT contracts use the phrase 'interim valuations'. Assessment dates are established in contracts to suit all parties, perhaps at the end of each month. Employers and contractors will agree the value of work done at assessment dates. On larger projects, both will likely have their own quantity surveyors to undertake this work. Very often, judgements will be made about percentage completion for trades e.g. reinforcement and concrete foundations to wing 'A' of a building 50% complete. If there is a bills of quantities and separate prices are identified for reinforcement and concrete, it may be possible to have different percentages for each e.g. reinforcement 60%, concrete 40%. Work is usually assessed on a cumulative basis, such that the total value of work completed is determined, and the amount of previous assessments is deducted e.g. assessment or valuation 2 = £100,000 minus assessment or valuation 1 of £40,000; amount completed in month 2 = £60,000. By always taking cumulative amounts, it will erase errors made in earlier assessments e.g. if an early assessment was made that work was 50% complete, but in actual fact it was only 45%, a subsequent accurate assessment of 70% would 'even things out'. It is important to note that precise accuracy for assessments is not usually required, since the objective is just to get some money flowing through supply chains. If a contractor is paid 50% for an element of work, and it is actually only 45% complete, that should not usually cause problems. The contractor may be underpaid for another element of work, and also when the contractor actually gets the money circa one month later, the work may be 60% or 70% complete. It is not good, though, to be too relaxed about assessments, because problems may arise if a member of a supply chain has been paid too much, and then goes into liquidation. Final account payments do need to be precise.

6.3.1 S curves

The empirical quarter/third ($^1/_4 - ^1/_3$) S curve is a long-established method of predicting assessment amounts or interim valuations on building projects. It is termed 'empirical' since it is based upon the experience of previous projects. Plotting on a graph, with the vertical scale in money and the horizontal scale in time, it has been noted that the relationship can be often expressed by a curve that approximates to a lazy 'S' shape. Whilst projects always need to progress at good speed at all stages, the premise is that in financial terms projects are likely to have slow starts and slow finishes. In the opening stages, work completed is of relatively low value and only few trades may be on-site initially e.g. site set-up for temporary cabins and services, with some excavation and groundworks. In closing stages, again work of relatively low value and few trades e.g. floor finishes, painting and decorating, test and commissioning and 'snagging'. Financially, this is expressed that in the first third of the project 'only' one quarter of work is completed. Similarly, in the last third of the project 'only' one quarter of work is completed. That leaves in the middle third, the need to complete one half of the project in financial terms. From a practical viewpoint, this can be substantiated by the possibility to have many trades working simultaneously on high-value work. On building projects this will particularly be the case when projects are watertight, thus enabling many first fix trades to be working together. Figure 6.2 illustrates a forecast 'value curve' for a £1.2m project with a duration of eight months. Four points are plotted: one-third time (8 months /3 = $2^2/_3$ months) with one-quarter value (£1.2m/4 = £300k), two-thirds time (8 months /2*3 = $5^2/_3$ months) with three-quarters value (£1.2m/4*3 = £900k); also the origin and completion points. The origin point to one-third time/one-quarter value is joined with a gentle curve to indicate that the value of work increases gradually as projects go through their first third stage. Similarly there is a curve to join the two-thirds time/three-quarters value point to the completion point to indicate that the value of work decreases gradually as projects move through final stages. The middle part of the 'S' curve is constructed using a straight line.

To forecast assessment amounts or valuations, the next step is to read up vertically on the graph at each assessment period (often monthly), and where time meets the curve, then read horizontally to determine the predicted amount. For example in Figure 6.2, reading up vertically at month 5, it meets the curve horizontally at a value of £825k.

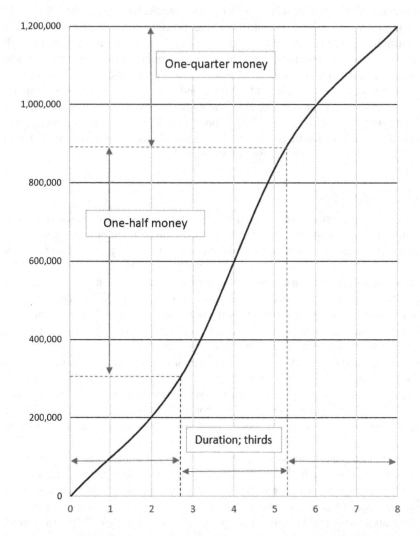

Figure 6.2 Empirical $^1/_4$–$^1/_3$ S curve.

Table 6.1 illustrates how figures at each month can be used to determine cash-flow requirements for the project.

Column B indicates the forecast cumulative gross valuations for months 1 to 8. Month 14 is included, since it is assumed that six months later, any defects will have been attended to and the second half of retention released. These are gross assessments or valuations. The word 'gross' is used to indicate the total amount of money due, before retention is deducted.

Table 6.1 Details of income and expenditure for a £1.2m project, assuming the ¼–⅓ rule (data in column B taken from Figure 6.2)

Valuation or assessment no ↓	Cumulative gross valuation	Cumulative net valuation after retention deducted	Monthly payment due	Cumulative cost after profit deducted	Monthly cost	Cashflow balance before payment received	Cashflow balance after payment received
Formulae→	Taken from S curve	Column B x 0.95, except cell C9 = B x 0.975	D3 = C3; D4 = C4- C3; D5 = C5-C4	E3 = BI/110 x 10	F3 = E3; F4 = E4- E3	G3 = F3*-1; G4 = H3- F4	H3 = D3 + C3; H4 = G4+ D4
1	800,00	76,000	76,000	72,727	72,727	-72,727	3,273
2	190,000	180,500	104,500	172,727	100,000	-96,727	7,773
3	375,000	356,250	4*	5*	168,182	8*	15,341
4	600,000	1*	213,750	545,455	7*	-189,205	10*
5	825,000	783,750	213,750	750,000	204,545	-180,000	33,750
6	1,010,000	959,500	175,750	918,182	168,182	-134,432	41,318
7	1,120,000	1,064,000	104,500	1,018,182	168,182	-58,682	45,818
8	1,200,000	2*	106,000	6*	100,000	9*	79,091
14	1,200,000	3*	30,000	1,090,909	72,727	79,091	109,091
Totals			120,0000	1,090,909	1,090,909	Profit in the bank at month 14 = ii*	

Note: £1.2m project value; eight-month duration; 5% retention; 10% profit; payment period from employer to main contractor one month; payment period from main contractor to supply chain one month.

One-third duration = 8/3 = 2 2/3 months; two-thirds duration = 8/2*3 = 5 1/3 months. One-quarter value = £1.2m / 4 = £0.30m. Three-quarters value = £1.2m / 4 *3= £0.90m.

Column C shows the cumulative net valuation after retention is deducted. Cell C2 indicates that figures in column B are multiplied by 0.95 to take off 5% retention. Cell C10 is multiplied by 0.975 since half retention is released on completion; only 2.5% is retained.

Column D indicates the month payments received. These are calculated by deducting the total of previous payments from the current cumulative net valuation.

Column E illustrates forecast cumulative costs; that is the figures in column B adjusted to remove profit.

Contractors bid for projects by estimating forecast actual costs they will incur if they are successful in winning projects; these costs may be termed 'net costs'. They will then make market judgements about the amount they will add for profit. Chapter 3.16 indicates that a benchmark norm is 2.5%, but in tough markets, contractors may bid with zero profit. In this example, it is assumed that work is plentiful and markets are very buoyant; the contractor adds 10% for profit. To determine 'net' costs from 'gross costs' the arithmetic calculation is illustrated thus for valuation No 1:

£80,000 gross amount includes 10% profit
Consider £80,000 as being equal to 110%
1% = £80,000 / 100 = £727.27
100% = £72, 727
Check calculation: £72,727 + 10% = £72,727 + £7,273 = £80,000
The formula to move from the gross in column B to the net in column E, assuming 10% profit, is therefore: gross valuation / 110 x 100.

Column F indicates monthly costs. These are calculated by deducting the total of previous costs from the current cumulative cost.

Column G is the cashflow balance at the end of each month, just before payment is received. At the start of projects, contractors will need to make some payments to their supply chains before they get their first payment. For example as illustrated in Figure 6.1, weekly paid operatives will be paid wages, and monthly paid staff their salaries at the end of month 1. Monthly paid site staff may also be paid salary before projects start, since some will be involved in planning and operational activities for mobilisation. Utility bodies that provide connections for temporary electrical and water supplies may require advance payments before they will schedule works. Contractors may not receive their first payment until towards the end of month 2, and therefore at the start of projects they will be in negative cashflow.

Column H is the cashflow balance at the end of each month, just at the point that payment is received.

Task 6.1: Eleven figures in Table 6.1 are substituted with asterisks thus: 1*, 2*, etc. What are those figures?

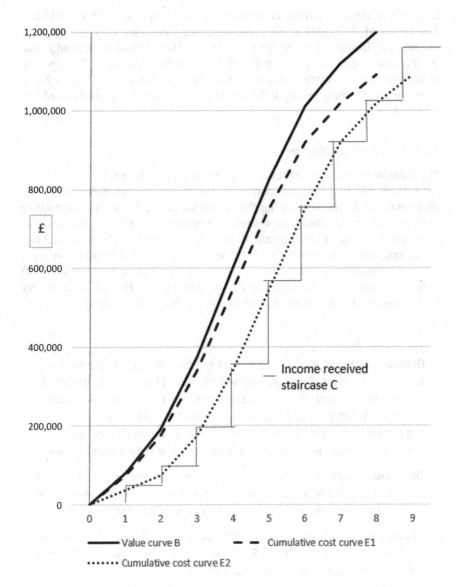

Figure 6.3 Empirical $^1/_4$–$^1/_3$ S curve, including cost curves and income staircase.

Figure 6.3 is a development of Figure 6.2. Added to the 'value curve' (labelled curve B) is the cumulative 'cost curve' based on the data in column E of Table 6.1 (labelled curve E1). This cost curve is then moved horizontally one month to denote a one-month period between receiving invoices and making payment; a one-month or 30-day payment period (labelled curve E2). The fourth 'curve' (labelled curve C) is more analogous to a staircase. The height of each riser is monthly payments received, as shown in column D of Table 6.1. The height of the treads is the cumulative net valuations after deduction of retention, as illustrated in column C Table 6.1. The width of each tread is one month, based upon the period between each payment received. The position of the first riser is month 2, when the first payment is received. That is based upon the first valuation being at the end of month 1, and payment received one month later.

6.3.2 The saw tooth diagram

The relationship between staircase C and curve E2 in Figure 6.3 is key. At each point where curve E2 is below staircase C, the project is in negative cashflow. At each point that the staircase nosing rises above the curve, the project is in positive cashflow. For greater clarity, the relationship between the staircase and the curve is presented in a saw tooth diagram, illustrated in Figure 6.4. The horizontal line on the saw tooth is zero in the bank account; areas above the zero line are positive and areas below are negative cashflow. If the contractor needs an overdraft to fund the project, the maximum facility is shown to be £189,205; alternatively the contractor will need to use its own internal cash balances to fund the project.

Discussion point 6.3: 'S curve' and programme-based forecasts are based on a 'model'. They give some indication of peaks and troughs in contractors' cashflow. At an operational level, how accurate is the model? In Table 6.1, will the contractor really be £189,205 negative cashflow on this project? What will the effect be of contractors receiving their payments after 30 days, and negotiating 60-day payments with their supply chains?

Discussion point 6.4: Some contractors may not need overdrafts from their banks, and when individual projects are in negative cashflow, contractors may finance them from cash reserves. Is this appropriate?

Discussion point 6.5: Why is it that contracting has a reputation for positive cashflow when Figure 6.4 shows that for the majority of the project, the contractor is in negative cashflow?

6.3.3 Cashflow based on programmes

The empirical S curve is for building projects constructed 'traditionally'; that may mean concrete foundations, brickwork and blockwork superstructure, tiled roofs and simpler mechanical and electrical systems and finishes. It is not suitable for projects that may have high-value items that can be completed in a short period of time. A multi-storey city centre office development with heavy concrete piles, steel frame, glass cladding and air conditioning is unlikely to have a cashflow model

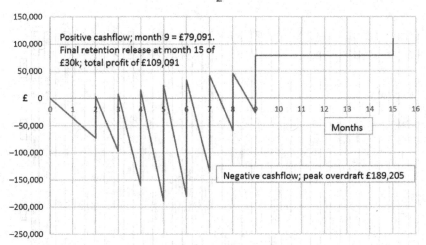

Figure 6.4 The saw tooth diagram.

approximating to an empirical S curve. The delivery for example of air conditioning equipment will significantly increase the value of work overnight. Similarly,
civil engineering project cashflows are also unlikely to resemble an S curve. Therefore, for 'non-traditional' or one-off projects, forecasts of cashflow need to be based
on construction programmes. The value of each activity on a master programme is
determined, and it is allocated on the programme as illustrated in Figure 6.5. For
those projects which have bills of quantities, all measured items can be allocated to
programme activities. Projects that have prices based upon elements or trades can
similarly allocate costs. Some approximations may be necessary; for example, an
electrical installation may have 30% of its cost allocated to first fix work, 30% to
second fix and 40% to final fix and test and commission. Contractors' preliminary
costs may be allocated on a straight-line basis, or alternatively it may be possible to
separate out fixed charges (at the beginning and end of projects), and time-related
charges. The New Rules of Measurement (NRM2, 2012, pp. 11 and 13) define fixed
charges as 'work, the cost of which is to be considered independent of duration'
(e.g. costs to initially establish, but not maintain, site compounds) and time-related
charges as 'work, the cost of which is to be considered dependent on duration' (e.g.
weekly hire charges for site cabins, hire rates for tower cranes or staff salaries).

6.4 Multiple project cashflows

Contractors are likely to have many projects running simultaneously. Whilst each
project has its own cashflow projection, at head office, there is a need to bring
all projects together to give one forecast of cashflow for perhaps a region or a
whole company. Figure 6.6 illustrates a case for one financial year with just three

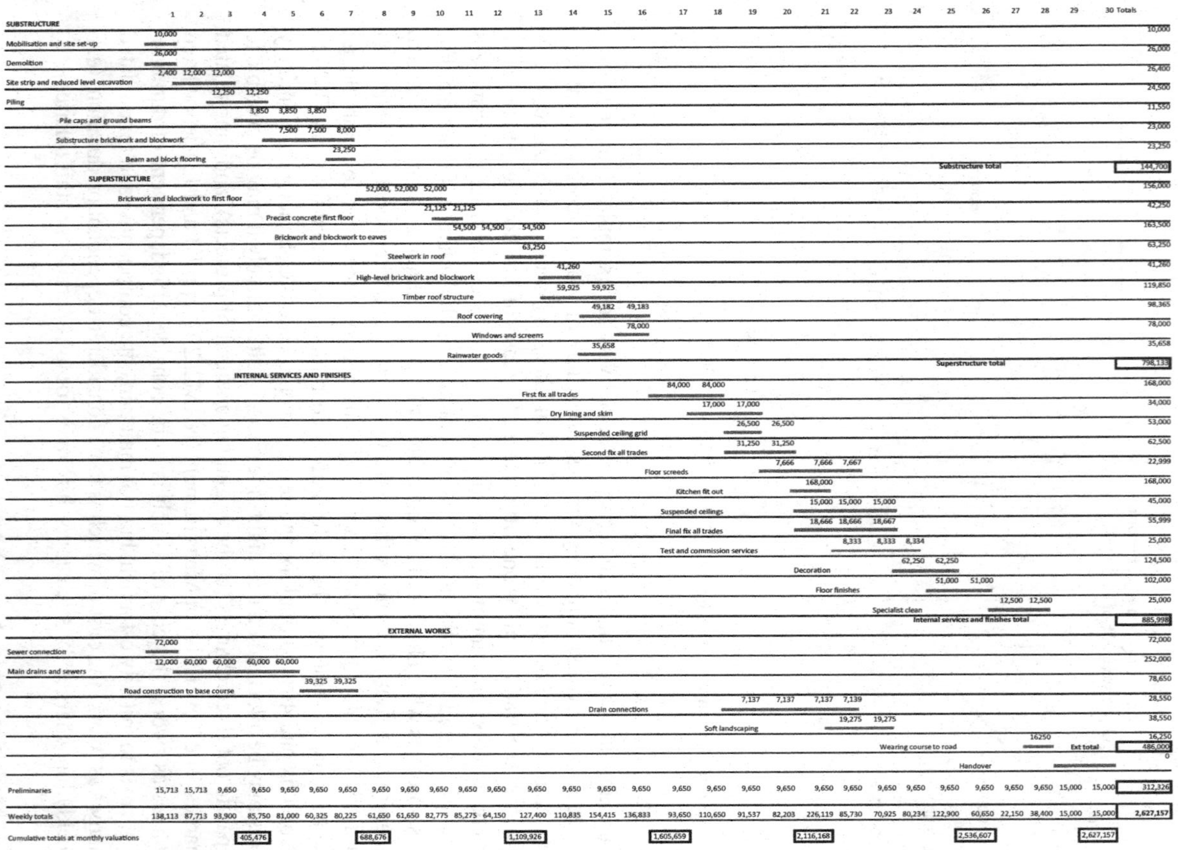

Figure 6.5 Cashflow based on a construction programme.

	Contractor's previous financial year			Contractor's financial year; 1st April to 31st March												Contractor's next financial year		
End of month	Jan	Feb	March	April	May	June	July	Aug	Sept	Oct	Nov	Dec	Jan	Feb	March	Apr	May	June
Month No	-3	-2	-1	1	2	3	4	5	6	7	8	9	10	11	12	1	2	3
Project 'A' Income				0.124	0.190	0.190	0.190	0.123	0.086	0.106								
Value £1.2M; nine months' duration. Three months in previous financial year																		
Project 'A' Cost				0.118	0.182	0.182	0.182	0.118	0.082	0.072								
Project 'B' Income					0.047	0.077	0.086	0.142	0.152	0.133	0.095	0.067	0.079					
Value £0.9M; nine months' duration. All in forthcoming financial year																		
Project 'B' Cost					0.045	0.073	0.082	0.136	0.145	0.128	0.091	0.063	0.055					
Project 'C' Income											0.047	0.077	0.086	0.142	0.152			
Value £0.9M; nine months' duration. Three months in next financial year																		
Project 'C' Cost											0.045	0.073	0.082	0.136	0.145			
Total income; rows 5, 9 and 13				0.124	*1	0.267	0.276	0.265	0.238	0.239	0.142	0.144	0.165	0.142	0.152			
Total cost; rows 7, 11 and 15 x -1 (to show cost as negative figures)				-0.118	-0.227	-0.255	-0.264	-0.254	-0.227	-0.2	*4	-0.136	-0.137	-0.136	-0.145			
Monthly cashflow; row 16 + row 17				0.006	0.01	*2	0.012	0.011	0.011	0.039	0.006	0.008	0.028	0.006	0.007			
Cumulative cashflow; cell F19 = cell E19 + cell F18				0.006	0.016	0.028	0.04	*3	0.062	0.101	0.107	0.115	0.143	*5	0.156			
Forecast of monthly income and expenditure for forthcoming financial year. Assume JCT 2011. 5% retention, 10% contractor's profit, one-month payment delay, one-month delay in meeting cost.																		

Figure 6.6 Cashflow prediction for multiple projects.

projects A, B and C. For simplicity, all projects are of nine months' duration. Project A starts three months before the given financial year and completes during it. Project B starts at the commencement of the financial year and Project C starts after six months, completing in the next year. Using an empirical S curve, forecasts of monthly valuations are calculated, and indicated above bar lines after deduction of retention; for example Project B valuation No 1 is forecast to be £47,000. Its cost calculated at £45,000 is indicated below the bar line. Both these sums are shown in month 2 to take account of a one-month delay in receiving and making payment. At the end of the financial year, the contractor has a cash surplus of £156,000.

Task 6.2: Five numbers in Figure 6.6 are substituted with asterisks thus: 1*, 2*, etc. What are those numbers?

6.5 Cashflow in private housebuilding

Private housebuilders build homes for private sales to the general public, not homes that are for sale to government, local authorities or housing associations. They are colloquially known as speculative builders or 'spec builders'. They commit funds to build projects speculating that they may achieve sales on completion, or better still, sales agreed before completion. They do indeed speculate, since on some occasions after completing properties they may have difficulty in selling. Many houses may lay empty for many months awaiting sale. Builders will speculate they wish to sell at a given price, but because sales are difficult they may have to reduce prices substantially or introduce incentives to achieve sales, such as free carpets and curtains or such like. The profits they had speculated they would make may be less than anticipated, or there may be losses involved. In the economic crisis in the UK of 2008/09, many large builders started developments speculating that they could sell at a given price, but as house prices fell rapidly between project starts and completions, they were then forced to reduce prices substantially and accept losses, just so they could get some cash into their businesses.

The reverse can happen too. Demand for homes can increase prices, and private housebuilders may be able to make higher profits than anticipated. This can be especially so in the context that the period of time between planning developments and completions can be many years. Planning starts at land purchase. Local authority planning permissions are often argued to delay commencement of much needed housing projects, sometimes by many years. Large private housebuilders feel secure if they have land banks of circa five years or more; that is if they build 1,000 units per year, they will want to own sufficient land to build 5,000 units. At the point they purchase land, they will budget for sales values many years ahead. If potential sales values are declining, they will not be so keen to start construction as if they are increasing.

In cashflow terms, private housing is the opposite of contracting. In contracting, work starts in month 1, and payment is often received at the end of month 2. Contractors receive regular monthly payments, and may often be in positive cashflow positions for most of project durations. In private housebuilding, substantial monies are spent purchasing land several years before construction starts. Land purchase costs need to include legal fees and staff cost of land buyers. When purchase is completed, there are more costs developing designs and seeking planning permission and building regulation approvals. A contribution needs to be made to local authorities to comply with Section 106 requirements; that is a sum of money to support infrastructure development outside the boundary of sites, to support the influx of people to an area. When construction starts, it may be necessary to spend money constructing all the roads and sewers to give good access and ensure that rainwater is able to drain away and not 'bog down' the site. Site compounds need to be established too; and also perhaps permanent fencing or boundary walls constructed to the whole perimeter of the site to give security. The construction of houses can then begin, and that will include a show house; it will not be sold until the end of the project. The first income that housebuilders receive is when the first houses are sold.

Example 6.2

A private housebuilder purchases land sufficient for 100 homes; there will be a mixture of four-bedroom detached, three-bedroom semi-detached and two-bedroom quasi-semi or terrace. The cost of the land purchase including all overheads is £4m, equivalent to an average of £40k per home. The cost of completing designs and achieving local authority permissions over three years is £240k; assumed as £80k per year. There is a Section 106 charge at project commencement of £200k. Additionally, there is a finance charge including interest fees and facility fees equivalent to 5% of cash outstanding.

It takes three months, costing £450k, to establish the site compound and construct the roads and sewers. The show house, which is fully furnished, costs

£150k to build. The builder then constructs 12 houses per month. The first 12 houses are completed in three months; thereafter 12 houses per month for the remaining 87 houses takes an additional 7.25 months; say 8 months including holidays. The overall time on-site is thus 3 months (site establishment) + 3 months (first 12 houses) + 8 months (last 87 houses) = 14 months. The first income the builder receives is 3 years 6 months after purchase of the land. The cost of constructing each of the remaining 99 houses, including external works, is an average £125k each. The average selling price is £200k per house x 100 = £20m total income.

Table 6.2 illustrates the cashflow position during each quarter of the life cycle of the development. All costs given are assumed to be on a straight-line basis, and sales are achieved in good time as houses are completed. The housebuilder makes an average of just over £20k per house, equivalent to 10.14% profit.

Figure 6.7 illustrates the data from row 21 of Table 6.2 in histogram form; positive cashflow for the project is only achieved in quarter 1 of year 4, when £3m income is received from the sale of the last 24 houses.

Task 6.3: Ten numbers in Table 6.2 are substituted with asterisks thus: 1*, 2*, etc. What are those numbers?

6.6 Using cashflow projections to forecast turnover

It is important for companies to forecast the value of work ahead over a number of years. Companies that are doing well will often proudly announce in the construction press the value of their order book. That may include large projects that extend

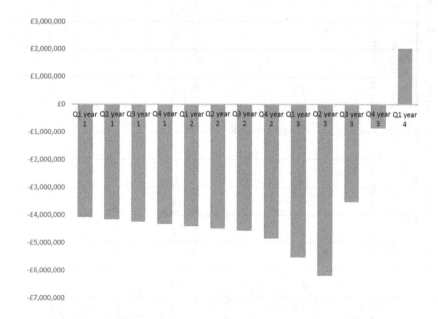

Figure 6.7 Cashflow prediction for a housing project.

Table 6.2 Private housebuilder's cashflow and profit projection on one development of 100 houses, three year+ duration

A	B — Q1 year 1	C — Q2 year 1	D — Q3 year 1	E — Q4 year 1	F — Q1 year 2	G — Q2 year 2	H — Q3 year 2	I — Q4 year 2	J — Q1 year 3	K — Q2 year 3	L — Q3 year 3	M — Q4 year 3	N — Q1 year 4	□ Totals
1 Land purchase	−4,000,000													−4,000,000
2 Design costs and LA permissions	−30,000	−30,000	−30,000	−30,000	−30,000	−30,000	−30,000	−30,000						−240,000
3 Section 106 fee								−200,000						−200,000
4 Site establishment and roads and sewers									−450,000					−450,000
5 Show house									−150,000					−150,000
6 Construction of first 12 houses £125k each										−1,500,000				−1,500,000
7 Construction of remaining 87 houses										−1,500,000	−4,500,000	−4,500,000	−375,000	−10,875,000
10 Total monthly costs	−4,030,000	−30,000	−30,000	−30,000	−30,000	−30,000	−30,000	−230,000	−600,000	−3,000,000 1*	−4,500,000	−4,500,000	−375,000	−17,415,000
11/12 Total cumulative costs	−4,030,000	−4,060,000 2*	−4,090,000	−4,120,000	−4,150,000	−4,180,000	−4,210,000	−4,440,000	−5,040,000	−8,040,000	−12,540,000 3*	−17,040,000	−17,415,000	
13 Sale of 99 houses										2,400,000	7,200,000 4*	7,200,000	3,000,000	19,800,000
14 Sale of show house													200,000	200,000
15/17 Sub-total monthly income										2,400,000	7,200,000	7,200,000	3,200,000	20,000,000
Overall total cumulative income										2,400,000	9,600,000	16,800,000 5*	20,000,000	20,000,000

18	Overall cashflow position; row 11 + row 27	-4,030,000	-4,060,000	-4,090,000	-4,120,000	8* -4,180,000	-4,210,000	-4,440,000	-5,040,000	-5,640,000	-2,940,000	-240,000	2,585,000	-556,937.5
19	Finance charges or credit 5% of row 20	-50,375	6* -51,125	-51,500	-51,875	-52,250	-52,625	-55,500	-63,000	-70,500	-36,750	-3,000	32,313	
20	Cumulative finance charges	-50,375	-101,125	7* -203,750	-255,625	-307,875	-360,500	-416,000	-479,000	-549,500	-586,250	-589,250	-556,938	
21	Cumulative cashflow	-4,080,375	-4,161,125	-4,242,250	-4,323,750	9* -4,405,625	-4,487,875	-4,856,000	-5,519,000	-6,189,500	-3,526,250	10* 2,028,063		
22	Overall profit											2,028,063		
23	Overall profit percentage cell N22 / cell N16 * 100 =											2,028,063 10.14%		

over several years, and therefore the value of order books may exceed the value of one year's turnover. Looking just one year ahead, companies need to achieve a set turnover to (i) support fixed costs such as head office overheads and staff salaries, and (ii) to generate an appropriate profit on money invested. Companies that are in danger of operating at under-capacity will often look to reduce fixed costs; that may involve staff redundancies. Staff positions can be vulnerable, since on the one hand whilst they are considered fixed costs and people are in 'permanent' positions, on the other hand, they can be less fixed than costs such as office premises which may be owned or occupied on long-term leases. On some occasions, it is in the public domain that companies close down completely regional offices, and re-deploy staff where possible or enforce redundancies. Alternatively, companies may plan for increased turnover and may hope to meet that with existing fixed costs e.g. not increasing resources at head office. Other companies may spot market opportunities, and plan for substantive increased turnover, with increased resources by opening up new regional offices and employing more people. On some occasions the best way to expand rapidly may be to buy other companies.

Example 6.3

A contracting company feels it is currently operating at under-capacity, with a turnover of £5.2m. To avoid compulsory redundancies for staff, it is targeting 10% growth in turnover to £5.72m in its next financial year, which starts in two months. It has nine projects running currently. Three are to finish in this financial year. Six of the remaining projects will run into next year, and indeed one of them also into the year after; they are shown as projects A to F in Table 6.3.

It has secured four new projects with values of £688k (project G), £332k (project H), £968k (project I) and £523k (project J). The first will start immediately, with a duration of nine months. The second one month from now, six months' duration. The third and fourth on the first day of the new financial year, both with

Table 6.3 Values of six projects A to F to be completed in next financial year; project F also two months into the following year

Project ID	Total value; £000s	Total value of work to be completed this financial year; £000s	Total value of work remaining; £000s	No of months into next financial year
A	600	200	400	4
B	780	460	320	4
C	125	45	80	6
D	338	138	200	5
E	97	37	60	10
F	569	59	510	12 plus two months into the following financial year
Totals	**2,509**	**939**	**1,570**	

15-month durations. The contractor is negotiating a one-year project K valued at £600k due to start in month 4 of its financial year; a letter of intent has been received. Further, the contractor has a variety of small works and on-going maintenance contracts that turnover circa £100k per month.

Table 6.4 illustrates the projected turnover of all projects A to J, including small works. For simplicity, it is assumed value and cost will be on a 'straight-line' basis (not the ¼–⅓ empirical curve). Retention is ignored, since this is value of work completed, not cashflow. Also, whilst the contractor will have retention deducted from payments, it should be getting equivalent amounts paid to it arising from retention released in previous years. Figures are shown to the nearest £100.

Task 6.4: Five figures in Table 6.4 are substituted with asterisks thus: 1*, 2*, etc. What are those figures?

Task 6.5: What value of work does the contractor still need to secure to meet its turnover target for next year?

6.7 Employment of labour; direct, indirect (self-employed) and subcontracting

The cost differences between employing labour directly or indirectly can be huge. In recent decades many, though not all, employers have preferred to use self-employed labour; also known as indirect employment. Some companies may have had the bulk of their workforce engaged individually as self-employed workers. However, there have been many recent government initiatives to deter self-employment, on the basis that companies who use this classification do so to avoid the payment of employers national insurance contributions (NIC). For earnings between £156.01 and £827.00, the NIC rate is currently 13.8%; on large wage bills, this is a significant sum of money. Government argues that individuals who work regularly for the same company, often classed as self-employed or indirect, are really 'employed'.

Direct labour are employed 'on the books' of companies and are paid through the Pay-As-You-Earn (PAYE) Inland Revenue system. The Employment Rights Act (1996) requires that after being employed for one month, employees should be given a 'written statement of particulars' detailing what is included in their contract. Additionally in construction, employers usually commit to compliance with the National Working Rule Agreement (NWRA) published by the Construction Joint Industry Council (CJIC, 2013). This agreement is negotiated between construction employers and trade unions, and details the rates of pay and terms and conditions of employment. Employees may be entitled to additional payment for overtime, fare and travel, shift and night working and subsistence. Also detailed are requirements for holiday pay, sick pay, accident and death benefit and pensions. Direct labour are entitled to statutory provisions for maternity, paternity and adoption leave. Occasionally, employers may negotiate exemptions from NWRA; for example where employees are regularly working night shifts, employers may find it difficult to remain competitive and pay premium rates. The wages element of the NWRA is renegotiated periodically to ensure wage levels are kept up-to-date. Many employers pay contributions to the benefits scheme operated

Table 6.4 Contractor's projected turnover for one financial year; projects A to J and small works

Project ID	Total value of work £000s	Calculation to determine amount in new financial year; £000s	Value of work in the new financial year; £000s	Duration this financial year; months	Month 1	Month 2	Month 3	Month 4	Month 5	Month 6	Month 7	Month 8	Month 9	Month 10	Month 11	Month 12	Totals £000S
Existing projects																	
A	600		400	4	100	100	100	100									*3
B	780		320	4	80	80	80	80									320
C	125		80	6	13.3	13.3	13.3	13.3	13.3	13.3							79.8
D	338		200	5	40	40	40	40	40								200
E	97		60	10	6	6	6	6	6	6	6	6	6	6			60
F	569	569–59/14 * 12	437.1	12	36.4	36.4	36.4	36.4	36.4	36.4	36.4	36.4	36.4	36.4	36.4	36.4	436.8
New projects																	
G	688	688 / 9 * 7	535.1	7	76.4	76.4	76.4	76.4	76.4	76.4	76.4						534.8
H	332	332 / 6 * 5	276.7	5	55.3	55.3	55.3	55.3	55.3								276.5
I	968	968 / 15 * 12 *1	*2	12	64.5	64.5	64.5	64.5	64.5	64.5	64.5	64.5	64.5	64.5	64.5	64.5	774
J	523	*1	418.4	12	34.9	34.9	34.9	34.9	34.9	34.9	34.9	34.9	34.9	34.9	34.9	34.9	418.8
In negotiation																	
K	600	600 / 12 * 9	450	9				50	50	50	50	50	50	50	50	50	450
Small works and maintenance	1,200	–	1,200	12	100	100	100	100	100	100	100	100	100	100	100	100	1,200
Totals	6,820		5,151.7	*4	606.8	606.8	606.8	656.8	476.8	381.5	368.2	291.8	291.8	291.8	285.8	285.8	*5

Note: difference between some figures due to rounding errors

by the Building and Civil Engineering Trust (B&CE, 2016), who then administer benefits on behalf of employees. The Trust will also take additional contributions from individuals to increase potential benefits; it will also take contributions from and provide benefits for the self-employed.

Other direct labour costs include redundancy pay and the requirement to give periods of notice. Those employed between two and twelve years for example, must be given one week's notice of termination for every year of service. Though costly, employers may find it better to give full pay in advance for a period of notice, and however unfortunate, ask employees to leave immediately. There can also be other complications employing people directly, dealing for example with tribunals in the event of disputes. A smaller cost for employers is the provision of personal protective equipment (PPE); self-employed labour provide their own. Additionally, there is a levy to be paid to the Construction Industry Training Board of 0.5% of payroll, though money can be claimed back to support the training of apprenticeships. From April 2017, employers in all industries will be required to pay the Apprenticeship Levy of 0.5%, though currently there are proposals for some exemptions for smaller companies.

Alongside employers' NIC, payment of wages during inclement weather can amount to significant sums of money. It can be the case that whole weeks or longer can be inclement, and in such cases even if no alternative work is available, the basic weekly wage must be paid in full in return for zero productivity.

Whilst employment of direct labour may incur significant sums of money and risk, it would seem correct that employers pay those sums and carry those risks, and merely include appropriate sums of money in their tenders.

Indirect or self-employed labour eliminates many of the above costs and passes risks to individuals. Arguably, the self-employed can manage these risks better and at lower costs than employers. However, Government set criteria that need to be met for a person to be truly considered self-employed, although as the House of Commons Briefing Paper 'Self-employment in the Construction Industry' Seely (2015) states, distinguishing between employment and self-employment is complex. The paper identifies there is a prevalence of 'false' self-employment, whereby intermediary companies are set up to give the appearance of the employed being self-employed, thus avoiding the need by employers to pay to the Treasury, Employers National Insurance Contribution. The criteria used to distinguish between employment and self-employment include whether the person provides plant, equipment or materials, and whether the person may employ others. It may be acceptable to employ a person who does not provide their own plant, equipment or materials, for a short number of weeks, but if the work is over a longer period, that person should arguably be classified as employed. In return for employers not having to pay many of the costs related to employment (holiday pay, travel, sick pay, pension contributions, inclement weather pay), it is normal practice that the self-employed will expect higher hourly rates of pay or higher rates if remuneration is based upon prices. Market rates will decide how much higher, but as a guide it may be around 25% or more. Individually, self-employed people need to be sure they manage this extra 25% carefully, since they may have periods without work, particularly for example bricklayers, who may find it difficult to secure inside work in winter months. They also need to provide for their own national insurance contributions, private pensions, sick pay, holidays, travel, PPE *et al.* There is a danger that self-employed people may not provide adequately for their retirement.

Some contractors consider there are too many complexities in employing labour directly or indirectly. Both bring many risks, and it may be better to pass those risks in full to bona-fide subcontractors. Therefore, there has been a move by some large contractors to subcontract only employment. Significant trades such as groundworks, brickwork, carpentry and joinery, which have been traditionally been directly employed may be subcontracted for all labour, plant and material. The only people that such contractors employ directly are professional staff.

6.8 Incentive schemes or fixed pay?

There are no authoritative publications to indicate the extent to which incentive schemes are used in construction. Payment based upon output or productivity is pervasive in new build works; perhaps less so in maintenance and refurbishment where assessing how long it may take to complete tasks at an operational level can be difficult. Incentive schemes should not be used in high-risk work, where care and attention is key to health and safety or environmental protection, and may not be suitable where there is great attention needed to quality of workmanship. Across the construction industry as a whole, perhaps 50% of operatives are paid based upon incentive schemes.

Trade unions support incentive schemes where they give tradespeople and operatives the opportunity to increase earnings; many employees may prefer them too for the same reason. Incentive schemes can be a psychological driver for individuals to attain optimum productivity.

Employers support incentive schemes, since employees who are motivated to work quickly are supporting a key goal of completing projects in good time. Employers are also able to link their incomes to their costs, since employers are usually being paid based upon a price, and business is more predictable if those prices are passed to employees. For example, if an employer is being paid £11.78/ m^3 (BCIS, 2016) for pouring concrete to foundation trenches, it gives predictability if the same or a lesser amount is paid to employees.

Alternatively, employees are paid an hourly rate. The national agreed rate for tradespeople for 2015/16 by the Building and Allied Trades Joint Industrial Council (BATJIC) is £11.50/hour. Based on the standard 39 hours week, that is £448.50/ week. Irrespective of the volume of productivity each week, tradespeople are paid this same £448.50. A survey by Hubbard (2016) for Building found that average weekly earnings across all skill levels was £601.00 per week in December 2015. For the skilled trades the average rate will be higher, with some reports of bricklayers being paid £1,000.00 per week. In fluctuating labour markets, with periods of high and lower demand, the wages of tradespeople (and indeed professional staff), may change rapidly. Construction is fragmented and tradespeople can be very transient, moving between employers after relatively short periods, highly motivated to increase earnings. Employers need to respond rapidly and increase if necessary rates of pay to attract and retain tradespeople. To make up the difference between £448.50 and £601.00 or £1,000.00 or any other 'market' figure, directly employed labour on fixed rates of pay will be paid a fixed or 'spot' bonus. Paying a fixed wage can constitute extra financial risk for employers. For example using concrete again, contractors may have contracted with employers to pour concrete in foundation trenches at a labour rate of £11.78/m³, no matter how long it takes. BCIS rates (2016) show that rate is based on 1 hour/m³ at a labour rate of £11.78/hour. Whilst there is the possibility that work may be completed in less time than estimated, there is also the risk that it will take a labourer perhaps 2 hrs/m³, and losses will result. However, many companies are happy to carry the risks of quality of work. If the workforce is being paid based upon a fixed wage, it is likely that individuals will take time to ensure each task is completed well, and the supervisory task is to ensure reasonable productivity and time keeping e.g. unreasonable amounts of time not lost to excessive break times.

6.8.1 Types of incentive scheme

Traditionally, many incentive schemes were set on a time basis, and were built up from time and motion studies. For example, if time and motion indicated that a bricklayer can lay 50 bricks in one hour, market conditions may indicate that employers will want bricklayers to earn an extra 50% of their pay in bonus, such that for basic pay of £448.50, an additional £224.25 would be earned in bonus; total £672.75. A target is set (to be beaten) of 33.33 bricks per hour; in 39 hours the bricklayer actually lays 50 bricks x 39 hours = 1,950 bricks. Given a target of 33.33, the time paid is then 1950 / 33 = 58.50 hours' pay, comprising 39 hours' basic pay and 19.50 hours' bonus. If the bricklayer works quickly and actually lays 66.66 bricks per hour, the hours achieved will be 78 total, 39 hours paid as basic pay and 39 as bonus.

Task 6.6: Based upon the above figures, what will be the wage of a bricklayer who one week lays an average of 33.33 bricks/hour, and another week 40 bricks per hour?

However, traditional time and motion studies are now rare in construction. Most incentive schemes are based around money, such that instead of 33.33 bricks per hour, the rate for laying bricks may be £400.00/1,000. Employers and tradespeople can associate quickly with these money rates, and movement takes place rapidly between sites because perhaps for example, one employer is paying £10.00 per thousand more than another. Employers can move rapidly to adjust money-based rates as markets dictate, mindful also that on some occasions where work is less plentiful and employers need to reduce costs, market rates may go down.

The choice of incentive or fixed schemes applies to directly and indirectly employed labour. It may be most appropriate to employ indirect labour a fixed wage, by agreeing an hourly rate, perhaps £15 or £20 or £25 per hour depending on market conditions. Table 6.5 illustrates calculation of pay for two directly employed carpenters, who are working as a team. Table 6.6 illustrates pay for a 2+1 gang of plasterers who are self-employed. The assumption is that both are working on incentive schemes.

Table 6.5 Calculation of weekly pay for two carpenters employed on a financial incentive scheme

Two joiners; one week's work	Quantity	Unit	Rate (£)	Amount (£)
Roof carpentry timbers				
225 x 75 purlins	30	m	10.82	324.60
200 x 25 ridge	15	m	3.18	47.70
300 x 75 hip rafters	10	m	14.30	143.00
125 x 50 rafters	100	m	3.18	318.00
100 x 25 hangers	25	m	1.81	45.25
100 x 25 runners	15	m	1.81	27.15
100 x 50 dragon tie	2	m	2.67	5.34
150 x 50 ceiling joists	90	m	3.18	286.20
Extra time allowance since brickwork out of square	5	hours	11.00	55.00
Total wages (£) to be shared between two				**1,252.24**

Table 6.6 Calculation of weekly pay for two and one gang of plasterers employed on a financial incentive scheme

Two and one gang of plasterers; one week's work	Quantity	Unit	Rate (£)	Amount (£)
Plastering				
12mm plasterboard dot and dab to walls	85	m²	2.28	*1
2mm finish to ditto	85	m²	5.42	460.70
9mm plasterboard screwed to ceilings	60	m²	*2	381.60
2mm finish to ditto	60	m²	6.77	406.20
9mm plasterboard to side and soffits of beams	*3	m²	7.95	119.25
Angle beads	40	m	3.76	150.40
Total wages (£) to be shared between three				**1,711.95**

Task 6.7: Three figures in Table 6.6 are substituted with asterisks thus: 1*, 2* and 3*. What are those figures?

6.9 Cost control

6.9.1 Introduction

Cost control is pervasive in all sections of the economy. Governments, other public bodies such as local authorities, and private businesses are constantly looking to reduce costs. In construction, leading reports often call for a reduction in costs. Latham (1994) for example, called for a reduction in real costs of 30% by year 2000. At the time many authoritative sources scoffed at the possibility, but then six years later many employers argued they had achieved it. Whilst cashflow is key for businesses, it is not so often a subject of debate on an every-day basis;

controlling costs is always on the agenda. There can be weekly meetings devoted to cutting and controlling costs. Organisations employ many specialist staff to help them to control costs. In construction, a key element in the role of employer, contractor and subcontractor quantity surveyors is cost control. Accountants, who are not specialist in construction, are also employed at head offices in these organisations. Cost control is also key lower down the supply chain, in builders merchants, suppliers and manufacturers. Companies design cost controls systems to suit their own internal preferences. A system for example in one private house-building company may be radically different to the system in another.

The basis of cost control systems is to compare budgets with actual costs. Where costs exceed budgets, corrective action should be taken. On an individual element of work, if 20% of work is completed and losses are being incurred, corrective action may be possible to ensure losses are not incurred on the remaining 80%. On whole projects, if work is similarly 20% complete and losses are being incurred, particularly close attention may be needed to ensure they are successful at completion. Perhaps for a site activity, the method of pouring concrete can be changed from hoisting by mobile crane and skip to pumping?

A key goal is that at the end of individual operations, or on projects totally, costs are the same as, or better, lower than budgets. Employers desire that final accounts for projects are less than or equal to tender bids. Consultants desire that design costs are less than or equal to fee income. Contractors seek to have lower costs than final accounts. Systems are designed such that costs are often compared against net income figures; net income has 'budgeted profit' taken off. Companies will bid for work including a profit, though sometimes they may do so without, just to win work and maintain turnover. Companies estimate actual costs they will incur if they were to win a project, for example £100k. They may submit a tender of £110k to include 10% profit. Assuming there are no changes or variations as work proceeds, and the final account is £110k, net income is £100k and gross income £110k. Arguably, the £10k profit belongs to the company, not to the project. The project has only £100k to complete the work, and its actual cost performance is monitored against the £100k. Companies will usually be happy if actual costs at the end of the project are £100k, but they will always aim for 'net gains'; that is actual costs less than £100k. Some companies would proactively set this project a cost target of £95k, and as work progresses compare actual costs against £95k. Main contractors may bid for work with 2.5% added for profit, but then try to achieve more; perhaps 4 or 5%. A company would be very unhappy if actual costs on this project were in the range of £100k+ to £110k, since it would be making less profit than it had budgeted for. It can be catastrophic if costs are over £110k. In the public domain, it is often reported that large companies lose money on flagship projects. In such cases it is not clear, but likely that loss estimates are against gross figures, in our case £110k, not the net. A reported estimate of loss of £10k announced in public may actually represent costs of £120k; total costs thus £20k more than expected.

Discussion point 6.6: A project with estimated net costs of £100k and bid amount of £110k completes successfully with costs of £95k. The project leader claims £15k has been made. Company directors argue that profit was only £5k, since £10k of the £15k belongs to the company. Who is correct? Another project with the same values completes with costs of £105k. Is this a £5k gain or £5k loss?

Whilst a primary function of cost control systems is to monitor the effectiveness and efficiency of site operations, it is also important to provide feedback to estimating teams for future projects. If money is being lost on individual elements of work, it may not be the fault of project teams who are working as well as can be reasonably expected. It may be just that estimates are too low; arguably on the next project, estimates for this type of work should be increased. It may also be the case that individual elements of work can easily make much higher profits than budgeted because estimates in these areas are generous. Estimators need to know that too, since higher prices make companies less competitive, and they may consequently not win projects that they should. It is not the function of estimators to estimate highly so companies make good profits; it is their job to estimate accurately. Profit addition is a judgement to be made after the estimate of costs and before the submission of bids. Estimators will be part of teams that make that decision. Some estimators may argue they are reluctant to change future prices based on feedback from project teams, since teams are quick to 'complain' about estimates that are low and keep quiet about estimates that are generous.

One other major function of cost control reports is to minimise the possibility of fraud and corruption. The CIOB (2013, p. 1) document 'A Report Exploring Corruption in the Construction Industry' found in a survey of members that "49% of respondents believe corruption is common with the UK construction industry". Corruption may arise is a multiplicity of ways, such as money passing inappropriately, or theft or labour and materials being diverted off-site to other activities *et al.*

For all organisations, it is inevitable that some elements of work or trades within projects will incur losses. That is the nature of business. For that single project, it is hoped that net gains on other elements or trades will outweigh net losses. However, that may not always be possible, and all experienced contractors incur losses on some projects. In such cases it is hoped that within each accounting year, any net losses on individual projects will be more than offset by net gains on other projects. Again, however, at the end of an accounting year, many experienced contractors incur losses across the full portfolio of their work; net losses outweigh net gains. If this is a repeating phenomena over several years, it may lead to business failure. Contractors who have strong cash balances, or who are able to gain the support of banks, will try to 'ride' some bad years in the expectation that profits will return in the future.

6.9.2 Frequency of cost reporting

The frequency of cost reports is key. The more frequent reports are produced, the more quickly potential problems can be identified, and the more quickly corrective action can be taken. If an activity were scheduled to last for 10 weeks, any corrective action needed would be more effective if implemented after one week, rather than four weeks. Some companies may reflect on costs on a weekly basis. It is likely that for Monday to Friday work, costs calculations will be performed in the early part of the following week, and figures available for management inspection towards the end of the week. Not all companies will monitor costs weekly. Most SMEs and large companies will likely monitor costs monthly, since that period is used traditionally for the majority of payments in and out of organisations. Since there can be some complexity in cost systems, it may be two or three weeks into a new month before figures are available for a preceding month. Small and micro companies may only reflect on costs at the end of projects, or indeed not monitor cost on individual projects at all. They will alternatively produce accounts at the end of the year for income tax purposes, and as work progresses make judgements about profitability based on cash balances. If balances at the bank are increasing, 'they must be making money'?

6.9.3 Final cost forecasts and prudence in reporting

Budgets are fixed at project commencement, and since very often contracts are signed, those budgets will not usually change, unless there are variations, as work proceeds. However, forecasts of costs can continually change. It is as work progresses that more details about projects become available, more detailed planning takes place, markets conditions develop and thus project teams develop their knowledge and are able to make more accurate predictions of costs. For example in the £100k project, if the contractor's cost control systems showed that 50% or work was complete and actual costs were £52k, it may be possible to forecast more accurately that the remaining 50% of work can be completed for £48k, or some other figure. That higher confidence may stem for example from a budget for a subcontract package at tender stage which is £2k higher than the value for which an order has been subsequently placed. On larger projects, labour and material prices may be volatile, and partway through projects new, more accurate predictions are possible. Therefore in cost control systems, reports detail figures up to a given moment in time, and also compare future cost predictions with likely final budgets.

In cost control systems, there should be no surprises ahead. If projects teams know that in the future losses are likely, those losses should be reported promptly. It may be that an order is placed for a subcontract package which is higher than that allowed at tender stage, or that labour and material to be used in the future are higher in price than envisaged at tender stages.

There should also be a measure of 'prudence' in cost control systems; that is not to report profits that are not secured. There may, for example, be extra work to be carried out, but it is not absolutely clear which party takes responsibility for payment. This could be a designer involved in more detailing; the cost of the designer's salary will show in a cost control report, but should a budget be allowed against it if there is only a 50% possibility the design consultant will be paid? If it is a contractor carrying out some remedial work, it may not be absolutely clear if it is necessary because of a specification problem which the employer may pay for, or a workmanship problem for which it carries responsibility. The emphasis may be to get the task done, so not to delay progress, and negotiate later about who pays. In cost control reports it is best to include such costs but not budgets. A comment can be added to the report that the company is hopeful of securing a payment, but as a measure of prudence, a budget has not been included until agreement is reached.

6.9.4 Manipulation of internal cost reports

Manipulation should not happen, but it does pervasively in all parts of the supply chain. It should be discouraged. In contracting it can happen at all reporting levels, from site to senior managers, and from senior managers to directors. Figures for projects may change twice as they pass through these reporting levels. There may be a wish not to report losses to superiors, since that may prompt some unwelcome investigations. Also, in cases where higher than anticipated profits are being made, there may be a wish at early stages of projects not to show these profits; better to hold some money back in case unexpected losses occur later. These losses can then be disguised as they arise. On a single project, if some elements of work are making losses and others higher than expected profits, figures may be changed to show that all elements are operating 'okay'; arguably this is not so serious. More problematic is if projects are making losses in overall terms, and figures are changed to show profits or losses as not being so severe. That may be done at early stages in projects in the hope that losses will be recovered on later work. If subsequently there is no recovery of losses, that may leave people involved in very difficult positions. In the context of a regional contracting operation reporting to its head office, it can reasonably be expected that some projects will be doing better than expected, some on schedule and some worse than expected. There may be a temptation to use money from the better than expected projects to 'cover' losses on projects where results are worse than expected. This can be facilitated internally by allocating salaries of staff on the 'worse projects' to the 'better projects'. Thus the costs on these 'worse projects' will be reported lower than is actually the case. Some companies may have zero internal tolerance of manipulation of cost control report figures.

6.9.5 Employer cost control

The headline figure for employers is often tender sums. However, there is a need to also control costs on elements of work other than just amounts paid to contractors. They need to have oversight of their own internal costs in overseeing projects and design team costs. Employers also have a plethora of other costs in the early stages of projects, for example legal and planning fees. At the end of projects they will have budgets for occupation, or if they are speculative development companies, budgets for marketing, advertising and selling. In all cases, the objective is to monitor costs as work progresses, and take corrective action if there are some early signs that costs may exceed budgets. Figure 6.8 illustrates an employer's budget for the construction phase of a project. It is assumed it is a construction management form of procurement, as described in section 2.11. Some elements of the project have higher costs than forecast at tender stage; other elements show savings. The overall position is that the project is forecast to complete at lower cost than budget.

Work package item	Tender Amount (£)	Order values (£): * = not yet placed	Paid to date (£)	Budget to date (£)	Net gain or loss to date (£)	Final cost forecast (£)	Forecast net gain or loss (£)
Main contractor's preliminaries	902,500	902,500	463,000	463,000	0	902,500	0
Intrusive investigations	10,000	12,525	12,525	10,000	−2,525	12,525	−2,525
Demolition works	53,000	45,000	46,821	53,000	6,179	46,821	6,179
Groundworks	262,000	250,560	283,264	262,000	−21,264	290,525	−28,525
Piling	0	0	0	0	0		
Concrete works	425,000	502,350	480,600	406,599	−74,001	508,650	−83,650
Roof coverings and roof drainage	165,000	158,625	82,500	85,816	3,316	162,500	2,500
External and internal structural walls	975,000	943,560	48,650	50,271	1,621	943,650	31,350
Cladding	260,000	278,325	138,600	129,475	−9,125	278,325	−18,325
Windows and external doors	245,000	240,010	15,000	15,312	312	240,010	4,990
Mastic	35,000	36,250	0	0	0	36,250	−1,250
Non-structural walls and partitions	184,000	161,265	12,000	13,692	1,692	161,265	22,735
Joinery	202,500	210,625	15,200	14,614	−586	220,625	−18,125
Suspended ceilings	165,000	173,000	0	0	0	173,000	−8,000
Architectural metalwork	125,000	156,000	0	0	0	156,000	−31,000
Tiling	75,000	72,385	0	0	0	72,385	2,615
Painting and decorating*	180,000	180,000	0	0	0	180,000	0
Floor coverings*	276,000	276,000	0	0	0	276,000	0
Fittings, furnishings and equipment	0	0	0	0	0	0	0
Combined mechanical and electrical engineering services	3,330,000	3,150,000	72,000	76,114	4,114	3,150,000	180,000

Figure 6.8 Employer cost control report for a project procured using construction management.

Work package item	Tender Amount (£)	Order values (£): * = not yet placed	Paid to date (£)	Budget to date (£)	Net gain or loss to date (£)	Final cost forecast (£)	Forecast net gain or loss (£)
Lifts and escalators	95,000	85,325	6,500	7,237	737	85,325	9,675
Facade access equipment	0	0	0	0	0	0	0
External works and drainage	445,000	440,360	200,600	202,714	2,114	455,325	−10,325
Risks	100,000	0	48,635	48,635	0	100,000	0
Provisional sums*	1,000,000	525,325	256,335	256,335	0	902,000	98,000
Credits	0	0	0	0	0	0	0
Dayworks (Provisional)	25,000	0	8,658	8,658	0	25,000	0
Totals	**9,535,000**	**8,799,990**	**2,190,888**		**−87,418**	**9,378,681**	**156,319**

Figure 6.8 (Continued)

6.9.6 Consultant cost control

Figure 6.9 illustrates a cost control report for a consultant. Estimates prepared by consultants for projects are based on the time required by designers. Reconciliations can be undertaken by a comparison of actual hours to those budgeted, and a conversion of hours to money. Hours are allocated to the project according to seniority in the company, and thus level of salary.

6.9.7 Contractor cost control

Figure 6.10 is developed from Table 6.5 and illustrates a weekly cost control report for joiners/carpenters fixing roof carpentry timbers. They are paid a fixed weekly wage, and overall the work is completed £153.50 lower than the net budget; a net gain is made. This net gain is valuable, in the case that net losses are incurred elsewhere on the project.

Figure 6.11 illustrates a monthly cost control report. Figure 6.12 shows a breakdown for the preliminary element of the project. Figure 6.13 details a breakdown for the subcontractors' part of the project.

6.10 Dayworks

Dayworks occur throughout the supply chain, from employers to contractors, contractors to subcontractors, and subcontractors to sub-subcontractors. Also employers of people (contractors, subcontractors and sub-subcontractors) may pay tradespeople on daywork. Colloquially, daywork is often disliked, especially from the viewpoint of the party that is paying, but sometimes also by the party being paid. It is often for work that arises that was not foreseen in earlier stages of projects. It could occur frequently on refurbishment-type projects, where unexpected tasks arise as work is opened up. Daywork indicates that work is executed and paid

Time reconciliation hours

	A Total hours in bid	B Percentage of work completed to date	C Budget time to date; column C x col D / 100	D Actual time taken to date from timesheets	E Net gain or loss to date; col E – col F	F Estimated time to complete	G Final time forecast; col F + col Fi	H Final net gain or loss forecast; col C – col 1. Flors H
Partner/director	80	90	72	60	12	6	66	14
Senior designer	450	75	338	382	–45	113	495	–45
Designer	600	75	450	440	10	150	590	10
Assistant	300	50	150	200	–50	200	400	–100
Trainee	100	50	50	95	–45	95	190	–90
Totals	1,530		1,060	1,177	–118	564	1,741	–211

Net money reconciliation (£)

Net hourly rate (£)		A Total money in bid; cell A12 x cell C3, etc.	B Percentage of work completed to date	C Budget cost to date; column C x col D / 100	D Actual cost to date; cell A12 x cell F3, etc.	E Net gain or loss to date; col E – col F	F Estimated cost to complete; cell A12 x cell H3, etc.	G Final cost forecast; col F + col FI	H Final net gain or loss forecast; col C – col 1. (£)
49.68	Partner/director	3,974	90	3,577	2981	596	298	3,279	696
31.05	Senior designer	13,973	75	10,479	11,861	–1,382	3,493	15,354	–1,382
21.73	Designer	13,038	75	9,779	9,561	217	3,260	12,821	217
17.39	Assistant	5,217	50	2,609	3,478	–870	3,478	6,956	–1,739
13.66	Trainee	1,366	50	683	1,298	–615	1,298	2,595	–1,229
	Totals	37,568		27,126	29,179	–2,052	11,826	41,005	–3,437

Figure 6.9 Consultant designer cost control report.

Item	Quantity	Unit	Cost (£)			Budget (£)		Net gain or loss (£)	Comments
			Rate	Amount	Including overhead addition say 25%	Rate	Amount		
Roof carpentry timbers									
225 x 75 purlins	30	m	10.82	324.60	405.75	15.62	468.60	62.85	
200 x 25 ridge	15	m	3.18	47.70	59.63	5.26	78.90	19.28	
300 x 75 hip rafters	10	m	14.30	143.00	178.75	19.65	196.50	17.75	
125 x 50 rafters	100	m	3.18	318.00	397.50	4.56	456.00	58.50	
100 x 25 hangers	25	m	1.81	45.25	56.56	2.89	72.25	15.69	
100 x 25 runners	15	m	1.81	27.15	33.94	2.89	43.35	9.41	
100 x 50 dragon tie	2	m	2.67	5.34	6.68	3.6	7.20	0.53	
150 x 50 ceiling joists	90	m	3.18	286.20	357.75	4.4	396.00	38.25	
Extra time allowance since brickwork out of square	5	hours	11.00	55.00	68.75		0.00	−68.75	Examine potential to charge to bricklayers?
Totals				**1,252.24**	**1,565.30**		**1,718.80**	**153.50**	

Figure 6.10 Contractor cost control report for two joiners fixing roof carpentry.

Monthly cost value reconciliation (£)		Conciliation up to date:	01.05.2017.
Contract period	50	Number of weeks in progress	17
Extension of time	0	Percentage period expired	34
Total period	50	Number of weeks to run	16
		Ahead or behind programme	2
Total contract value	2,627,157	Gross valuation to date	981,467
Less contingencies	0	Less materials on-site	22,500
Balance	2,627,157	Over or under claim	16,250
Subcontractors at tender	1,703,444	Balance	942,717
Contractor's work	923,713	
% value of work complete; cells C23 / F23	35.56	

Notes: at tender stage the project was forecast to make £59,708 profit. The tender bid was £2,627,157. After variations and extra works, the forecast final account is £2,651,161. The employer is forecast to exceed its budget by £24,004. The profit should be £57,624; it is actually forecast to be £54,797. That is a net loss of £2,816.

Figure 6.11 Contractor's monthly cost control reconciliation.

Monthly cost value reconciliation (£)		Conciliation up to date:			01.05.2017.		
Item	Total budget at tender	Budget to date	Cost to date	Net gain or loss	Final budget forecast	Final cost forecast	Forecast net gain or loss
Labour	140,331	68,252	85,219	−16,967	143,250	162,000	−18,750
Plant	55,896	48,926	56,231	−7,305	58,000	64,000	−6,000
Materials	162,691	125,623	127,522	−1,899	168,222	162,691	5,531
Subcontractors	1,703,444	493,144	477,411	15,733	1,717,159	1,700,727	16,432
Preliminaries	312,326	121,070	111,452	9,618	312,326	312,168	158
Provisional sums	13,636	0	0	0	13,636	13,636	0
Sub-total	2,388,324	857,015	857,835	−820	2,412,593	2,415,222	−2,629
Company overheads 7.5%	179,125	64,276	64,338	−62	180,944	181,142	−197
Company finance charges	0	0	0	0	0	0	0
Profit 2.5%	59,708	21,425	0	21,425	57,624	54,797	−2,826
Overall totals	**2,627,157**	**942,717**	**922,173**	**20,544**	**2,651,161**	**2,651,161**	**−24,004**

Figure 6.11 (Continued)

for on the basis of time (plus materials and plant) taken to do tasks. If tasks take a short time, payment is low, and if they take a long time, payment is high. The party paying has two fears: (i) the party being paid has no incentive to complete the work at reasonable speed, and (ii) the final cost of completing the work is unknown – authorising work to be completed on daywork is like 'writing a blank cheque'. There is also the fear that the party being paid may allocate more time in claims for payment than was actually used e.g. six hours of daywork may be claimed for an item of work that actually took four hours to complete. However, especially for smaller tasks, or when a payment is due because one party is 'standing' because of the fault of another party, daywork can be a very sensible and amicable method of agreeing amounts of money due. Contractors' employees may charge 'standing' time if they are waiting for material deliveries that are arranged by the contractor.

From the viewpoint of the party being paid, daywork might be attractive. For tradespeople, it would be acceptable to work at a 'steady pace', although it is hoped not at 'snail pace'. However, key is the actual amount of money being paid, and if the hourly rate per hour of time is not attractive, there can be a disincentive to work on daywork. This can be the case in daywork situations between contractors and tradespeople. For example, a tradesperson capable of earning £20.00 per hour whilst being paid on an incentive scheme will not want to work on daywork for long periods at a rate of £15.00 per hour. Contractors, subcontractors and sub-subcontractors may be happy to work for higher-tier parties on daywork, since in the context that contracting can involve a loss of money, agreed daywork rates are usually high enough to ensure they do make some profit.

The alternative to working on daywork is that a payment is made based upon a price. If for example there is some extra foundation trench excavation arising from ground conditions being worse than envisaged, and a bills of quantities is available, it is accepted practice that payment will be made based upon rates in the

Budget release (br) and cost (c)	Item	\	Months; actuals 1	2	3	4	Totals (£)	Net gain or loss	Months; forecast 5	6	7	8	9	10	11	12	Total forecasts budgets and costs	Total forecast net gain or loss	Comments
BR	Employer's requirements	Site cabins	1,166	1,166	1,166	1,166	4,664	−1,134	1,166	1,166	1,166	1,166	1,166	1,166	1,166	1,174	14,000	−1,398	
C			1,506	1,362	1,528	1,402	5,798		1,200	1,200	1,200	1,200	1,200	1,200	1,200	1,200	15,398		
BR		Mobilisation	1,500				1,500	−1,362									1,500	−1,362	
C			2,862				2,862										2,862		
BR		Demobilisation					0	0								600	600	0	
C							0								600	600		
BR	Supervision	Construction manager and foreman	5,633	5,633	5,633	5,633	22,532	−68	5,633	5,633	5,633	5,633	5,633	5,633	5,633	5,637	67,600	−696	
C			5,555	5,621	5,712	5,712	22,600		5,712	5,712	5,712	5,712	5,712	5,712	5,712	5,712	68,296		
BR		Site engineer and foreman	3,958	3,958	3,958	3,958	15,832	−1,268	3,958	3,958	3,958	3,958	3,958	3,958	3,958	3,962	47,500	−15,452	Extra foreman support to finish
C			4,500	4,250	4,250	4,100	17,100		3,852	6,000	6,000	6,000	6,000	6,000	6,000	6,000	62,952		
BR	General labour	Planner and QS	2,375	2,375	2,375	2,375	9,500	1,254	2,375	2,375	2,375	2,375	2,375	2,375	2,375	2,375	28,500	7,454	
C			3,522	1,562	1,562	1,600	8,246		1,600	1,600	1,600	1,600	1,600	1,600	1,600	1,600	21,046		
BR		Attendant labour	2,426	2,426	2,426	2,426	9,704	3,068	2,426	2,426	2,426	2,426	2,426	2,426	2,426	2,434	29,120	6,484	
C			1,628	1,952	1,856	1,200	6,636		2,000	2,000	2,000	2,000	2,000	2,000	2,000	2,000	22,636		
BR		General attendance			200	200	400	344	200	200	200	200	200	200	200	200	2,000	344	
C			0	0	0	56	56		200	200	200	200	200	200	200	200	1,656		
BR	Site facilities	Administration	1,041	1,041	1,041	1,041	4,164	−680	1,041	1,041	1,041	1,041	1,041	1,041	1,041	1,049	12,500	−672	
C			962	1,058	1,623	1,201	4,844		1,041	1,041	1,041	1,041	1,041	1,041	1,041	1,041	13,172		
BR		Services	2,200	680	680	680	4,240	−1,896	680	680	680	680	680	680	680	700	9,700	−2,836	
C			3,986	750	775	625	6,136		800	800	800	800	800	800	800	800	12,536		
BR		Mobilisation	2,500				2,500	438									2,500	438	
C			1,862	200			2,062										2,062		
BR	Temporary works	Maintenance	5,000				5,000	1,414	1,000								6,000	1,414	
C			3,586				3,586		1,000								4,586		
BR		Mobilisation/removal	2,500				2,500	−1,062	200								2,700	−1,062	
C			3,562				3,562		200								3,762		
BR		Temporary work					0	0					1,250	1,250	1,250	1,250	5,000	0	
C			0				0						1,250	1,250	1,250	1,250	5,000		
BR	Mechanical plant	Lifting	5,000	5,000	5,000	5,000	5,000	−10,949	5,000								25,000	4,051	
C			2,626	3,856	6,211	3,256	15,949		5,000								20,949		

Figure 6.12 is a wide numeric reconciliation table (rotated on the page). Reconstructed below with 4 elapsed periods, a "to date" reconciliation (budget/cost/variance), 8 forecast periods, and the project total (budget/cost/variance). "BR" = budget, "C" = cost.

Category	Item		1	2	3	4	To date	Var	5	6	7	8	9	10	11	12	Total	Var
	Transporting	BR	333	333	333	333	1,332	525	333	333	333	333	333	333	333	337	4,000	525
		C	0	222	300	285	807										3,475	
	Concreting	BR					0	0									0	0
		C					0										0	
Non-mechanical plant	Scaffold	BR				3,000	3,000	0	4,000	5,000	2,326	1,000					15,326	0
		C				3,000	3,000		4,000	5,000	2,326	1,000					15,326	
	Instruments	BR	180	180	180	180	720	157	180	180							1,080	157
		C	95	156	156	156	563		180	180							923	
	Miscellaneous	BR			1,000	1,000	2,000	1,206	1,000	1,000	1,000	1,000	1,000	1,000	1,000	1,000	10,000	1,206
		C			262	532	794		1,000	1,000	1,000	1,000	1,000	1,000	1,000	1,000	8,794	
	Small tools	BR	2,800	200	200	200	3,400	−458	200	200	200	200	200	200	200	200	5,000	1,142
		C	3,256	200	200	200	3,858										3,858	
Contract conditions	Insurances	BR	716	716	716	716	2,864	380	716	716	716	716	716	716	716	724	8,600	1,148
		C	621	621	621	621	2,484		621	621	621	621	621	621	621	621	7,452	
	Bonds	BR	9,600				9,600	1,088									9,600	1,088
		C	8,512				8,512										8,512	
Miscellaneous	Winter working	BR					0	0									0	0
		C					0										0	
	Quality assurance	BR	60	60	60	70	250	−67									250	−67
		C	100	62	57	98	317										317	
	Safety	BR	100	100	100	50	350	−398	50	50	50	50	50	50	50	50	750	−1,598
		C	21	156	321	250	748		200	200	200	200	200	200	200	200	2,348	
	Setting out consumables	BR	100	100	100	100	400	−150	100								500	−150
		C	321	53	65	111	550		100								650	
	Specialist clean on completion	BR					0	0								3,000	3,000	0
		C					0									3,000	3,000	
Total budget							**111,452**	**−9,618**									**312,326**	**158**
Total cost							121,070										312,168	

Figure 6.12 Contractor's preliminary reconciliation.

Sub name	Trade	Tender net value	Order net vale	Factor: columns c/d	Cost to date (£) Measured work	Recoverable variations	Unrecoverable ditto	Miscellaneous	Total: columns f to i	Budget to date (£) Measured work: columns f* columns e	Recoverable variations	Unrecoverable ditto	Miscellaneous	Total: columns k to n	Net gain or loss	Final cost forecast	Final budget forecast	Forecast net gain or loss
	Measured work																	
ABC Ltd	Demolition	25,350	22,265	1.14	22,265	2,000			24,265	25,350	2,000			27,350	3,085	24,265	27,350	3,085
DEF	Piling	23,888	23,888	1.00	23,888	5,200			29,088	23,888	5,200			29,088	0	29,088	29,088	0
GHI	Groundworks	4,57,421	432,655	1.06	302,126	6,250	1,565	265	310,206	319,420	6,250	0	265	325,935	15,729	440,735	463,936	23,201
JKL	Precast floors	63,863	68,352	0.93	68,352				68,352	63,863				63,863	-4,490	68,352	63,863	-4,490
MNO	Steel work	61,669	59,000	1.05	22,500				22,500	23,518				23,518	1,018	59,000	61,669	2,669
PQR	Roofing	95,906	92,326	1.04	6,000				6,000	6,233				6,233	233	92,326	95,906	3,580
STU	Windows and screens	76,050	76,050	1.00	15,000				15,000	15,000				15,000	0	76,050	76,050	0
VWX	Plumbing	34,767	32,225	1.08	2,000				2,000	2,158				2,158	158	32,225	34,767	2,542
YZA	Mechanical installation	131,147	143,147	0.92												143,147	131,147	-12,000
100 Ltd	Electrical	119,818	128,623	0.93												128,623	119,818	-8,805
101	Suspended ceilings	95,550	94,563	1.01												94,563	95,550	987
102	Plaster	33,150	32,150	1.03												32,150	33,150	1,000
103	Floor screeds	22,424	21,500	1.04												21,500	22,424	924
104	Kitchen equipment	163,800	163,800	1.00												163,800	163,800	0
105	Decoration	121,388	119,562	1.02												119,562	121,388	1,826
106	Floor finishes	99,450	101,250	0.98												101,250	99,450	-1,800
107	Specialist clean	24,375	22,000	1.11												22,000	24,375	2,375
108	Macadam	15,844	17,869	0.89												17,869	15,844	-2,025
109	Soft landscaping	37,586	34,222	1.10												34,222	37,586	3,364
Sub-totals: rows 4 to 22		1,703,444		—	462,131	13,450	1,565	265	477,411	479,429	13,450	0	265	493,144	15,733	1,700,727	1,717,159	16,432
	Prelims																	
110	Scaffolding	15,326	16,123	0.95	13,525		250		13,775	12,856				12,856	-919	16,373	15,326	-1,047
111	Temporary water connection	2,500	3,000	0.83	2,500				2,500	2,083				2,083	-417	2,500	2,083	-417
112	Temporary electric conn	2,000	1,500	1.33	2,000				2,000	2,667				2,667	667	2,000	2,667	667
113	Temporary electrics fit out	1,500	1,456	1.03	1,650				1,650	1,700				1,700	50	1,650	1,700	50
Sub-totals; rows 25 to 28		21,326	22,079	—	19,675	0	250	0	19,925	19,306	0	0	0	19,306	-619	22,523	21,776	-747
Overall totals; rows 23 and 28		1,724,770	1,724,770	—	481,806	13,450	1,815	265	497,336	498,735	13,450	0	265	512,450	15,114	1,723,250	1,738,935	15,685

Notes: Recoverable variations are those changes made by the employer for which payment is made to subcontractors and will also be received from the employer. Unrecoverable variations are those where the contractor makes payment to the subcontractor arising from a mistake or error of the contractor. Factors in column E that exceed 1.0 are instances where the contractor hopes to make a net gain. Factors less than 1.0 are where the contractor forecasts net losses.

bill. However, if a machine and a gang of operatives are delayed because excavations reveal some ancient artefacts, time lost will likely be paid for on a daywork basis. The party being paid in such cases will always have a responsibility to minimise delays and look for alternative work whilst decisions are made.

It is standard practice that time taken and the amount of materials used are signed-off or agreed on a weekly basis. Very often, representatives of the paying party that sign daywork sheets will do so on the basis of 'record purposes only'. These representatives in JCT contracts may be architects or clerk of works, and in NEC contracts project managers or supervisors. It is not the task of professionals based on-site to agree to payment, since that is usually a function for quantity surveyors. It may be the case that a subsequent investigation reveals that daywork that has been signed for on record purposes only, as though payment should be made from an employer to a contractor, is in fact a contractor risk in contract documentation.

If some substantial unexpected work arises, it may be possible before work starts to agree a price, no matter how difficult judgements may be about the amount of time that may be necessary. In that way everybody is clear about payment in advance; the higher-tier party has cost certainty and the lower-tier party is motivated to work quickly. For example a task executed on daywork may have employer cost of £10,000 and contractor cost of £9,000; the contractor quite reasonably makes £1,000 profit. If a price is agreed in advance, maybe that could be £9,000 and since the contractor is motivated to work at good speed, perhaps the cost is only £7,000. The employer might actually be quite happy to agree a price of £11,000 and let contractor cost be £7,000, since again cost certainty in advance of work execution is important, and if the contractor completes at less cost, that will likely mean less time. There is then reduced possibility of consequential costs arising from delays to programme that extra work may have involved. Though windfall profits (excessively high) may not be desirable, arguably employers should not 'bemoan' that contractors make a little higher profit than normal on extra work whether executed on a daywork or price basis, since there are lots of occasions when contractors lose money. Employers need contractors to make money, otherwise contractors will fail.

It could be argued that professional staff paid salaries and tradespeople who are paid the same wages each week irrespective of work output are paid on daywork. Employers may commission consultants to carry out some feasibility or design work on a fixed hourly rate because perhaps there is too much complexity involved to assess how long it may take. Freelance self-employed professional staff who may change employers on a regular basis are often paid on an agreed hourly rate. However, on all these occasions, it is not thought of as payment on a daywork basis. There is usually sufficient trust in construction that people will work at good speed, and if they do not, there is always the option that employment can be terminated.

Employers may wish to include a sum of money in tenders in the eventuality that daywork will arise. Chapter 3's Table 3.5 illustrates typical presentation if a bills of quantities is used. The RICS through BCIS (2016) publish standard daywork rates. For example in 2016, the building craft rate is £14.28/hour. This is based on wages of £11.33/hour and an addition for costs incurred by contractors in employing labour directly e.g. employers national insurance contribution, holiday pay *et al.*

As section 6.8 explains, contractors may need to pay craftspeople more than the basic rate of £11.33 in order to attract them. Therefore, in bills of quantities, contractors are given the opportunity to add a percentage to the RICS rates. BCIS (2016) reports, that based upon a sample of 32 projects, the mean percentage addition by contractors was 64.8%. Contractors will be mindful they are adding this percentage in competition, since as Figure 6.5 illustrates, higher percentages will increase bid amounts. The RICS also periodically publishes its 'Schedule of Basic Plant Charges'; the last edition currently available is dated 2010. In this schedule, a wheeled hydraulic excavator up to 11 tonnes (typically a JCB 3CX) has a rate of £25.86/hour. Based upon a sample of 16 projects, BCIS (2016) reports that the mean percentage addition to plant rates in 2015 was 17.4%. Prices for materials are based upon their 'prime cost'; that is cost of invoice after deduction of discounts. Mean percentage addition to materials prices based on a sample of 38 projects was 13.2%. Table 6.7 illustrates a daywork sheet for an item of extra work. The total cost amounts to £3,856.52.

Table 6.7 A daywork sheet costed to show price to be charged to the employer

				Break out mass concrete foundation due to change in dimensions required by employer. Excavate new trench and place concrete; all by hand since machines unable to gain access		
	Quantity	Unit	Rate (£)	Amount (£)	Percentage addition	Total (£)
Labour					75%	
General operative	62	hours	13.59	*1	631.94	1,474.52
Engineer setting out	4	hours	30.00	120.00	90.00	210.00
First line supervisor	10	hours	30.00	300.00	*2	525.00
Sub-total labour				1,262.58	946.94	2,209.52
Plant					20%	
Compressor 250cfm	80	hours	5.54	443.20	88.64	531.84
Heavy breakers2	160	hours	2.41	385.60	77.12	*3
Hoses	inc					
Points and chisels	inc					
Skips; disposal of waste concrete and excavated material	2	No	95.00	190.00	38.00	228.00
Compacting plate	16	hours	2.44	39.04	7.81	46.85
Small tools	lump sum			100.00	20.00	120.00
Sub-total plant				1,157.84	231.57	1,389.41
Material					15%	
Concrete	2	m³	92.00	184.00	27.60	211.60
Part load charges	4	m³	10.00	40.00	6.00	46.00
Sub-total material				224.00	33.60	*4
Overall total				**2,644.42**	**1,212.10**	**3,856.52**

Task 6.8: Four numbers in Table 6.7 are substituted with asterisks thus: 1*, 2*, etc. What are those numbers?

6.11 Combining the effect of cashflow and profits/losses

Cashflow and making profits or incurring losses are two different concepts, but they are related. Ideally companies should be well capitalised. They should have sufficient cash balances to enable them to pay wages, salaries and invoices from supply chains without having to rely on bank overdrafts or other forms of debt. Good cashflow arises, *inter alia*, by getting paid by customers 'before' having to pay wages, salaries and invoices. Better still, it also arises if companies make good profits which are held in businesses to pay supply chains promptly. Companies that pay promptly may get better prices and excellent service, thus allowing them to complete projects on time, hassle-free and profitably.

Companies that are not well capitalised will always need to pay wages and salaries on time, but they may delay payments to external supply chains. Companies that do not pay promptly may be quoted higher prices and receive less good service from supply chains. Thus they have difficulty completing projects on time, there is lots of hassle and projects are not profitable; all things could spiral down into liquidation.

Negative cashflow can arise if companies have to pay wages, salaries and invoices before they are paid by customers. A cashflow crisis can arise if excellent and efficient work is completed for customers – made excellent 'paper' profits, but then payment is received late, or not at all – a bad debt. Bad debts are common in industry and commerce. If companies need to exceed their agreed overdraft limits to pay supply chains, banks may refuse to support, and force companies into liquidation. Cashflow positions can be made even worse by loss-making projects. Such projects may drain any cash reserves companies may have, or if they are already in a negative cashflow position, may force them into liquidation.

Figure 6.14 illustrates the position of four companies. Company A starts its financial year with good cash reserves, and each month it is consistently making profits on its projects. Company B similarly starts off in a good position, and whilst it may have many profitable projects, it has some projects which are causing considerable difficulty and are forcing it into a negative cashflow position. Company C starts the year with negative cashflow, but through excellent work is consistently making profits and elevates itself into positive territory; it has a better outlook for next year. Company D is in trouble, starting the year in negative cashflow and continually making losses. Unless that company has some other means of raising cash, it may fail. Companies that do well through a combination of profit and cashflow may give good dividends to owners or shareholders, and if they are publicly quoted companies, their share price might increase. Companies that do not do well have limited opportunity to give good dividends to owners or shareholders, and if they are publicly quoted companies, their share price might decrease.

In summary, good profits support good cashflow. Heavy losses can impact on cashflow by taking money out of businesses and cause them to fail. Companies that have built up good cash balances may be able to survive some loss-making projects; those without good cash balances may fail with few loss-making projects.

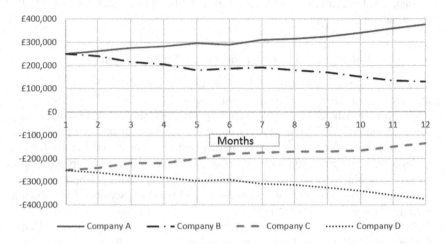

Figure 6.14 The impacts of profit on cashflow.

Example 6.4

The following example illustrates how cashflow and profits, though different, are related. John is an entrepreneur. He started a small construction company five years ago. His company had a turnover in the last financial year of £3m. The profit was £185k, upon which tax was paid. That is 6.16% of the gross turnover, or based on costs of £2.815m (£3m – £185k), £185k as a percentage of £2.815m is 6.57%. Some of the profit was re-invested in the business and buying some mechanical plant. The rest was shared between himself and his spouse as dividend payments; substantial monies were paid into personal pension funds to limit income tax liability.

The profit figure of £185k is after he has paid himself a salary of £100k, and his spouse a notional salary of £35k to do bookwork.

The following is a record of performance in the current financial year and is represented in Table 6.8.

The year started with a cash balance of £50k and an agreed bank overdraft of £100k. An increase in turnover was targeted of 20%. That would be £3.6m, with an even more ambitious profit target of 8% of the gross turnover: £3.6m x 0.08 = £288k or £24k per month.

The financial year starts as planned. Arising from work in month 1, John notes that the client on project A has accidentally paid £10k too much for a major variation. The payment is received at the end of month 2, and John accepts it happily. Project B declares £12k more profit than anticipated at the end of month 3. A major variation arises on project C; John completes this work on a daywork basis to the client's complete satisfaction. From this a further unanticipated £15k profit arises at the end of month 4. Also in month 4, John settles a dispute with a client from the previous financial year on project D; he receives £8k.

As a consequence of this better-than-expected start to the year, John decides to purchase a new company car for himself costing £40k at the end of month 4.

John contracts to build a new house, project E, for a client on a lump sum fixed price basis. At the start, ground conditions are worse than envisaged in one corner of the plot. That costs an extra £15k paid out at the end of month 5. Otherwise the project goes to plan.

On project F, an electrical subcontractor goes into liquidation. The project is delayed by one month whilst he secures a new specialist. An enhanced upfront payment of £10k is made to the specialist at the end of month 5 to get the work done. The client levies liquidated damages of £30k at the end of month 9.

Two projects G and H run into a delay because of weather conditions. John is able to secure an extension of time on both to avoid paying liquidated and ascertained damages of potentially up to £50k. He pays a consultant QS £10k at the end of month 7 to secure the extensions. He had to pay site manager salaries and other costs to stay on-site longer than envisaged. He had budgeted to make a profit on both projects. It transpired they both lost money. An anticipated combined profit for both of £27k turned into an actual loss of £8k (a net loss of £35k). This 'leak' of £35k impacted on profits at the end of months 6 to 10.

At the end of month 6 a reliable client on project I was due to pay £50k for an interim valuation; the money was received one month late. Also, at the end of month 6 another client did not pay £20k that was due on project J. Nor did a payment for £15k arise at the end of month 7. After much hassle, John completed the project with another payment of £15k due, making a final payment of £50k at the end of month 8. John received a letter from the client's solicitor refusing to pay until a long list of defects were attended to. John was aggrieved that these were not defects, but an excuse not to pay. He secured the services of a construction lawyer to seek legal recovery; he had to make an upfront payment of £15k at the end of month before the lawyer would start work. John recognised it would be many months before he would be able to recover this £50k.

The client on project A that had paid John £10k too much at the end of month 2 noted this mistake in an audit. John was forced to refund this amount at the end of month 9. Whilst John was suffering many costs that he had not budgeted for at this time, other projects were making profits as planned. John took a three-week holiday in month 10 and employed a colleague freelancer for four weeks as cover; that cost £8,000. He was particularly busy in month 11, and a little unwell after his holiday. He retained the consultant at the same cost for the next month.

John was threatened with an unfair dismissal claim from the first site manager he employed when the company started five years ago. John was informally advised that the dismissal was indeed unfair and decided he did not have the time nor the inclination to fight the case. He was also very reluctant to pay more legal fees. He paid the individual off with £10k at the end of month 12.

A client due to pay John £75k on project K at the end of month 11 goes into liquidation. A personal problem suddenly arises at home, and John is forced to take a dividend payment of £50k to help resolve the issue at the end of month 12. John is mindful his cash balance at the start of the year was £75k, and at the end

Table 6.8 Effect on cashflow over one year of several unexpected losses and delayed payments

	This financial year												Next year
End of month →	1	2	3	4	5	6	7	8	9	10	11	12	1
Cash balance	50,000												
Profit each month	24,000	24,000	24,000	24,000	24,000	24,000	24,000	24,000	24,000	24,000	24,000	24,000	24000
Overpayment Project (Proj) A		10,000	*1										
Positive cashflow — More profit than anticipated Proj B			12,000										
Major variation Proj C				*2									
Dispute payment Proj D				8,000									
Late payment received Proj 1							50,000						
Total of positives; rows 3 to 9	74,000	34,000	36,000	*3	24,000	24,000	74,000	24,000	24,000	24,000	24,000	24,000	24,000
Total of negatives; rows 16 to 29 x −1 (to make negative)			−40,000	−22,000	*4	−92,000	−32,000	−22,000	−62,000	−15,000	−93,000	−50,000	
Balance of positives and negatives; row 11 + row 12	74,000	34,000	−4,000	47,000	14,000	−68,000	42,000	*8	−38,000	9,000	−69,000	−26,000	
Cashflow position, end of the month; cell E14 = cell D14 + E13	74,000	108,000	104,000	151,000	165,000	*5	139,000	141,000	103,000	112,000	43,000	17,000	*10
New car			40,000										
Ground conditions Project (Proj) E						*6							

Electrical contractor Proj F					10,000
Damages payment Proj G & H			30,000		
Consultant QS fee Proj G & H		10,000		7,000	
Net loss on two projects Proj G & H	7,000	7,000	7,000	7,000	
Negative cashflow Payment delayed Proj 1	50,000	*7 15,000	15,000		
Client did not pay Proj J		15,000	15,000		
Legal fee Proj J					
Refund overpayment Proj A			15,000		
Unfair dismissal payment			10,000		*9
Client liquidation payment Proj K					
Freelance colleague	8,000				8,000
Dividend					50,000
Payment not received Proj L					50,000

of the year it is only £17k. He is however confident of a good year next year and has the bank overdraft upon which to rely if necessary.

The bank notes the fall in cash balances during the course of the year, and at the same time the financial sector generally became uneasy about the construction industry. The bank makes a policy decision not to fund small businesses in this sector, and his £100k overdraft facility is removed in month 1 of the next financial year. Also in month 1, a client does not pay £50k on time on project L.

Task 6.9: Complete the 10 missing figures in Table 6.8.

Discussion point 6.7: What are the possibilities for John given that he is forecast to have a £9k negative balance in month 1 of the next financial year, and he has no overdraft facility?

Discussion point 6.8: What item or items of expenditure could have been avoided, thus not putting the company into a position of having a negative balance in month 1 of the next financial year?

6.12 Model answers to discussion points

Discussion point 6.1: What may the financial advantages be of reverse factoring?

Larger tier 1 contractors are more likely to be able to negotiate large competitive 'overdraft' facilities to support payments, with just one bank facility fee, rather than many lower-tier suppliers individually negotiating finance facilities with likely higher interest charges and fees. When lower-tier suppliers bid for projects, they are able to take account of the discount they are required to offer for earlier payments (in effect add a sum of money to their bid, to allow contractors to take it off), but they will also have lower internal finance costs themselves; it is hoped overall, they can offer lower bids.

Discussion point 6.2: If contractors are forced to pay supply chains in 30 days, what consequences may this have for those who have business models that depend upon holding the cash of others to fund their business? Is it viable that contractors can bid higher prices, and thus make employers pay instead?

One consequence of being forced may put some companies at risk of liquidation or promote mergers with other companies. Another potential outcome is that in order to pay on time, contractors will need to generate more profits and retain those profits in their businesses; perhaps markets will shift contractor profitability from traditional 2.5% norms to 5%. However, it is not for employers to pay more, since contractors that pay on time will benefit from lower prices from their supply chains, who will not have to include in bids for their costs of supporting late payment. Table 6.9 assumes that the cost of finance in projects is 5%; that is for bank costs and the employment of professional staff chasing payments. In this example, overall the bid level to the employer is £22,250 lower, or circa 2% of the project value. The government document Guide to Best 'Fair Payment' Practice (Office of Government Commerce, OGC, 2007) estimated that 'public sector clients could expect to save up to 2.5% on construction costs from the introduction of better payment processes'.

Table 6.9 Comparison of bids to an employer based upon contractors paying supply chains in 60 or 30 days

	Present Position; many contractors paying supply chains in 60 days		*Future Position; all contractors paying supply chains in 30 days*
Estimated cost of contractor's own labour and staff salaries	100,000		100,000
Estimated cost of materials purchased from suppliers, plant hired or purchased, and of subcontractor's work based on 60-day payments	900,000	Estimated cost of materials purchased from suppliers, plant hired or purchased, and of subcontractor's work based on 30-day payments; assume 5% less	855,000
Total cost	1,000,000		955,000
Add profit 2.5%	25,000	Add profit 5%	47,750
Total bid to employer	**1,025,000**		**1,002,750**

Discussion point 6.3: 'S curve' and programme-based forecasts are based on a 'model'. They give some indication of peaks and troughs in contractors' cashflow. At an operational level, how accurate is the model? In Table 6.1, will the contractor really be £189,205 negative cashflow on this project? What will the effect be of contractors receiving their payments after 30 days, and negotiating 60-day payments with their supply chains?

The cost side of the model shows that contractors make payment on a straight-line basis. Whilst contractors do have to make some payments in advance of being paid themselves, very often the largest payments that contractors make are to suppliers and subcontractors at the end of the month; money going out may be like an inverted staircase, with the largest riser at the end of the month. Also, the income side of the model shows the contractor is paid at the end of the month, but on many projects the payment arrangements from employer to contractor may mean contractors are paid within 21 days of valuation dates. Contractors may therefore be paid circa one week before they make payments to supply chains. Figure 6.15 is adapted from Figure 6.1, and illustrates an alternative to the straight-line money out or cost assumption, with payment received in 21 days.

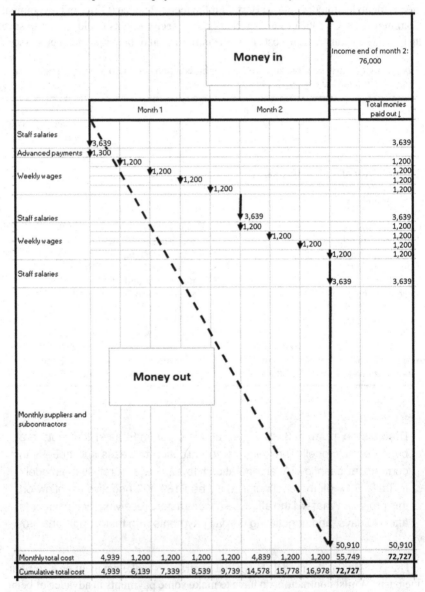

Figure 6.15 Adapted from Figure 6.1, a pragmatic illustration of cash outflow in contracting companies.

Discussion point 6.4: Some contractors may not need overdrafts from their banks, and when individual projects are in negative cashflow, contractors may finance them from cash reserves. Is this appropriate?

Cash has its interest value. Contractors are not in business to finance projects for employers freely. Many contractors will impose internal finance charges on projects that are in negative cashflow. When bidding for projects, contractors have the opportunity to include sums of money in tenders for internal (also external if necessary) finance charges. On those projects where there is a risk that employers may not pay on time, the sum of money included may be higher than for projects where employers are known to have good payment records.

Discussion point 6.5: Why is it that contracting has a reputation for positive cashflow when Figure 6.4 shows that for the majority of the project, the contractor is in negative cashflow?

Figure 6.16 illustrates the effect of paying supply chains one month after the contractor gets paid. The contractor gets paid after one month, and the supply chain is paid after 60 days. Whilst best practice initiatives are moving towards 30-day payments, Figure 6.16 is a reflection of historic and perhaps even current practice.

Discussion point 6.6: A project with estimated net costs of £100k and bid amount of £110k completes successfully with costs of £95k. The project leader claims £15k has been made. Company directors argue that profit was only £5k, since £10k of the £15k belongs to the company. Who is correct? Another project with the same values completes with costs of £105k. Is this a £5k gain or £5k loss?

For the project completing with £95k costs, both are correct, but perhaps it is better that systems in companies and the mindset of staff are directed towards comparing against the net, and thus the profit is £5k. Often project teams can do nothing about the amount of profit (or lack of profit) added to projects. That has been decided by companies before the involvement of project teams and will likely have been governed by market conditions; higher profits in buoyant markets compared to lower or no profits in tough markets. However, we live in a world of spin, and if actual costs are £100k, that can be a successful project, and it is unlikely the project leader would want to share in formal or informal internal discussions with colleagues that no profit has been made. For the project completing with £105k costs, it is just a matter of semantics about whether this is a £5k

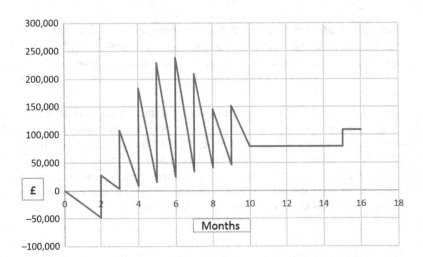

Figure 6.16 Saw tooth diagram illustrating the impact on cashflow of contractor's extending payment periods to supply chains to 60 days.

gain or loss. Arguably a loss, but in accountancy terms a profit will indeed show in the company annual report, albeit a profit less than was hoped for.

Discussion point 6.7: What are the possibilities for John given that he is forecast to have a £9k negative balance in month 1 of the next financial year, and he has no overdraft facility?

He may reflect he should have stopped work on project J when he did not receive the second payment; too late now though. Also, it may have been better to lease a car rather than take £40k out of the business to purchase it. John may be able to negotiate a gradual reduction in his bank overdraft. He may alternatively accept a lower payment of perhaps £30k or £35k on project J, just to get money in to keep the business going. He may have some assets in the business that he can sell. Perhaps he might consider the possibility to take a personal loan, perhaps secured against his home. Many businesses in this situation may unfortunately delay payments to their supply chains, and John may explain, particularly to those working on project L, that since he has not been paid, he cannot pay them. He needs to chase the payment on project L vigorously and diplomatically, avoiding if possible legal costs.

Discussion point 6.8: What item or items of expenditure could have been avoided, thus not putting the company into a position of having a negative balance in month 1 of the next financial year?

The purchase of the new car at £40k was a significant outlay. It is often better to keep money in businesses such that cash balances are readily available if unexpected costs arise. It would be very unfortunate if the business was not able to stay 'afloat' and continue to make its £24k profit each month, just because of the outlay for the car. Small entrepreneurs may best be advised to build up cash reserves; using the colloquial words of many, 'stash the cash'.

6.13 Model answer to exercise

Exercise 6.1

A tier 2 subcontractor submits an invoice for £100,000.00 due for payment in 120 days. The tier 1 contractor approves the invoice quickly, and the subcontractor decides to take the payment after 30 days; that is 90 days early. What is the 'discount or 'charge' for taking early payment? Assume an interest rate of 4%.

Model answer

4% annual interest charge / 365 = 0.0109589041% per day
0.0109589041% per day x £100,000.00 / 100 = £10.95 per day
£10.95 per day x 90 days = £986.30 'discount'
Actual payment after 30 days = £100,000.00 – £986.30 = £99,013.70

6.14 Model answers to tasks

Task 6.1: Eleven figures in Table 6.1 are substituted with asterisks thus: 1*, 2*, etc. What are those figures?
Task 6.1 answers: 1* = 570,000; 2* = 1,170,000; 3* = 30,000 (release of retention); 4* = 175,750; 5* = 340,909; 6* = 1,090,909; 7* = 204,545; 8* = 160,409; 9* = 26,909; 10* = 24,545; 11* = 109,091
Task 6.2: Five numbers in Figure 6.6 are substituted with asterisks thus: 1*, 2*, etc. What are those numbers?
Task 6.2 answers: 1* = 0.2367; 2* = 0.012; 3* = 0.051; 4* = –0.136; 5* = 0.149
Task 6.3: Ten numbers in Table 6.2 are substituted with asterisks thus: 1*, 2*, etc. What are those numbers?
Task 6.3 answers: 1* = –600,000; 2* = –4,090,000; 3* = –17,040,000; 4* = 7,200,000; 5* = 20,000,000; 6* = –50,750; 7* = –152,250; 8* = –4,150,000; 9* = 4,570,500; 10* = 2,028,063
Task 6.4: Five figures in Table 6.4 are substituted with asterisks thus: 1*, 2*, etc. What are those figures?
Task 6.4 answers: 1* = 535 / 15 * 12; 2* = 774.4; 3* = 400; 4* = 606.8; 5* = 5150.7
Task 6.5: What value of work does the contractor still need to secure to meet its turnover target for next year?

Task 6.5 answer: £5,720,000 – £5,150,700 = £569,300

Task 6.6: Based upon the above figures, what will be the wage of a bricklayer who one week lays an average of 33.33 bricks/hour, and another week 40 bricks per hour?

Task 6.6 answers: 33.33 bricks/hour laid. Total over 39 hours is 1,300. Target over 39 hours is 1,300. Payment 39 hours x £11.50 = £448.50. 40 brcks/hour laid. Total over 39 hours is 1,560. Target over 39 hours is 1,300. 260 extra brick laid / 33.33 = 7.8 hours made. Payment wages 39 hours x £11.50 = £448.50, and bonus 7.8 hours x £11.50 = £89.70; total pay £538.20. Or 1,560 / 33.33 = 46.8 hours pay for 39 hours' work = £538.20.

Task 6.7: Four figures in Table 6.6 are substituted with asterisks thus: 1*, 2*, etc. What are those figures?

Task 6.7 answers: *1 = 193.80; *2 = 6.36; *3 = 15; *4 = 1,711.95

Task 6.8: Five numbers in Table 6.8 are substituted with asterisks thus: 1*, 2*, etc. What are those numbers?

Task 6.8 answers: *1 = 842.58; *2 = 225.00; *3 = 462.72; *4 = 257.60

Task 6.9: Complete the 10 missing figures in Table 6.5.

Task 6.9 answers: 1* = 24,000; 2* = 15,000; 3* = 47,000; 4* = –10,000; 5* = 97,000; 6* = 15,000; 7* = 20,000; 8* = 2,000; 9* = 75,000; 10* = –9,000

References

B&CE (2016) Building and Civil Engineering Trust. Available at: bandce.co.uk/ Accessed 25.04.16.

BIS (2013) Construction 2025. Industrial Strategy: Government and Industry in Partnership. Available at: www.gov.uk/government/uploads/system/uploads/attachment_data/file/210099/bis-13–955-construction-2025-industrial-strategy.pdf Accessed 01.05.16.

Cabinet Office (2012) A Guide to the Implementation of Project Bank Accounts (PBAs) in Construction for Government Clients. Available at: www.gov.uk/government/publications/project-bank-accounts Accessed 31.03.16.

CIC (2014) Construction Supply Chain Charter. Construction Industry Council. Available at: www.gov.uk/government/uploads/system/uploads/attachment_data/file/306906/construction-supply-chain-payment-charter.pdf Accessed 31.03.16.

CIOB (2013) 'A Report Exploring Corruption in the Construction Industry'. The Chartered Institute of Building. London. Available at: www.giaccentre.org/documents/CIOB.CORRUPTIONSURVEY.2013.pdf Accessed 25.08.16.

CJIC (2013) Working Rule Agreement for the Construction Industry. Construction Joint Industry Council. Available at: www.ucatt.org.uk/files/publications/2013cjcagreement.pdf

Fitzpatrick, T. (2016) Build UK to track payment. *Construction News*. 4 March 2016, p. 5.

HMSO (1996) The Employment Rights Act. Available at: www.legislation.gov.uk/ukpga/1996/18/data.pdf Accessed 03.05.16.

Hubbard, M. (2016) Cost update Q4 2015. *Building*. 15 March. Available at: www.building.co.uk/data/cost-data/cost-update/cost-update-q4–2015/5080709.article Accessed 26.04.16.

OGC (2007) Guide to Best 'Fair Payment' Practices. Office of Government Commerce. Available at: www.bipsolutions.com/docstore/pdf/18463.pdf Accessed 17.04.16.

RICS (2012) *New Rules of Measurement 2: NRM2.* Coventry: Royal Institution of Chartered Surveyors.

Seely, A. (2015) Self-Employment in the Construction Industry. House of Commons Briefing Paper. Available at: researchbriefings.files.parliament.uk/documents/SN00196/SN00196.pdf Accessed 25.04.16.

7 Financial management post practical completion

7.1 Introduction

Chapter 7 looks at the role of quantity surveyors once all work on sites has been completed, and practical completion has been achieved. On traditional contracts, where contractors have been tasked with either constructing, or designing and constructing, buildings or other assets for employers, the role of quantity surveyors at this stage will be limited. It will be typically focused on preparing final accounts and final valuations for payment of remaining retentions and other funds outstanding. As a result, the first part of the chapter introduces readers to the final account process and explains how each element of final accounts are developed. This includes how final accounts will be agreed if termination has taken place during the course of a contract.

The second part of the chapter looks at the growing importance of life cycle cost analysis in view of the longer-term interests contractors are now taking in buildings. Contractors are often no longer appointed to simply construct, or design and construct, buildings. They can also be tasked with funding projects and taking a long-term interest in assets by providing core services to employers over concessionary periods of 25 to 40 years, in return for annual unitary charges, rather than simply construction phase interim payments. Consequently, construction organisations are asking quantity surveyors to forecast and value engineer operational costs of buildings over these longer periods. Contractors look to design teams to achieve a *value for money* balance between capital and operational costs of buildings. This can be with a view to reduce the long-term financial risk of maintaining assets to no more than 2% of capital costs per annum. Given the increase in focus towards operational (life cycle) expenditure, the second part of this chapter introduces the principles life cycle costing.

7.2 Final accounts

The term 'final account' refers to the process of calculating and agreeing adjustments to contract sums (tender price received or the amount originally set out in contracts to be paid to contractors for completing work) so that the amount of the final payment to bring contracts to an end can be determined (Designing Buildings, 2016).

Once contracts have been completed and defects liability periods have expired, contractors are entitled to the final payment and the issue of the final completion certificate. In order to make this final payment, however, it is first necessary for quantity surveyors acting for employers and main contractors to negotiate and settle final accounts. Although this section discusses this as a process between employers and main contractors, the process will filter down through supply chains, such that contractors also agree final accounts with subcontractors, and subcontractors with sub-subcontractors and so on. There is no set order for settlement, whether you work from the top (employer) to the bottom tier of supply chains, or in reverse from the bottom to the top; both present benefits and drawbacks for the parties involved.

Manufactured outputs are often mass produced in factory settings, following extensive prototyping to resolve any potential problems. In comparison, the majority of construction output is one-off bespoke designs. Construction projects often necessitate a large number of changes or adjustments as problems such as unexpected ground conditions arise, or employers make changes as they are able to visualise their buildings better as work develops. In some sectors of the construction industry such as volume housebuilding, standardised schools, factory warehouse units *et al.*, there is the possibility to use 'pattern book' models, or designs produced in large volumes on different sites. Even then, adjustments are common due to unforeseen problems such as ground conditions or purchasers of buildings ordering extra works. Construction projects often encounter a large number of changes, which translate into a large number of adjustments to contract sums; work that is changed, removed or added needs to be valued. These adjustments require that quantity surveyors prepare detailed documents setting out how much employers are due to pay contractors and showing how this has been calculated. As Table 7.1 illustrates, whilst there are two ways this can be achieved, it is ultimately for employers and their professional teams to select the most suitable option.

As this is a negotiated process, it is important that basis of the calculation is acceptable to and agreed with contractors. The outcome of these negotiations must be approved by contract administrators (architects or project engineers) and employers before settlement is finalised and sums agreed 'in full and final settlement', so the final payment can be issued and the final certificate agreed. Final accounts may be subject to internal or external audit to ensure there are no arithmetic errors or such like, and also to assure parties that no inappropriate payments have been made. It is therefore important that amounts agreed are clearly justified. A final account statement is illustrated in Figure 7.1.

Discussion point 7.1: How would you go about negotiating the final account for a project?

Table 7.1 Available methods for producing final accounts

Method	For	Against
Bill of omissions and additions are listed separately under each Architect's Instruction (AI).	Cost of each variation is obvious showing total values for omissions and additions.	The account document is likely to be thicker due to repetition of items in different variations and space needed for separate totals.
All omissions can be collected together, and similarly all additions, with each group identified under the appropriate work section or trade heading.	Less items, resulting in a compact document that is quicker to prepare and price.	The cost of each variation is not individually available.

Statement of Final Account

Dated: 16 May 2016
Between: AZAZ Social Housing and ZZZ Construction Group
For: The design and construction of: 105 dwelling houses.
At: Smith Street Estate, Smiths Town.

1	Contract Sum	31,349,000
2	LESS risk allowance	0
	SUB-TOTAL	**31,349,000**
3	Net omissions/additions (compensation events, early warning notices and adjustments to provisional sums up to and including the project manager's instruction (PM132)	2,345,000
	Final account TOTAL (exclusive of VAT)	**33,694,000**

We hereby agree to accept the sum of £33,694,000 (thirty-three million, six hundred and ninety-four thousand pounds exactly) (excluding VAT) in full and final settlement of the final account for the above contract.

This sum is in full and final settlement of the amount claimable under the final account including all sums claimable by the main contractor ZZZ Construction Group or by any subcontractor engaged by a contractor or any suppliers to ZZZ Construction Group or their subcontractors.

This settlement does not in any way affect the contractual obligations of either party in relation to other matters that might arise under the terms of the contract including, but not limited to defects, warranties and retention.

Signed:. . . . .
Position:. . . . .
Dated:. . . . .

Figure 7.1 Example final account statement.

Source: RICS (2015).

Final account documents produced by quantity surveyors should include the following information, where appropriate:

- Summary of account;
- Adjustments to prime cost sums;
- Adjustment to provisional sums;
- Variation accounts;
- Fluctuations;
- Settled value for contractor claims.

Before each section of the final account is explained in detail, it is important to state that the process of agreement depends on the pricing model used. For most projects, final accounts are agreed based on changes to lump sum prices presented by contractors at tender stage, and which have subsequently been used as the *contract sum* in the contract documents. As a result final accounts are presented as a series of additions or deductions from those initial sums.

In the public sector, however, as part of a move to more trust-based collaborative procurement processes, there has been a move away from lump sum prices, towards cost reimbursement models based on open book accounting principles. Under these models, parties may agree target prices and methods of shared pain (for overspending) or shared gain (for underspending). Final accounts are agreed based on expenditure and an adjustment for any overspend (penalty) or underspend (bonus). In other situations, a ceiling price, referred to as a *guaranteed maximum price* (GMP), may be agreed, resulting in contractors taking risks associated with projects over-running GMPs. It is also possible contractors are incentivised by payments based on performance, as measured against a pre-determined range of key performance indicators. Final accounts are based on assessments of actual expenditure and performance against pre-agreed benchmarks. An example showing the operation of shared pain/shared gain mechanisms, based on the NEC3 Standard Form of Contract Option C, is provided in Figure 7.2.

AZAZ Social Housing agreed a contract for the renewal of front doors on 80 properties in Lancashire. The work was let using the NEC3 Option C (Target Contract with Activity Schedule) with a target cost of £93,000 agreed along with the following share percentages:

Share range	Contractor's share percentage
Less than 80%	15%
From 80% to 90%	30%
From 90% to 110%	50%
Greater than 110%	20%

Figure 7.2 Operation of a shared pain/shared gain mechanism.

Now let's assume at completion after the effects of compensation events (changes authorised by the employer) have been added or subtracted from the target cost stipulated in the contract, the revised target cost for the project stood at £100,000. To show the full effects of this clause, we will model two situations, the first is an underspend against target of £25,000, whereas the second is an overspend of £30,000.

Scenario 1: The final cost (Price for Work Done to Date or PWDD) of the project was £75,000, due to clever value engineering by the contractor. Savings made are therefore £25,000. Based on the pre-determined share ranges the contractor would receive a bonus of £8,750.

Share range	Final PWDD	Contractor's share	Calculation
Less than 80%	£80,000 or less	15%	£5,000 x 0.15 = £750
From 80% to 90%	£80,000–£90,000	30%	£10,000 x 0.30 = £3,000
From 90% to 110%	£90,000–£110,000	50%	£10,000 x 0.50 = £5,000
Greater than 110%	£110,000 or more	20%	0
		Total:	**£8,750**

Scenario 2: The final PWDD for the project was £130,000. As a result, the final account exceeded the target by £30,000. Again, based on the pre-determined share ranges the contractor would pay the client £.

Share range	Final PWDD	Contractors share	Calculation
Less than 80%	£80,000 or less	15%	0
From 80% to 90%	£80,000–£90,000	30%	£0
From 90% to 110%	£90,000–£110,000	50%	£10,000 x 0.50 = £5,000
Greater than 110%	£110,000 or more	20%	£20,000 x 0.20 = 4,000
		Total:	**£9,000**

Figure 7.2 (Continued)

Discussion point 7.2: What are the benefits of using a target cost mechanism and open book accounting rather than a lump sum model with final account?

7.2.1 Summary account

Summary accounts are normally located towards the rear of final account documents. A summary sheet is the place where parties sign the document to indicate agreement to the final settlement sum. An example of a summary account sheet is shown in Figure 7.3. A summary is useful, since some employers may not have the expertise or interest to examine accounts in detail. It is a simplified summary statement of the account outlining the original sum, the total value of omissions (deductions) from it and the value of subsequent additions to the original contract sum.

7.2.2 Adjustment to prime cost sums

Chapter 3 explains that prime cost sums are not now standard items in JCT Standard Forms of Contract or indeed recognised in NRM2. However, this does not prevent employers inserting a series of bespoke clauses as amendments to JCT contracts or Z clauses as part of NEC standard forms to facilitate the use of nominated subcontractors and suppliers. Incorporation of such clauses is likely to be accompanied by the inclusion of prime cost sums in tender documents to allow contractors to include a percentage addition for overheads, profit and attendance. Given that PC sums are usually approximate estimates of order values for nominated subcontractors and suppliers, they need to be adjusted and settled as part of final accounts.

Agreed Contract Sum		**£31,349,000**
	Omission	*Addition*
Provisional Sum Adjustment	100,000	NIL
Prime Cost Sum Adjustment	NIL	NIL
Variations/Compensation events Summary	123,000	2,568,000
Additional works Summary	NIL	NIL
Contractor's claims summary	NIL	NIL
	223,000	2,568,000
Total:		
Total OMISSIONS		(£223,000)
Total ADDITIONS		£2,568,000
Total Final Account		**£33,694,000**
LESS Retention (Post-PC retention @1.5%)		£505,410
LESS Interim Payment Certificate Numbers 1–23		£32,789,000
Amount Due to Contractor		**£399,590**
ALL FIGURES EXCLUSIVE OF VAT		

Figure 7.3 Final Account Summary Form: Smith Street Estate.

7.2.3 *Adjustment to provisional sums*

Whether or not provisional sums are used will depend on the contract selected. For instance whilst JCT makes express provision for their inclusion, they are not recognised as part of the NEC3 standard forms. In the latter they are risks and are dealt with as compensation events. Under JCT standard forms there are two types of provisional sums used in contract documents as specified in the NRM2, thus:

- *Defined provisional sums* – the work cannot be quantified, but sufficient detail can be provided to allow contractors to take account of them in pricing and programmes.
- *Undefined provisional sums* – the work cannot be explained in any detail; as a result contractors are unable to make provision for them in pricing and programmes.

For most construction projects defined provisional sums are favoured by employers, since that means that when the work is executed, there is merely the issue of agreeing the amount of payments. Undefined provisional sums create the possibility that contractors will claim for extra time in programmes to complete work, with consequential extension and time, and loss and expense claims. As design teams are able to detail and quantify work to be executed in drawings or schedules, instructions are issued to omit provisional sums, and add back work into contracts as a variation.

7.2.4 *Valuing change to the works*

Most, if not all, standard forms of contract make provision for changes to be made. The terminology used by JCT contracts is *variations*, whereas NEC3 standard forms describe *compensation events*. Either way, their purpose is the same; they allow employers to make changes to work without having to agree new contracts. Amounts of money are included in interim valuations as work progresses, but these amounts can be provisional. As part of final account processes, the value of changes is finally agreed. NEC3 (Cl 63.1) stipulates that the "*assessment of compensation events as they affect prices is based on their effect on defined cost plus the fee*", which effectively means they are based at cost plus overheads, profit and attendance. JCT provides a hierarchical series of valuation rules for changes to the scope of the works, thus:

1. *Bill rates* – work which is of similar character and executed under similar conditions to the work set out in contract bills.
2. *Pro-rata rates* – where work is not quite the same or executed under different conditions or the quantity is significantly different, the work is valued based on the bill rates with adjustments made for differences.

3   *Fair rates* – where the work is different than that included in the bill, or in procurements methods where there is no bill nor pricing information available, the value of variations will be based on fair rates or agreed rates. These are usually established from commercial price books.

4   *Daywork* – an extensive discussion of daywork is provided in chapters 5 and 6. Daywork rates are used for work that cannot be measured.

7.2.5 Fluctuations

A comprehensive discussion of fluctuations can be found in chapter 3. Fluctuations refer to compensation paid (or recovered from in the case of price decreases) to contractors by the employers for the impact of price changes from the time of the submission of tenders to the work being carried out on-site. Similar to the case of valuing changes, provisional amounts of money are included in interim valuations as work progresses. The value of compensation is finally agreed and included as part of the final account.

Discussion point 7.3: Why do you think most contracts are based on a fixed price basis without allowance for fluctuations?

7.2.6 Contractors' claims

Claims made by contractors may be agreed in principle as work proceeds and provisional sums of money are included in interim valuations. Alternatively claims may be disputed by employers, who may refuse to pay any money. Final accounts can only be concluded when agreement has been reached on claims, whether that be negotiated agreement between parties, or by perhaps dispute resolution processes.

Whether for provisional sums, changes or claims, including provisional amounts in interim valuations can be preferred by employers, since only at the end of projects are all liabilities on the table, and quantity surveyors can negotiate values on behalf of employers mindful of any project cost limits.

Discussion point 7.4: What claims could a contractor make under a JCT Standard Form?

7.2.7 Dealing with final accounts following termination

In some unfortunate situations it may be necessary for quantity surveyors to deal with final accounts following the termination of contracts before work is complete. Although termination does not materially change processes used for agreeing final accounts, it does require quantity surveyors to develop several separate accounts, thus:

- *Notional final account* – this is a theoretical final account, developed based on the assumption that the project would have proceeded as originally intended. As a result, the value of variations and functions would be added or omitted from the contract sum and a theoretical final figure established. This allows the employer to develop a claim for additional costs against the original contractor.
- *Final account – contractor 1* – this will be the account for works completed up to the point of termination.
- *Final account – contractor 2* – the final account for the newly appointed contractor. This will include all work completed from the commencement of its contract.
- *Final settlement* – this presents the overall effect of the termination for the employer. It includes the two final accounts, along with additional costs (losses) incurred by the employer such as security, additional fees *et al.*

Exercise 7.1

On a bus station project the employer terminated the employment of the appointed contractor (contractor A) arguing that the delays suffered on the project were unacceptable and amounted to a failure to work regularly and diligently. In addition the employer's agents also citied repeated failure to act on architect's instructions related to inferior design and material use. The contract was then let on a negotiated arrangement with contractor B, who in addition to completing the bus station also corrected defective work undertaken by contractor A. Three months after termination, the first contractor, contractor A, entered into insolvency. How would you agree the final account for this project?

7.3 Budgets for life cycle and maintenance

The costs of building do not end at practical completion, although that marks the end of the capital costs associated with initial construction. Practical completion marks the start of a far more expensive phase in the life of buildings: occupancy and operation. Regrettably many employers, especially public sector organisations, routinely separate the two phases in building life cycles for accountancy purposes. Organisations may have two budgets for their buildings: a capital account (budget) for initial construction and a revenue account (budget) for maintenance and operational costs. As a result of this budget structure, it is arguable that organisations do not receive value for money from their asset portfolios, since those controlling individual budgets may consider the potential to reduce maintenance expenditure as having a negative effect on capital accounts, so they will not proceed with interventions. Consequently capital expenditure (CAPEX) and operational expenditure (OPEX) are considered to be two independent and, to an extent, competing budgets.

This is unfortunate when it is considered that the occupancy phase of a building life cycle is often far more expensive than initial construction costs, and slightly more construction phase expenditure could give significant financial benefits for employers during building occupancy. This is illustrated in Table 7.2 and graphically in Figure 7.4 where the annual occupancy costs for a school dining hall measuring 763m² are explored.

In this example, it can be seen that cleaning costs account for the majority of the annual expenditure or future occupancy cost, accounting for 26% of the annual expenditure. When the costs of constructing the school dining room are considered against occupancy costs of the building over its 20-year expected life cycle, the report found that the dining room costs 1.98 times more to 'occupy' as it did to build.

Discussion point 7.5: To what extent do you think separating CAPEX and OPEX has prevented the delivery of true value for money?

Discussion point 7.6: Do you think the adoption of BIM will overcome these issues?

Despite a lack of commitment to the use of life cycle costing, in the public sector, government policy is shifting away from a CAPEX view of projects. This shift can be seen through guidelines for the appraisal of potential projects published by

Table 7.2 Occupancy costs for a 763m² school dining room

Element	Initial cost (£)	Future occupancy costs (£)
Total initial capital costs	**881,865**	
Annual costs		
2.1 Major replacement		928
2.2 Redecorations		294
2.3 Minor repairs, etc.		1,723
2.4 Unscheduled repairs, etc.		173
2.5 Grounds maintenance and external works		137
3.1 Cleaning		3,023
3.2 Utilities		1,931
3.3 Administrative costs		2,695
3.4 Overheads		547
Total average annual costs (£/100m²)		**11,452**
Projected total (over 20 years)		**1,747,647**
Ratio (Running costs: Capital costs)		**1.98**

Source: BCIS (2011).

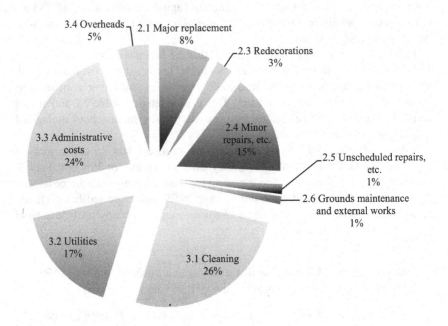

Figure 7.4 Occupancy costs for a 763m² school dining room.

Source: BCIS (2011).

HM Treasury and in the latest construction policies published by government. HM Treasury (2013) published the latest edition of *The Green Book*: *Appraisal and Evaluation in Central Government*. As part of the guidelines, the Treasury advocates value for money, which it asserts is measured through the ratio between functionality and whole-life cost of built assets. Subsequently this has started to feed into public sector tender evaluation criteria. The importance of life cycle costing was again reinforced through the publication of the government's vision for the evolution of the construction industry captured in its strategic policy document *Construction 2025: Industry Strategy – Government and Industry in Partnership* (HM Government, 2013). The policy links to other government policies such as the *Government Digital Strategy* (HM Government, 2012) and the development of digital information management in construction through maturity level 2 and 3 BIM models, which run for the full life cycle of assets. There is also the government's commitment to an eased transition from construction to occupancy through a policy called *Government Soft Landings* (Cabinet Office, 2012), where there will be a three-year transition between construction and asset management teams. At the core of the industrial strategy, however, is the government's commitment to "a 33% reduction in the initial cost of construction and the whole-life cost of built assets" (HM Government, 2013).

Collectively these policies reinforce the government's commitment to whole life value, and align with the view proposed by Wolstenholme (2009), as part of his industry review for Constructing Excellence, when he opined that construction should be re-branded as the built environment to reflect our long-term interests in buildings. Clearly this collective policy framework advocates the need for a shift in viewpoint away from a short-sighted CAPEX and OPEX view of buildings and the separate management of capital and revenue budgets to an integrated cost model based on total expenditure (TOTEX) and one budget.

The importance of life cycle costing is not, however, limited to public sector clients. The evolution of procurement practices within the public sector over the last two decades has seen a worldwide increase in the adoption of long-term service-based procurement models founded on some form of public–private partnership (PPP) arrangement. As described in chapter 2, in the UK, PPP projects are normally delivered under the policy frameworks of PFI or more recently PF2. The accumulated value of these projects at March 2016 was £57.7 billon spread over 722 projects (Office of Budgetary Responsibility, 2016).

As a result, construction organisations are now finding themselves tasked with not simply designing and/or constructing buildings, but also fulfilling traditional employer roles of sourcing finance and providing asset management services for the length of concessions (25–40 years). During these periods contractors are paid a service charge, from which they need to recover the capital cost of the building, finance costs and the costs of providing facilities management services such as cleaning and maintenance. With many contractors seeking to spend no more than 2% of CAPEX on the operation of buildings per annum, the adoption of a TOTEX view becomes critical. For contractors, developing detailed life cycle cost and whole life cost models at the commencement of projects becomes critical to ensuring the 'best value' balance is achieved between capital expenditure and annual operation costs, and to ensure the 2% maximum is preserved. Consequently, it is important for quantity surveyors and construction managers to have a robust understanding of life cycle costing and the operational and maintenance phases of the life cycle of buildings.

7.3.1 *Life cycle cost (LCC) analysis or whole life cost (WLC)?*

A variety of terms can be used interchangeably to describe life cycle concepts. Often people talk of LCC and WLC as though one concept, with two different titles. Unfortunately some view definitions of LCC and WLC incorrectly; authoritative guidance is provided in NRM3 (RICS, 2014, pp. 15–17), thus:

* *Life cycle cost* – is the cost of an asset, or its parts throughout its life cycle, whilst fulfilling the performance requirements.
* *Life cycle costing* – a methodology for the systematic economic evaluation of life cycle costs over a period of analysis, as defined in the agreed scope of assessment.
* *Whole life cost* – all significant and relevant initial and future costs and benefits of a building, facility or an asset, throughout its life cycle, whilst fulfilling performance requirements.

- *Whole life costing* – a methodology for the systematic economic evaluation used to establish the total cost of ownership, or the whole life costing of option appraisals. It is a structured approach addressing all costs in connection with a building or facility (including construction, maintenance, renewals, operation, occupancy, environmental and end of life). It can be used to produce expenditure profiles of a building or facility over its anticipated life span or defined period of analysis.

As can be seen from the definitions above, WLC is a far more comprehensive level of analysis than LCC. As illustrated in Figure 7.5, the focus of LCC is limited to considering the expenditure profile of the building, whereas WLC considers not only expenditure profiles but also potential income streams, the impact of externalities and other non-construction-related expenditure associated with buildings.

Figure 7.5 identifies four principal cost centres associated with developing whole life cost models for buildings; these cost centres are each explained below to give a full appreciation of what would typically be included in WLC analyses.

1 Non-construction costs

Non-construction costs cover any additional costs building owners are likely to incur over the lifetime of assets, or over periods of study being evaluated. Non-construction costs is a broad cost centre that includes any expenditure incurred by employers that falls outside the scope of construction costs under contracts managed by quantity surveyors. Examples of non-construction costs therefore include:

- Site costs – the costs associated with purchasing land or existing buildings. Transactional costs will also be included (e.g. legal fees, auction costs).
- Finance costs – any interest and bank charges or other fees employers pay in relation to funding projects.
- Rental costs – if employers seek to lease out buildings, for example retail development. Employer costs including legal and agent fees on lettings, void costs and other associated outlays need to be included.

2 Life cycle costs

The life cycle cost of projects covers all standard expenditure associated with buildings which can include:

- Construction costs;
- Renewal costs;
- Operation and occupancy costs;
- Maintenance costs;
- End-of-life costs (demolition costs and the like).

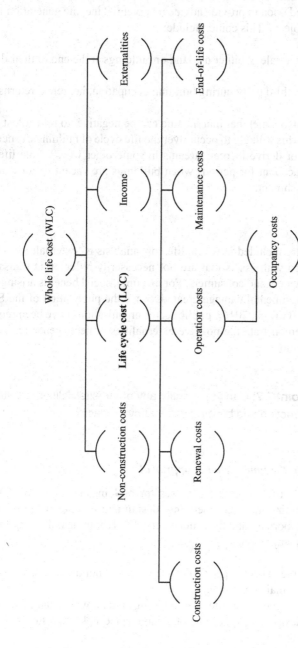

Figure 7.5 Cost categories for LCC and WLC analysis.

Source: RICS, 2014, p. 23.

3 Income

Income is usually the responsibility of development surveyors who provide detailed advice to employers about overall development budgets. Quantity surveyors can be called upon to provide advice on potential income generation likely to emerge from projects. This could include:

- Income from the sale of either the land or buildings at the end of their design life;
- Income from a third party during building occupation, i.e. rental returns.

It is important to also remember that income can be negative, to reflect lost earning opportunities. This is likely to occur over the life cycle of buildings, especially if they are rented or derive income streams in some other way. Whole life cost analysis needs to account for periods when buildings are vacant or for example undergoing refurbishment.

4 Externalities

The final cost centre included in whole life cost analysis is 'externalities'. These are costs associated with assets that are not necessarily reflected in transaction costs between providers and consumers. For example, social benefits arising from buildings, or environmental damage created. Since the publication of the Social Value Act in 2012 (Gov.uk, 2016), public sector organisations have been required to measure and demonstrate the positive externalities projects generate, by way of a social value ratio.

Discussion point 7.7: Can you identify any other externalities resulting from the construction and building occupancy process?

7.3.2 Determining the time period of appraisal

Before undertaking life cycle cost analyses for buildings (LCC or WLC), it is important to determine intended operating, design and functional life of assets. These terms are important, and they mean very different things; defining them is therefore critically important:

- *Design life* – the expected period designers expect buildings to last based on the life span of main components;
- *Operating life* – the actual life span buildings attain, which can be longer or shorter than design life, depending on maintenance, upkeep, refurbishment or rehabilitation;
- *Functional life* – the time span before buildings will cease to function in their current use; usually this is related to problems of obsolescence.

Whilst it is possible for quantity surveyors to analyse data relating to when buildings are demolished to accurately predict the operating life of assets, functional life is usually far shorter than true operating life of assets. In general, buildings fail for one of two major reasons as illustrated in Figure 7.6; these are:

- Deterioration resulting from age and component failure; or
- Obsolescence or changes not related to the structure, but to functionality.

It is important that a distinction is drawn between these two primary causes of building failure. Deterioration is often due to age and use of buildings, although as Table 7.3 shows there are other common causes of failure. As with cars, as buildings age, they develop defects, which can be corrected through routine maintenance. Similarly as parts of buildings wear, such as roof coverings, they can be replaced to preserve building usability and functionality. The levels and speed of deterioration can be to an extent controlled during the design of buildings or refurbishment schemes. For instance, designers can select products with predicted short or long life expectancies. When managing assets during occupancy cycles, components can be controlled by proactively managing replacement e.g. rather

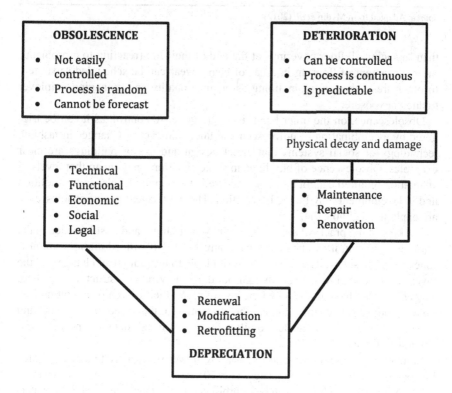

Figure 7.6 Obsolescence, deterioration and depreciation.
Source: Adapted from Flanagan et al. (1989).

Table 7.3 Causes of building deterioration

Main factors	Causes of deterioration
Weathering factors	Radiation (solar, infrared and ultraviolet)
	Temperate
	Water
	Other chemical agents such as sulphates
	Normal air constituents
	Freeze-thaw
	Wind
Biological factors	Insect vermin
	Surface growths (algae, bacteria, fungi)
	Animal vermin
	Plant agents (tree routes, ivy)
Stress factors	
Incompatibility factors	Chemical
	Physical
Use factors	Manufacture and construction
	Design
	Building occupants and other users
	Maintenance
Other natural factors	Ground movement (soil movements)

Source: Adapted from Mydin *et al.* (2012).

than replacing all floor coverings at the end of their life (reacting to a problem), the replacement of flooring in areas of heavy wear can be scheduled separately to other areas, thereby maintaining asset functionality and managing employer financial exposure.

Obsolescence, on the other hand, is much harder to control, as it can be triggered by a combination of unforeseen and uncertain events. Changes in fashion, technology or social systems that affect design and use of buildings are clear examples. Obsolescence of buildings in some situations may mean that levels of adaptation required cannot be accommodated at economic levels, and demolition and redevelopment may be the best option. The main categories of obsolescence are explained in Table 7.4.

Looking at the practical application of physical failure and obsolescence in the built environment, it can be noted that many 1960s UK office buildings are now functionally obsolete. This is largely attributable to technological change and the invention of computers. Even though buildings may remain structurally sound, suggesting they have not reached the end of either their design or operating life, floor to ceiling heights may prevent the integration of the raised access floors and false ceilings which are needed to allow owners to install data and power infrastructure cabling.

Similarly, a house may be expected to last well over 100 years, although the design life for most modern volume housing is only 60 years. However, as very little changes with the way we live, houses can be refurbished to keep them functioning as owners desire. Yet, by the same token, many high-rise tower blocks constructed during the

Table 7.4 Categories, definitions and basis of assessment for obsolescence

Type of obsolescence	Definition	Basis for assessment of building life	Examples of factors leading to obsolescence
Physical (Deterioration)	Life of the building to when physical collapse is possible	How long will the building meet human desires (with the exclusion of economic considerations)?	Deterioration of the external brickwork; deterioration of the concrete frame
Economic	Life of the building to when occupation is not considered to be the least cost alternative of meeting a particular objective	How long will the building be economic for the employer to own or operate?	The value of land on which the building stands is more than the capitalised full rental value that could be recovered from the building
Functional	Life of the building to when the building ceases to function for the same purpose as it was constructed	How long will the building be used for the purpose for which it was built?	Train stations converted into industrial, retail and residential dwellings. Cinemas converted into bingo halls. Old mills converted into prestigious apartment buildings in city centres
Technological	Life of the building until the building is no longer technologically superior to alternatives	How long will the building be technologically superior to alternatives?	Prestige offices unable to accommodate introduction of high levels of computing facilities. Storage warehouse unable to accommodate the introduction of robotics in goods handling
Social	Changes in the needs of society result in a lack of use for certain buildings	How long will the building meet with social desire of the population (with the exclusion of economic considerations)?	Churches converted to restaurants, retail units and residential dwellings; 1960s high-rise tower blocks becoming sink estates leading to high levels of anti-social behaviour
Legal	Legislation resulting in the prohibitive use of buildings unless major changes are made	How long will the building meet legislative requirements?	Major changes to buildings following the introduction of asbestos controls
Aesthetic	Style of architecture no longer fashionable	Brutal architecture of the 1960s/1970s	Buildings clad with concrete or ceramic tiles; 1960s office buildings

Source: Adapted from RICS (1986 and 2014, p. 507).

1960s have been demolished. This was not because the way people live has changed, but as a result of a combination of deterioration and obsolescence, together with poor designs and low build quality. Socio-economic change has also been a factor leading to many tower blocks experiencing unpopularity and low demand amongst residents in the social housing sector, known as social obsolescence. It is also the case that some buildings are demolished neither due to deterioration or obsolescence, but because the land they occupy has more value in alternative use, or the land is required for the delivery of major infrastructure schemes such as HS2.

Quantity surveyors need to forecast the operative life of buildings whilst trying to foretell social, technical and other changes. For this reason life cycle cost studies are typically undertaken based on duration of the owner's interest in buildings; so for PF2 projects, this would be the length of concessionary periods. For some grant funders in the public sector, such as Sport England, a 20-year analysis will often suffice.

7.3.3 Component life considerations

Once quantity surveyors have considered the overall design life of buildings, they need to consider the life span of individual components. Given buildings consist of a large number of individual components, all with different maintenance requirements and design life spans, this is not an easy task. This is especially so, given trade literature can give vastly different life expectancies for similar materials. Added to this is the complication that the majority of data is based on controlled experiments and will not necessarily reflect the realities of building use. It is, however, possible to access real component life data as part of the Building Cost Information Service (BCIS) subscription. Data included in *Component Life* represents the findings of surveys issued to a sample of 80 building surveyors, who report on life expectancies of a pre-determined list of common building components. This information is typically presented in the format shown in Figure 7.7, which relates to the life expectancy of uPVC windows with double-glazed units. The information provided includes sample size on which the data is based, the full range of estimated life expectancies for components in years and descriptive statistics which identify measures of central tendency including range, mean, median and mode data.

However, it has to be remembered that all data relating to component durability is affected by a range of building-specific factors including, but not limited to:

- Level of specification of materials;
- Quality of initial installations;
- Interaction with other materials;
- Levels of use and abuse;
- Frequency and standards of maintenance;
- Use (heavy or light);
- Local conditions (weathering for instance);
- Acceptable level of performance by users.

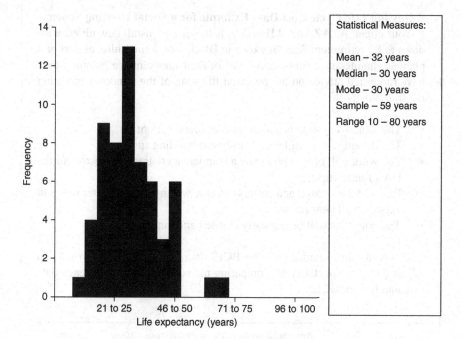

Figure 7.7 Typical life expectancy for uPVC double-glazed windows

Source: BCIS (2016).

Quantity surveyors should use their professional judgements to adjust life expectancy data, by considering the impact of these factors. This is shown in the example in Figure 7.8, which considers the life expectancy for uPVC windows installed in a social housing scheme. When specifics of the project are considered, the life span of the component is significantly reduced.

Analysis at component level is important to employers with large property portfolios, as they translate life cycle analysis into maintenance plans for buildings with a view to controlling exposure to component failure risks. For this reason most major organisations proactively manage components, so they are replaced under controlled conditions before they fail, therefore requiring reactive maintenance. Reactive maintenance is financially difficult to control and impacts negatively on building users, leading to potential disputes. Similarly, owners of large property portfolios should routinely undertake periodic inspections of between 5% and 10% of their stock to allow them to plan maintenance investment. Consequently, quantity surveyors may be tasked not with predicting component life expectancy, but the length of time components will be retained.

7.3.4 Discount rates, interest rates and inflation

Life cycle cost planning differs from the approach to cost planning discussed in chapter 1, mainly due to the timeframes involved. Whilst cost planning forecasts

Adjusting Life Cycle Cost Data Example for a Social Housing Scheme
Your client, AZAZ Social Housing, is looking to install new uPVC windows to 80 retirement flats they own in Blackpool, Lancashire, as part of a planned maintenance scheme. As part of their maintenance planning they have requested advice on the potential life span of the windows assuming the following:

- The window system is a high-market specification;
- The design was completed by in-house building surveyors;
- The work will be completed by a competent contractor selected for the HA's framework;
- Blackpool is coastal and suffers extreme weather in the winter with salt exposure all year round;
- The windows will be regularly cleaned and maintained.

Using this information and the BCIS data (the median value) we can adjust the life expectancy by comparing the scenario to the reference data (assumed) from BCIS.

Factors	Reference service life assumptions	Project conditions	Factors
Quality of components	Average	High-quality spec	1.1
Design level	Not known	Good	1.1
Work standard	Good practice	Good practice	1
Indoor environment	Not relevant	Higher than average temperature	0.7
Exterior environment	UK average	Blackpool (coastal)	0.7
In-use conditions	External envelope	External envelope	1
Maintenance level	Cleaned regularly	Cleaned regularly	1
Overall factor	1.00		0.593*
Reference service life	*30 years*	*Project service life*	*10.6 years*

*To calculate multiply the reference factor of 1.00 by the factors determined.

Figure 7.8 Adjusting life expectancy data
Source: RICS (2014, p. 9).

construction costs for buildings, usually focused on final accounts that will be agreed shortly after practical completion, life cycle costing needs to consider building occupancy, running and maintenance costs. Quantity surveyors therefore need to predict cost beyond practical completion, up to periods of between 20 and 60 years. Surveyors need to consider both current costs and future costs. The difficulty with future costs is the impact of *time value*, the effects of risk over and above a risk-free investment, inflation and interest. As a result, surveyors cannot simply add numbers together that represent different phases of expenditure. The method of calculation used is termed Net Present Value (NPV). The theoretical concept of this is discussed in chapter 8, 'Capital Investment Appraisal'. To calculate the NPV for each year in life cycle analysis, surveyors need to identify an appropriate discount rate, which ensures the time value of money is taken into account. It is important not to mistake the *discount rate* for the *interest rate*. Both these concepts are explored in chapter 8.

When selecting discount rates, BSI (2008) and the NRM3 (RICS, 2014) guide surveyors to consider selecting one of the following:

- The opportunity cost of capital;
- The client's selected discount rate;
- The societal rate of time preference (for public sector projects);
- The cost of borrowing the funds;
- Returns lost on investments elsewhere (bonds or equities *et al.*);
- The cost of generating equity;
- The WACC Rate (Weighted average cost of capital).

For most private sector projects, surveyors will typically adopt employer target discount rates developed by others as part of initial investment appraisals. However, if developing life cycle cost plans for public or third-sector employers (e.g. charities, voluntary or community groups), it is likely the societal rate of time preference would be used. The current rate published by HM Treasury (2013) in the *Green Book* is 3.5%.

The societal rate of time preference, as defined in NRM3 (RICS, 2014, p. 505), is a discount rate that reflects a government's judgement about the relative value society as a whole assigns (or which the government feels it ought to assign) to present versus future consumption. The societal time preference rate is not observed in the market and bears no relation to the rates of return in the private sector, which are developed based on individual investors' perceptions of consumption, risk preference and investment preference.

The final consideration surveyors face is whether to include or exclude inflation; in other words should they select *nominal* or *real* discount rates. If a nominal discount rate is selected, the assumption is that inflation is an impossible economic variable to forecast at an acceptable degree of accuracy; therefore life cycle cost analyses and discount rates selected do not include provision for inflation. At the core of this argument is the assertion that there is often only a small change in the relative values of the various items within life cycle cost plans. Thus, a future increase in the values of the cost of building components is likely to be matched by a similar increase in terms of other goods and services (Ashworth and Perera, 2015). This is complicated by the significant range of building components, which incur non-uniform price changes due to inflation and other market factors. The alternative, less common approach is to make some provision for inflation within real discount rates. Surveyors could, with some apprehension, try to use market expectations, short- and long-term inflation forecasts and judgements of prevailing economic situations. Since inflation prediction is both an art and science in its own right, requiring high-level math and economics knowledge, it is generally advisable for surveyors to avoid trying to make predictions about inflation.

The concluding point on the selection of discount rates is to consider risks these present to surveyors; when different discount rates are used, the effects on the present value of components can be clearly seen. Figure 7.9 illustrates the variations on present value costs using three different discount rates: 2%, 8% and 12%. The data shows the higher the discount rate and the longer into the future the predictions, the more variability there is to whole life costs.

Table 7.5 further illustrates the effects of this by using the three percentage values (2%, 8% and 12%) to calculate the present value for a tile roof covering (150m²) with a capital cost of £21,750. The table suggests in year 5, using a discount rate of 2% yields a present value of £19,698.98 (£21,750 x 0.9057). However, if this was increased to 3%, the value inserted into the life cycle cost analysis would change significantly to a present value of £18,761.55 (£21,750 x 0.8626). That simple adjustment has reduced the component replacement cost by £937.43.

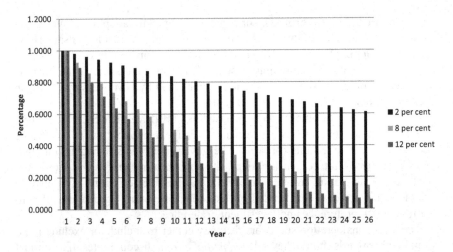

Figure 7.9 Effects of different discount rates.

Table 7.5 Discounting data

Discount Rate	Year 5	Year 10	Year 15	Year 20	Year 25
2%	0.9057	0.8203	0.7430	0.6730	0.6095
3%	0.8626	0.7441	0.6419	0.5537	0.4776
8%	0.6806	0.4632	0.3152	0.2145	0.1460
12%	0.5674	0.3220	0.1827	0.1037	0.0588

Similarly, if life expectancy of the roof increased to say 25 years, with a discount rate of 8%, then the present value would be worth only £1,278.90 (£21,750 x 0.0588). It is therefore important that quantity surveyors apply sensitivity testing (as described in chapter 5) to the outcome of the life cycle cost analyses to ensure the effects of change are considered.

Discussion point 7.8: How would you apply sensitivity testing to a life cycle cost analysis?

7.3.5 Building in-use considerations

Surveyors should take great care to select appropriate cost data, adjust life cycle components calculations to align them with employer anticipated maintenance cycles and finally select the most appropriate discount rates. However, life cycle cost modelling can, and will be, affected by building end users and the way buildings are used. If buildings are used and maintained as surveyors assume, life cycle analyses should provide a reasonably close resemblance to employer maintenance cost profiles; however, this is not always the situation. It is often necessary for employer asset management teams to replace components several times during the life cycle of buildings, and perhaps more regularly than quantity surveyors predict in life cycle analyses. BCIS (2006, p. 1) advocate, *"All buildings deteriorate naturally but deterioration may be accelerated as a result of human endeavours. It is the human factor that makes determining component and material life expectancy difficult".*

The life of buildings as a whole may depend on whether or not efficient and regular repair and maintenance are undertaken. Maintenance is often seen as a low-priority expenditure when other financial pressures are considered. For example, consider the case of school head teachers who experienced the consequences of public sector austerity and reductions in annual payments for provision of compulsory education. Despite the drop in funding, pressures to return high GCSE grades in core topics remained. In such circumstances, head teachers may see reductions in proactive long-term maintenance budgets preferable to decreasing short-term spending on teaching. This is especially in the context that the consequences of under-spending on maintenance may not become apparent for some years ahead, whilst reduced budgets for teaching can have immediate effect. As a result of reduced maintenance, both life cycle costing models and life expectancies of assets are compromised.

It is therefore important to remember that life cycle cost analyses tends to reflect ideal situations. However, as with life in general, what happens during the future life of buildings is impossible to predict with any level of certainty. Unexpected situations may arise which cannot be accounted for within the life cycle costing forecasts prepared during project design phases.

Discussion point 7.9: What do you think the benefits of life cycle costing are?

7.3.6 Life cycle costing – applications through the building life cycle

Life cycle costing can be undertaken at various points in the life cycle of buildings. Ideally studies will be undertaken early, perhaps during project appraisal or feasibility and option evaluation phases. This should ensure maximum value is gained from considering potential amendments to both the design of buildings and specification of components. There is a need to reduce employer exposure to high maintenance costs, or if that is not possible, to smooth or level costs by trying to design out peaks in component replacement expenditure. Whilst this is something of an ideal and maybe theoretical proposition, it is possible to make amendments during the early development of projects to deal with maintenance. For example by enhancing building *buildability*, it may be possible to reduce maintenance disturbance.

As projects move into design stages, further opportunities for life cycle costing will emerge, possibly as part of value engineering exercises or whilst designers consider different design features or specifications for major components. For example, in the case of the hotel used in earlier chapters, a life cycle costing exercise could be conducted to evaluate the whole life benefits of pod bathrooms and off-site manufacturing as compared to traditional bathroom installations. Whilst capital costs may initially appear prohibitive, value for money could be obtained through lower maintenance and operational expenditure. As with pre-contract cost control, discussed in chapter 1, cost limits at elemental or sub-elemental level could be determined, which take into consideration future costs, and thus ensuring employers receive best possible value for money from their projects.

As projects move through the construction phase, and near to practical completion, or if public sector and as part of the government's soft landings initiative, a life cycle cost analysis could be commissioned to prepare a cost-in-use (revenue) budget for buildings and to start phased handover processes by transforming the life cycle cost analyses into initial asset management and maintenance plans. Thereby initial maintenance and asset management plans are informed directly by design choices and construction methods adopted. This could then be inserted into level 3 BIM (Building Information Model) to further centralise and digitise main asset records. Employers are then able to receive advice throughout the life cycle of assets by reconciling actual costs and maintenance interventions against those forecast. For organisations with large property portfolios, such as supermarket chains or government organisations that are regular developers of buildings, this information could be used to further refine future buildings, thereby improving the accuracy of occupancy and maintenance cost predictions.

Discussion point 7.10: Life cycle cost analysis throughout the life cycle of the building is often linked to the need to deliver sustainability. Yet it is only worth 2 BREEAM credits so nobody really bothers. What do you think are the benefits of LCC when considering sustainability?

7.3.7 Developing life cycle cost plans

Chapter 1 provides a discussion of cost planning related to the capital costs of buildings. For the vast majority of employers this will be the key service they commission from quantity surveyors. However, those wishing to adopt a more strategic outlook will further commission quantity surveyors to undertake and produce whole life cost plans to forecast expenditure for construction projects over their entire life cycle, or a pre-determined number of years into the future. With this in mind, the RICS published the New Rules of Measurement in three volumes:

- NRM1 – Order of cost estimating and elemental cost planning – examines pre-construction cost forecasting;
- NRM2 – Detailed measurement for building works – replaces SMM7, and guides surveyors in producing bills of quantities;
- NRM3 – Order of cost estimating and cost planning for building maintenance works – supplements NRM1, and extends its scope to consider building costs in-use.

NRM3 therefore represents an extension of NRM1 and provides maintenance advice, and helps surveyors to forecast costs of maintaining and running buildings. For ease of use, NRM3 follows the exact same format as NRM1, however, NRM3 makes use of Net Present Value (NPV) and payback periods to model short-, medium- and long-term costs of buildings with the purpose of developing both life cycle cost and forward maintenance plans. This linkage between NRM1 and NRM3 is illustrated in Table 7.6; it shows that both documents include formal cost planning phases, but NRM3 requires quantity surveyors to extend the scope of coverage within cost plans to include renewal and maintenance costs for assets.

NRM3 includes provision for a fourth cost planning phase, formal cost plan 4. For new buildings this will be an update of formal cost plan 3 and is based on as-built information, thus taking account of any variations issued during construction stages. NRM3 also expands coverage of the earlier formal cost planning phases. As explained in chapter 1, NRM1 requires quantity surveyors to produce construction (capital expenditure) cost plans. NRM3 aligns with these and extends the scope of cost plans to include not only forecasts of capital costs, but to also include forecasts of costs for life cycle renewal works (replacement of elements, sub-elements and components) and annualised maintenance works.

Table 7.6 NRM Vol. 1 versus NRM Vol. 3 formal cost planning stages

RIBA stage	NRM1 – pre-contract cost advice	NRM3 – maintenance cost advice
0. Strategic definition		
1. Preparation and brief	Order of cost estimate: • Floor area method • Functional unit method • Elemental method	Order of cost estimate: • Floor area method • Functional unit method • Elemental method
2. Concept design	Formal cost plan 1	Formal cost plan 1 (Renewal/maintain)
3. Developed design	Formal cost plan 2	Formal cost plan 2 (Renewal/maintain)
4. Technical design	Formal cost plan 3	Formal cost plan 3 (Renewal/maintain)
5. Construction		
6. Handover and close-out		Formal cost plan 4 (Renewal/maintain)
7. Building in-use		

To explain this further, NRM3 (RICS, 2014, p. 23) adopts the acronym 'CROME' to describe the key constituents of life cycle costs of buildings and demonstrate broadly how these constituents (CROME) relate to construction costs and to other building maintenance costs. It is also important to point out that NRM3 does not aim to consider the full range of factors traditional life cycle cost models include; it only focuses on renewal and maintenance costs. CROME stands for:

- C = Construction costs;
- R = Renewal costs;
- O = Operation and occupancy costs (not covered by NRM3);
- M = Maintenance costs;
- E = Environmental and/or end-of-life costs (not covered by NRM3).

NRM3 and BSI (2008) provide guidance on what costs should be included in each of the CROME cost categories. This is illustrated in Figure 7.10. Unfortunately due to the size of a complete whole life cost plan it is not possible to provide a fully worked example here. However, guidance on this can be found in NRM3, pp. 67–68, where the basic framework is illustrated along with detailed guidance notes. It will be noted that each year is fully costed, with renewal costs separated from maintenance costs. All costs are provided at the current time and a discount rate is then applied within the cost plan to identify the Net Present Value of each stream of expenditure. Within the cost plan the costs should be sourced as follows:

- Renewal costs – these will be capital costs of construction and subsequent replacement of components (e.g. carpets) at capital cost; or refurbishment to

Construct	Renewal	Occupy	Maintain	End of Life
•**Capital building works**	•**Forward maintenance**	•**Operation and occupancy**	•**Annualised maintenance**	•**End-of-life costs**
•**Construction** works	•**Major repairs/replacement**—predicted scheduled actions	•**Cleaning** costs — internal and external	•**Planned**—schedule tasks	•**Disposal** inspections
•**Refurbishment** works			•**Reactive**—unscheduled tasks	•**Decommissioning**
•**Fit out and adaption** works	•**Refurbish** and upgrade works	•**Fuel** costs—energy usage, lighting, ventilation, etc.	•**Proactive**—inspect/monitor	•**Demolition**—if this is being carried out
•**End-of-life** works (demolition)	•**Redecorations** - (if separated)	•**Water and drainage** costs		•**Environmental** costs—landfill, recycling, disposal. Making the building environmentally safe.
•**Main contractor's**:	•**Maintenance contractor's**:	•**Administrative** costs—property management, insurance, staffing, waste management, etc.	•**Maintenance contractor's**:	•**Reinstatement** costs—for example, under a PPP agreement, where it has to be returned to an agreed standard of repair
•Preliminaries	•Management and admin costs		•Management and admin costs	
•Overheads and profit	•Overheads and profit	•**Occupancy** costs—additional costs including ICT, IT, telephones, catering, etc.	•Overheads and profit	
•**Other specific costs**:	•**Other specific costs**:		•**Other specific costs**:	
•Project/design team fees	•Consultant/specialist fees		•Consultant/specialist fees	
•Development/project costs	•Employer definable works		•Employer definable works	•**NOT COVERED BY NRM**
•**BASE COST ESTIMATE** (excluding risks/inflation/VAT)	•**BASE COST ESTIMATE** (excluding risks/inflation/VAT)	•**NOT COVERED BY NRM**	•**BASE COST ESTIMATE** (excluding risks/inflation/VAT)	

Figure 7.10 Costs included as part of CROME acronym.

major elements of a structure (e.g. roof) at capital cost. So for example, there may be a need to spend £170,000 re-roofing a building in year 50.

- Maintenance costs – the annual expenditure on maintenance should be included. This information can be sourced from the BCIS subscription service by selecting the 'life cycle costs tab'. The database will provide detailed life cycle cost models showing maintenance cost breakdowns for building types. Other sources of maintenance cost data include the *SFG20 Library of Maintenance Specifications for Engineering Services* (Building Engineering Services Association, BESA, 2016) and *Guide M: Maintenance Engineering and Management* (Chartered Institution of Building Service Engineers, CIBSE, 2014).

7.3.8 Example calculations

One of the difficulties of using life cycle cost analyses in practice is the mathematics associated with evaluation. The economics of time needs to take account of both present and future values for elements, sub-elements and components. Whilst chapter 8 provides a general overview of the different calculations and explains the theory behind their use, this section of chapter 7 shows how these methods can be applied and used in the context of life cycle analyses.

When undertaking life cycle analyses, surveyors will only really need to deal with three of the methods explored in chapter 8:

- Calculating the PV (present value) of a lump sum to be paid in the future (PV of £1);
- Calculating the PV of a regular annual payment for a number of years (PV of £1 payable at regular intervals);
- Calculating the AE (annual equivalent) of a lump sum to be paid now.

The example and exercise provided in this section of chapter 7 demonstrates typical scenarios quantity surveyors could encounter when producing life cycle cost plans. Further examples are provided at the rear of the chapter for you to attempt in your own time.

Example: Alternative building designs

An architect has provided an employer with two potential designs for a new office building. Both designs have similar layouts, but one has lower specification products and the other higher or medium-level specification. This is reflected in the capital cost values. The low specification building will cost £400,000 whilst the medium specification building will cost £550,000. However, it is anticipated the medium specification building will require significantly less maintenance. The full details of the building are provided in Table 7.7. Based on the information, the life cycle cost of the building can be ascertained; the calculations are illustrated in Table 7.8.

Table 7.7 Life cycle data for a low and medium specification building

	Low specification building	Medium specification building
Initial cost	£400,000	£650,000
Annual maintenance	£10,000	£6,000
Quinquennial maintenance	£14,000	£10,000
Cost of roof replacement	£40,000	£30,000
Life of roof	15 years	20 years
Life of building	50 years	50 years
Discount rate	10%	10%

Table 7.8 Life cycle cost analysis for low and medium specification levels

	Cost heading	R%	N (years)		Low spec building	Medium spec building
	Initial Cost				£400,000	£650,000
	Annual Maintenance					
LS	£10,000 x PV of £1 PA	10%	50	9.9148	£99,148	
MS	£6, 000 x PV of £1 PA	10%	50	9.9148		£59,489
	Quinquennial Maintenance					
LS	£14,000 x PV of £1	10%	5	0.6209		
	Year 5	10%	10	0.3855		
	Year 10	10%	15	0.2394		
	Year 15	10%	20	0.1486		
	Year 20	10%	25	0.0923		
	Year 25	10%	30	0.0573		
	Year 30	10%	35	0.0356		
	Year 35	10%	40	0.0221		
	Year 40	10%	45	0.0137		
	Year 45					
	£14,000	@	Σ	1.6155	£22,617	
MS	£10,000 x PV of £1					
	Year 5	10%	5	0.6209		
	Year 10	10%	10	0.3855		
	Year 15	10%	15	0.2394		
	Year 20	10%	20	0.1486		
	Year 25	10%	25	0.0923		
	Year 30	10%	30	0.0573		
	Year 35	10%	35	0.0356		
	Year 40	10%	40	0.0221		
	Year 45	10%	45	0.0137		
	£10,000	@	Σ	1.6155		£16,155
	Roof Replacement					
LS	£40,000 x PV of £1	10%	5	0.2394		
	Year 15	10%	10	0.0573		
	Year 30	10%	45	0.0137		
	Year 45					
	£40,000	@	Σ	0.3104	£12,417	
MS	£30,000 x PV of £1					
	Year 20	10%	20	0.1486		
	Year 40	10%	40	0.0221		
	£30,000	@	Σ	0.1707		£5,121
					£534,182	£730,765

In the calculations in Table 7.8 the sums from Table 7.7 have been reduced to a common timescale by calculating the Net Present Value for each item. In this example, valuation tables have been used for simplicity to determine the present value for each element. This calculation can also be undertaken using a calculator or Excel spreadsheet. To do this you would simply use the PV of £1 equation $(1/1+i)^n$ where i is the interest rate and n is the number of years. So for 5-year (quinquennial) maintenance, the PV cost would be calculated as follows:

Using the equation above, the PV of £1 is determined thus:

$$\frac{1}{(1+i)^n}$$

In this example, given a rate of interest of 10% (added as a decimal in the formula so 10% / 100 = 0.10), and the number of years as 10, the formula becomes:

$$\frac{1}{(1+0.10)^{10}}$$

This calculation determines the PV of £1 in 10 years, which equals 0.3855. The next step is to multiply this number by the cost of the maintenance works, which is £14,000. So £14,000 x 0.3855 = £5,397. That means at today's prices, £14,000 would be the equivalent of £5,397.

Although you will see capital costs have not been adjusted, these are what we call 'year zero costs', so they are already at current prices and do not require any amendment.

Whilst the example is simplistic and does not meet the rigours required for an NRM compliant study, the results of the analysis nonetheless suggest that the lower specification will present the client with better overall value for money; however, the analysis does not consider potential differences in rental return, nor does it consider occupancy costs which may differ significantly. Finally, the analysis does not allow for sensitivities such as changes in interest rates, target return rates or life expectancy and maintenance levels that will be achieved in practice.

7.4 Chapter summary

In this chapter the importance of final accounts has been introduced, discussed and evaluated. Examples have provided details of how final accounts are constructed, and the various elements that need to be broken down in documents, including provisional sums, prime cost sums, expenditure on variations and finally fluctuations and claims submitted by contractors. The chapter reviewed how quantity surveyors should deal with final accounts in the event of a termination during the construction of the works, or if a contractor unfortunately become insolvent and the employer is resolved to terminate the contract. The concept of the 'notional' final account was introduced and the process of reaching a resolution explained.

The major focus of the chapter concentrated on the need for quantity surveyors to examine building life cycles and to predict costs of maintenance, occupancy and replacement of elements, sub-elements or components. The concepts of CAPEX, OPEX and TOTEX were introduced and the way many organisations manage their asset budgets by having capital and revenue accounts has been critiqued. It is argued that only by adopting a TOTEX approach, and commissioning life cycle cost analyses at the outset of projects, can employers achieve maximum efficiency and best overall value from their investment in buildings. Finally the chapter introduced NRM3 and life cycle cost plans, although due to their scale, it was not possible to provide an example life cycle cost plan.

Exercise 7.2

Consider a choice between window specifications for a proposed building in a conservation area; only timber sliding sash windows are permitted. The architect provides the employer with a choice between cheaper Redwood sliding sash windows that will be painted with primer, undercoat and gloss, or more expensive Sapele (hardwood) windows which will receive two coats of wood stain. Both windows include 20mm double-glazed units and will achieve the same U-value. Full details for each option are provided in Table 7.9. Based on the information, calculate the life cycle cost of 10 windows over a 20-year period.

Exercise 7.3

In PF2 procurement, life cycle costing is widely used in order to inform the bidding strategies of consortia. In the case of a new PF2 school, identify some of the specific operational costs that the consortium would need to consider when evaluating the project.

Exercise 7.4

A common argument in the life cycle costing literature, and made by cost consultants in practice when advising clients about the benefit of life cycle costing, is the assertion that LCC will invariably result in a higher capital cost outlay in exchange for future long-term savings. Why would this not be seen as an attractive proposition for speculative developers?

Table 7.9 Life cycle data for softwood and hardwood windows

	Softwood	*Hardwood*
Initial cost of 10nr windows	£3,500	£6,200
Cleaning cost (per annum)	£120	£120
Renewal (R)	Every 15 years	Every 30 years
Maintenance (M) redecoration	Every 5 years	Every 5 years
Redecoration costs	£250.00	£150.00

7.5 Model answers to discussion points

Discussion point 7.1: How would you go about negotiating the final account for a project?

The typical way requires contractors or subcontractors to develop their view of the project final account. This will include a full breakdown for all the additional items of work, with the valuations based on the rules provided in the JCT SBC/Q or NEC3 compensation event clauses. This will be normally forwarded to the employer's quantity surveyor who will check and either agree or disagree with the valuations provided. At this point the quantity surveyor may return a counter-offer to the contractor, and the negotiation process will commence. Often this will be a discussion of initial final account items and the valuations provided by the contractor.

Discussion point 7.2: What are the benefits of using a target cost mechanism and open book accounting rather than a lump sum model with final account?

The benefits of this approach are supported by some in the public sector. Open book accounting may be seen initially as cost plus pricing, and therefore employers do not receive clear benefits. However, by adding additional features or incentives based on attainment of key deliverables monitored by KPIs such as levels of social benefit (social value) *et al.*, projects will invariably deliver more value. Also the addition of a target cost mechanism incentivises contractors to use their expertise to reduce final accounts for projects, as savings will be returned to contractors in additional bonus. Enhanced buildability and better value are possible.

Discussion point 7.3: Why do you think many contracts are based on a fixed price basis without allowance for fluctuations?

Essentially this becomes a question of risk. The reason employers often opt for lump sum based procurement routes (design-bid-build or design and build) is that they are seeking price certainty. As a result, price movement risk is passed to contractors along with other significant project risks. Whilst this may attract risk premiums, since projects are usually competitively tendered, most employers form the view that risks will be competitively priced, not over-priced.

Discussion point 7.4: What claims could a contractor make under a JCT Standard Form?

Claims by contractors are permitted under the loss and/or expense clauses. These claims cover the direct losses and related expenses incurred by contractors. The reason they are so complex is often due to the difficulties establishing the direct losses incurred. Additionally delay events are hardly often singular, and even more rarely simply allocated to one party or the other.

Discussion point 7.5: To what extent do you think separating CAPEX and OPEX has prevented the delivery of true value for money?

Unfortunately this is very much the situation facing most property managers. Many asset managers argue that only when property owners and their accounts adopt a TOTEX model (total cost, or one single account) will these issues be effectively resolved and true value for money delivered. Yet as the Green Book continues to define value for money, as the relationship between life cycle cost and functionality, pressure is not being applied to change budget systems.

Discussion point 7.6: Do you think the adoption of BIM will overcome these issues?

BIM will provide more information for asset managers to make maintenance and replacement of components more efficient. Regrettably, it will not change current budget systems, and capital and revenue accounts will continue to be widely used.

Discussion point 7.7: Can you identify any other externalities resulting from the construction and building occupancy process?

There are a significant number of potential externalities associated with construction. These can include:

- Pollution;
- Environmental damage;
- Social value;

- Employment creation (direct and indirect);
- Regeneration;
- Gentrification (social change);
- Improved wellbeing in the immediate local area;
- Site remediation and re-use;
- Crime and anti-social behaviour reductions;
- Education and training opportunities.

Discussion point 7.8: How would you apply sensitivity testing to a life cycle cost analysis?

Applying sensitivity testing to life cycle cost models requires that surveyors manipulate the various major elements of analyses to understand their impact on projects overall. This could include modifying discount rates, the life expectancy of components, and the duration of the study *et al.*

Discussion point 7.9: What do you think the benefits of life cycle costing are?

The benefits of life cycle costing are essentially linked to the argument that enhanced expenditure at the construction phase can have a positive effect on building asset management and maintenance cycle. The flip side of the head teacher example is to argue that if we design to minimise maintenance requirements of buildings, we could lessen the impact of decisions to reduce maintenance expenditure.

Discussion point 7.10: Life cycle cost analysis throughout the life cycle of the building is often linked to the need to deliver sustainability. Yet it is only worth 2 BREEAM credits so nobody really bothers. What do you think are the benefits of LCC when considering sustainability?

Sustainability is this context is often assumed to be environmental, and to an extent economic. As a result many argue the use of LCC within a sustainability context can ensure energy usage is fully modelled, and the costs of ensuring heating or air handling systems are sustainable can be examined from a cost viewpoint. The use of LCC can also help employers understand design features of buildings such as morphology by looking at impact of changes in life cycle costs.

Finally, the use of LCC alongside value engineering can ensure materials and natural resources are used in the most effective way possible by considering the design life and financial impact of the various alternatives.

7.6 Model answers to exercises

Exercise 7.1

On a bus station project the employer terminated the employment of the appointed contractor (contractor A) arguing that the delays suffered on the project were unacceptable and amounted to a failure to work regularly and diligently. In addition the employer's agents also citied repeated failure to act on architect's instructions related to inferior design and material use. The contract was then let on a negotiated arrangement with contractor B, who in addition to completing the bus station also corrected defective work undertaken by contractor A. Three months after termination, the first contractor, contractor A, entered into insolvency. How would you agree the final account for this project?

These are always challenging issues to resolve; however, as the contract has been terminated the quantity surveyor will need to negotiate and settle a final account with both contractor A, or in this case its receivers, and contractor B. This will result in the following accounts:

- Notional final account – a theoretical account for the project based on the continued employment of contractor A.
- Final account for contractor B – this will cover the period from its employment to practical completion and will include the additional costs associated with correcting defects from the first contractor.
- Final account with contractor A – this will cover the work it completed (in accordance with the contract), but no payment for defective work corrected by contractor B. There will be a deduction to cover the costs of appointing contractor B.

Exercise 7.2

Consider a choice between window specifications for a proposed building in a conservation area; only timber sliding sash windows are permitted. The architect provides the employer with a choice between cheaper Redwood sliding sash windows that will be painted with primer, undercoat and gloss, or more expensive Sapele (hardwood) windows which will receive two costs of wood stain. Both windows include 20mm double-glazed units and will achieve the same U-value. Full details for each option are provided in Table 7.9. Based on the information calculate the life cycle cost of 10 windows over a 20-year period.

Table 7.10 Life cycle cost analysis for softwood and hardwood windows

	Cost Heading	r%	n (years)		Hardwood	Softwood
	Initial cost				£6,200	£3,500
	Annual cleaning					
HW	£120 x PV of £1 PA	3%	20	14.88	£1,786	
SW	£120 x PV of £1 PA	3%	20	14.88		£1,786
	Redecoration					
HW	£150 x PV of £1					
	Year 5	3%	5	0.863		
	Year 10	3%	10	0.744		
	Year 15	3%	15	0.642		
	Year 20	3%	20	0.554		
	£14,000	@	Σ	2.803	£421	
SW	£250 x PV of £1					
	Year 5	3%	5	0.863		
	Year 10	3%	10	0.744		
	Year 20	3%	20	0.554		
	£10,000	@	Σ	2.161		£541
	Replacement windows					
SW	£3,650 x PV of £1					
	Year 15	3%	15	0.642		
	£3,650	@	Σ	0.642	£2,344	
					£8,407	£8,171

SOLUTION

In the calculations in Table 7.10 the sums from Table 7.9 have been reduced to a common timescale by calculating the Net Present Value for each item; this time a 3% discount rate has been applied. The softwood windows are not redecorated in year 15, as they will be replaced at this stage.

Exercise 7.3

In PF2 procurement, life cycle costing is widely used in order to inform the bidding strategies of consortia. In the case of a new PF2 school, identify some of the specific operational costs that the consortium would need to consider when evaluating the project.

There are a number of issues associated with operational expenditure, these could include:

• Obsolescence;
• Maintenance cycles;
• Changes in maintenance expenditure;
• Product life spans;

- Damage by pupils and other school building users;
- Discounting rate;
- Risk levels;
- Economic outlook.

Exercise 7.4

A common argument in the life cycle costing literature, and made by cost consultants in practice when advising clients about the benefit of life cycle costing, is the assertion that LCC will invariably result in a higher capital cost outlay in exchange for future long-term savings. Why would this not be seen as an attractive proposition for speculative developers?

Developers building on a speculative basis will not be driven by the desire to retain a long-term stake in buildings. They are normally organisations who are willing to accept high risks in the expectation of high returns, what economists would categorise as profit maximisers. They and their shareholders are driven by the need to provide high returns on capital employed to compensate for the risks taken. As this means they may take a short-term interest in buildings, they would see higher levels of expenditure at the construction phase as an erosion of ROCE, together with the demand for a higher fee from their professional advisors. Ultimately, LCC may contradict the business objectives of developers. As they take a short-term interest, OPEX is hardly something they would be concerned about. CAPEX and worth (resale value) are key motivators.

References

Ashworth, A. and Perera, S. (2015) *Cost Studies of Buildings*. 6th Edition. Oxon: Routledge.

BCIS (2006) *BMI Life Expectancy of Building Components: A Practical Guide to Surveyors Experiences of Buildings in Use*. London: Building Cost Information Service.

BCIS (2011) *BMI Special Report – Occupancy Costs of Specialist School Blocks*. London: Building Cost Information Service.

BCIS (2016) BCIS Independent Data for the Built Environment. Available by subscription at: service.bcis.co.uk/BCISOnline/ Accessed 02.05.16.

BESA (2016) SFG20 Library of Maintenance Specifications for Engineering Services. Building Engineering Services Association. Available at: www.thebesa.com/ Accessed 04.06.16.

BSI (2008) *BS ISO 15686–5:2008 Buildings and Constructed Assets. Service Life Planning. Life Cycle Costing*. London: British Standards Institute.

Cabinet Office (2012) *The Government Soft Landings Policy*. Available at: www.bimtaskgroup.org/wp-content/uploads/2013/02/The-Government-Soft-Landings-Policy-18022013.pdf Accessed 05.06.16.

CIBSE (2014) Guide M: Maintenance Engineering and Management. Chartered Institution of Building Service Engineers. Available at: www.cibse.org/knowledge/cibse-guide/cibse-guide-m-maintenance-engineering-management Accessed 04.06.16.

Designing Buildings Wiki (2016) Final Account. Available at: www.designingbuildings.co.uk/wiki/Final_account Accessed 03.04.16.

Flanagan, R. (1989) *Life Cycle Costing: Theory and Practice.* Oxford: BSP.

Gov.uk (2016) Social Value Act 2012 Updated 2016. Available at: www.gov.uk/government/publications/social-value-act-information-and-resources/social-value-act-information-and-resources Accessed 04.06.16.

HM Government (2012) Government Digital Strategy. Available at: www.gov.uk/government/uploads/system/uploads/attachment_data/file/296336/Government_Digital_Stratetegy_-_November_2012.pdf Accessed 05.06.15.

HM Government (2013) Construction 2025: Industry Strategy: Government and Industry in Partnership. Available at: www.gov.uk/government/uploads/system/uploads/attachment_data/file/210099/bis-13–955-construction-2025-industrial-strategy.pdf Accessed 05.04.16.

HM Treasury (2013) The Green Book: Appraisal and Evaluation in Central Government. Available at: www.gov.uk/government/publications/the-green-book-appraisal-and-evaluation-in-central-governent. Accessed 19.04.16.

JCT (2012) SBC/Q2011 Standard Form of Building Contract. The Joint Contracts Tribunal.

Mydin, M.A.O., Ramli, M. and Awang, H. (2012) Factors of deterioration in buildings and the principles of repair. *Eftimie Murgy ResltA.*, 19 (1), pp. 345–352.

NEC (2005a) *Engineering and Construction Contract.* London: Thomas Telford.

NEC (2005b) *NEC3 Engineering and Construction Contract Guidance Notes ECC.* London: ICE Publishing.

Office of Budgetary Responsibility (2016) Private Finance Initiative and Private Finance 2 Projects. 2015 Summary Data. Available at: www.gov.uk/government/uploads/system/uploads/attachment_data/file/504374/PFI_PF2_projects_2015_summary_data.pdf Accessed 13.04.16.

RICS (2012) *New Rules of Measurement 2: NRM2.* Coventry: Royal Institution of Chartered Surveyors.

RICS (2014) *New Rules of Measurement 3: NRM3.* Coventry: Royal Institution of Chartered Surveyors.

RICS (2015) *Final Account Procedures.* Coventry: Royal Institution of Chartered Surveyors.

Wolstenholme, A. (2009) Never Waste a Good Crisis. Available at: constructingexcellence.org.uk/resources/never-waste-a-good-crisis/ Accessed 03.05.16.

8 Capital investment appraisal

8.1 Introduction

A key type of decision facing an organisation's management is whether to make a capital investment, that is whether to invest in a project which will provide some benefit for the long term, at least for longer than the current year. A feature of this type of decision is that the cost may be incurred shortly after the decision is taken whilst the benefits – cost saving, additional revenues – occur over a period of years.

Five investment appraisal techniques are explained initially and then worked examples are illustrated towards the end of the chapter.

This is a large topic to cover for any student who wants a firm understanding of this key business decision-making technique. The various aspects are explained in chapters 4, 5 and 6.

8.1.1 The capital budgeting cycle

A common feature of business activity is the need to commit funds by purchasing land, buildings, machinery, etc. in anticipation of being able to earn, in the future, an income greater than the funds committed. This indicates the need for an assessment of the size of the outflows and inflows of funds, the life of the investment, the degree of risk attached (greater risk being justified perhaps by greater returns) and the cost of obtaining funds.

The key stages in the capital budgeting cycle may be identified as:

* Step 1   Needs for expenditure are forecast.
* Step 2   Projects to meet those needs are distinguished.
* Step 3   Alternatives are appraised.
* Step 4   Best alternatives are selected and approved.
* Step 5   Expenditure is made and monitored.
* Step 6   Deviations from forecasts are investigated.

Step 3 occupies a significant place in the theory and practice of long-term decision making.

8.1.2 Types of capital projects

Reasons for capital expenditure vary widely. Projects may be classified into the following categories:

- **Maintenance** – replacement of worn-out or obsolete assets, safety and security, etc.
- **Profitability** – cost savings, quality improvement, productivity, relocation, etc.
- **Expansion** – new products, new outlets, research and development, etc.
- **Indirect** – office buildings, welfare facilities, etc.

A particular investment project could combine any number or all of the above classifications.

8.1.3 Working capital

In most business and industrial projects, investment is required, both in fixed assets (*items which are purchased for long-term use and are not likely to be converted quickly into cash, such as land, buildings and equipment*) and in working capital (*the capital of a business which is used in its day-to-day trading operations, calculated as the current assets minus the current liabilities*), although the risk attached to working capital is less than that for fixed assets. Values of land may appreciate and so present less risk, but money invested in machinery is a sunk cost, which is unlikely to be recovered, save for perhaps minimal scrap values.

In capital investment and project appraisal, accurate estimates of working capital requirements are desirable, not only for assessment of project profitability, but also to facilitate forecasting of capital requirements.

8.1.4 Capital expenditure forecast

In preparing budgets, it is necessary to consider how much money can or must be allocated to capital expenditure (long-term use). Capital development schemes may be started because a surplus of cash resources is revealed by the long-term plan, but usually management decides on a capital development scheme and then seek the means to finance it.

Initially, the budget will be an expression of management's intention to allocate funds for certain broad purposes. In the budget period, money will be required for:

- Previously authorised existing projects; and
- New projects, full details of which may not yet be available.

The forecasts will indicate whether sufficient funds are available, and perhaps when additional funds will need to be obtained. It is advisable, therefore, for managers to submit long-term capital expenditure forecasts, say for two to five years ahead; consequently, the possibility of obsolescence and the direction of the future development of the business must be taken into account.

The capital budget is the outcome of a dual process:

• Higher management allocating funds to various areas in relation to the corporate plan, i.e. according to the long-term objectives of the company; and
• Individual managers seeking to utilise the funds for specific purposes.

The importance of this aspect of planning cannot be over-emphasised, because present capital investment will determine the structure and profitability of the company in the near future. Errors made in forecasting and planning will, therefore, have serious results, and may prove difficult to rectify.

8.1.5 Relevant cashflows for an investment appraisal

As mentioned in the introduction, a feature of an investment appraisal is that the cost may be incurred at the start of a project whilst the benefits occur over a period of years. An example is the purchase of equipment to make a new product which can then be sold, generating contribution, for a number of years. Similarly, investment in a new computer system now might produce cost savings in future years.

With one exception – accounting rate of return (which we will discuss later in this chapter) – investment appraisal methods are based on estimated future cashflows, not future profits. It is the timing of the cashflows which is important, for example when will the revenues from selling a new product be received, rather than the timing of profits, which are based on the matching principle.

The cashflows which are relevant to the decision are those which will be affected by the decision: if the investment goes ahead which of the organisation's cashflows, both inflows and outflows, will change? The cashflows affected by a decision are sometimes referred to as incremental cashflows.

Learning activity 8.1.6

Investment projects do not only include investment in plant and equipment or buildings. Think of some other types of capital projects that may be applicable to the construction industry.

Self-assessment question 8.1

A company is considering whether to launch a new product. Indicate whether each of the following cost and revenue items would be 'relevant' or 'irrelevant' to an investment appraisal decision:

(a) The cost of new equipment required to make the product;

(b) An increase in working capital (inventories, historically known as stock and debtors, now known as trade receivables) caused by the increases in revenue;

(c) The fixed cost of the factory building such as rent, which will not increase as a result of the new product;

(d) The variable costs, materials and labour, of making the product;

(e) The sales value of the new product;

(f) The estimated resale value of the equipment at the end of the new product's life cycle;

(g) The depreciation (assets charged to profit and loss over its estimated economic life) of the equipment;

(h) The loss of contribution from another product which will have to be discontinued to make way for the new product.

8.2 Non-discounting methods for simple projects

There are two non-discounting methods of investment appraisal, the payback period (PB) and the accounting rate of return (ARR).

The payback period simply shows how long it will take to recoup the initial investment. Although the timing of the cashflows is taken into account, the time value of money is not. Identical nominal cashflows in different years are treated as identical values when using the payback period calculation, which you will see is not the same with discounted methods.

The payback period is commonly used in practice, partly because it is easy to calculate and understand, but partly also because the length of time that it will take to get your money back is a useful piece of information. In an attempt to overcome the weakness that the method does not take into account the time value of money, the discounted payback method is sometimes used. It calculates how long it will take the discounted cashflows to recoup the initial investment.

The accounting rate of return takes into account neither the time value of money nor the timing of the cashflows. However, it is calculated in the same way as the return on investment (ROI) and therefore indicates what effect a project will have on a firm's ROI: if the accounting rate of return for a project is higher than the firm's current cost of capital, acceptance of the project should generate positive cashflows and increased profits into the business, and vice versa. It is the only investment appraisal technique that is based on accounting profits, all the others being based on cashflows.

8.3 Discounting methods of appraising capital investment projects

One of the most important steps in the capital budgeting cycle is determining whether the benefits from investing large capital sums outweigh the large initial costs of those investments. There is a range of methods that can be used in reaching these investment decisions; the most popular being discounted cashflow (DCF).

There are three basic discounted cashflow methods which will be discussed later in this chapter:

- Net Present Value (NPV)
- Internal rate of return (IRR)
- Profitability Index (PI)

8.4 The time value of money

A simple method of comparing two investment projects would be to compare the amount of cash generated from each. The project which generates the greater net cash inflow after taking into account all relevant revenues and costs would be the preferred project. However, such a simplistic view would fail to take into account the time value of money, the effect of which may be covered in the general rule below:

> *There will be a time preference for receiving the same sum of money sooner rather than later. Conversely, there will be a time preference in paying the same sum of money later rather than sooner. The reasons for this time preference could be:*

- **Consumption** – money received now can be spent immediately.
- **Risk preference** – risk disappears once money is received.
- **Investment decision** – money received can be invested in the business or externally.

If consideration is given to these factors it can be seen that inflation affects time preference but is not its only determinant. High levels of inflation for example will produce greater consumption preference and thus greater time preference.

Discounting analysis is usually based on an investment decision and in particular the ability for a business to invest or borrow and receive or pay interest.

It is important to note that the discounting process is fundamental to DCF calculations and is analogous to compound interest (*interest added to the principal of a deposit or loan so that the added interest also earns interest from then on*) calculations in reverse.

Simple interest arises when interest accruing on an investment is paid to the investor as it becomes due and is not added to the capital balance on which subsequent interest will be calculated. Compound interest, however, arises when the accrued interest is added to the capital outstanding and it is this revised balance on which interest is subsequently earned.

Self-assessment question 8.2

Marsh places £4,000 on deposit in a bank earning 5% compound interest per annum.

You are required to find the amount that would have accumulated:

(a) After one year;
(b) After two years;
(c) After three years;
(d) To find the amount that would have to be deposited if an amount of £5,000 has to be accumulated after one year, then after two years and finally after three years with interest available at 5% per annum.

The answers to part (d) form the mechanics behind discounted cashflow and the subsequent calculation of a present value through the principle of discounting.

8.4.1 Discounting

People often have a time preference for money and would prefer to receive money sooner rather than later. It is therefore inappropriate to give the same value to similar sums receivable at different times over the life of a project. This is what traditional methods of investment appraisal do such as the payback period. Because of investor's rates of time preference for money, a more suitable method of investment appraisal reduces the value (discounts) of later cashflows to find a value with which one would be equally happy now as a given receipt due in several years' time. This calculation of a present value (PV) is what was illustrated with the compound interest calculations.

8.4.2 Net Present Value

The NPV technique estimates future cashflows of a project to a present value at the interest rate prevalent at that time. If the sum of those discounted cashflows is positive (greater than the cost of the project), then it makes sense to proceed with the project. If they are negative, then the decision should be not to proceed with it.

8.4.3 Internal rate of return

The internal rate of return (IRR) is the interest rate at which the Net Present Value (NPV) of a project is zero. A manual calculation of the IRR requires the calculation of the NPV at two different interest rates. One interest rate produces a positive NPV and the other a negative NPV. An interpolation of the interest rate at which the NPV is zero could then be calculated.

8.4.4 Profitability Index

The Profitability Index (PI) is the ratio of the present value of the future cashflows to the initial investment. Projects with a higher PI are deemed to be more attractive to a business. This method is a variation on NPV investment decisions and is useful in situations where capital for investments is in short supply.

Practical example

A haulage company has three potential projects planned. Each will require investment in two refrigerated vehicles at a total cost of £240,000. The vehicle has a three-year life. The three projects are:

(a) Expected cash inflows, after deducting all expected cash outflows, are £120,000 per annum.
(b) Expected cash inflows, after deducting all expected cash outflows, are £90,000 per annum.
(c) Expected cash inflows, after deducting all expected cash outflows, are £80,000 in year 1, £140,000 in year 2 and £160,000 in year 3.

You are required to calculate for each project the payback period, the ARR, NPV and IRR, and to discuss why the results from NPV and IRR calculations may sometimes be incompatible with each other. The cost of capital is 10% for the business.

PAYBACK PERIOD

Cashflows	Project A	Project B	Project C
	£000s	£000s	£000s
Outlay	240	240	240
Net cashflows			
Year 1	120	90	80
Year 2	120	90	140
Year 3	120	90	160
Payback period	2 years	2.7 years	2.1 years
Workings	*120 + 120 = 240*	*90 + 90 + 60 / 90*	*80 + 140 + 20 / 160*

ACCOUNTING RATE OF RETURN

Cashflows	Project A	Project B	Project C
	£	£	£
Outlay (a)	240,000	240,000	240,000
Profits (cashflows minus depreciation)			
Year 1	40,000	10,000	nil
Year 2	40,000	10,000	60,000

Cashflows	Project A	Project B	Project C
	£	£	£
Year 3	40,000	10,000	80,000
Average annual profit (b)	40,000	10,000	46,667
Accounting rate of return (b × 100 / a)	16.7%	4.2%	19.4%

DISCOUNTED CASHFLOW

Project A

End of year	Cashflow	Discount factor	Present value
1	£120,000	0.9091	£109,092
2	£120,000	0.8264	£99,168
3	£120,000	0.7513	*£90,156*
			£298,416
Less Initial Outlay			*(£240,000)*
Net Present Value			£58,416

Project B

End of year	Cashflow	Discount factor	Present value
1	£90,000	0.9091	£81,819
2	£90,000	0.8264	£74,376
3	£90,000	0.7513	*£67,617*
			£223,812
Less Initial Outlay			*(£240,000)*
Net Present Value			(£16,188)

Project C

End of year	Cashflow	Discount factor	Present value
Year 1	£80,000	0.9091	£72,728
Year 2	£140,000	0.8264	£115,696

End of year	Cashflow	Discount factor	Present value
Year 3	£160,000	0.7513	*£120,208*
			£308,632
Less Initial Outlay			*(£240,000)*
Net Present Value			£68,632

Project C is the most desirable project.
Project A is next in rank.
Project B would be rejected as it gives a negative Net Present Value.

INTERNAL RATE OF RETURN

Project A
Find two values of NPV using discount rates lying either side of the actual IRR.
A first guess of 22% produces a NPV which is positive.
Higher discount rate of say 25% is used for the second guess which produces a negative value.

	Cashflows		Discount rate 22%		Discount rate 25%
End of year	£	£		£	
1	120,000	8,197	98,364	800	96,000
2	120,000	6,719	80,628	640	76,800
3	120,000	5,507	*66,084*	512	*61,440*
			245,076		234,240
Outlay			*(240,000)*		*(240,000)*
Net Present Value			*5,076*		*(5,760)*
22% + ((5,076 / 10,836) x 3) = 23.40%					

Project B
Find two values of NPV using discount rates lying either side of the actual IRR.
A first guess of 5% produces a NPV which is positive.
Higher discount rate of 10% is used for the second guess for a negative value.

5% + ((5,079 / 21,267) * 5)   =   6.19%

Project C
Find two values of NPV using discount rates lying either side of the actual IRR.
A first guess of 25% produces a NPV which is positive.

Higher discount rate of 30% is used for the second guess to produce a negative value.

25% + ((4,480 / 27,274) * 5)   =   25.82%

INTERNAL RATE OF RETURN DECISION RULE

- Where the IRR of the project is **greater than** the cost of capital, **accept** the project.
- Where the IRR of the project is **less than** the cost of capital, **reject** the project.
- Where the IRR of the project **equals** the cost of capital, the project is **acceptable** in meeting the required rate of return of those investing in the business but gives no surplus to its owners.

Project C is the most desirable project for IRR purposes.
Project A is next in rank.
Project B is the bottom-ranking project.
NPV accepts all projects with NPV > 0. Ranking of projects is by value of NPV.
IRR finds the value of the discount rate that makes NPV = 0. Project will be accepted if IRR > k (cost of capital).

Self-assessment question 8.3

Williams Ltd, a construction company, is considering the selection of one from two mutually exclusive investments projects, each with an estimated 5-year life. Project Y costs £1,616,000 and is forecast to generate annual cashflows of £500,000. Its estimated residual value after five years is £301,000. Project Z, costing £556,000 and with a scrap value of £56,000, should generate annual cashflows of £200,000. The company operates a straight-line depreciation policy and discounts cashflows at 15%.

Calculate the payback, ARR, NPV and IRR for each project and discuss which seems to be the better investment opportunity.

8.5 Conclusions

Capital investment appraisal attempts to determine whether the benefits from investing in large capital projects compensate for the initial outlay.

Traditional methods such as payback, profitability index and ARR are simplistic and easy to calculate, interpret and understand. They do, however, fail to appreciate the time value of money.

Two discounted cashflow techniques do take into account the time value of money, namely, NPV and IRR.

This chapter has focused on basic investment appraisal techniques, but this is not the full story when evaluating projects. Managers must take great care when identifying relevant cashflows alongside an understanding of the effects of tax, inflation, risk and qualitative issues when assessing investment proposals.

8.6 Present value table

Present value of $1 = 1 / (1 + r)^n$

Where r = discount rate and n = number of periods until payment

Table 8.1 Present value table

Periods (n)	Discount rate (r)									
	1%	2%	3%	4%	5%	6%	7%	8%	9%	10%
1	0.990	0.980	0.971	0.962	0.952	0.943	0.935	0.926	0.917	0.909
2	0.980	0.961	0.943	0.925	0.907	0.890	0.873	0.857	0.842	0.826
3	0.971	0.942	0.915	0.889	0.864	0.840	0.816	0.794	0.772	0.751
4	0.961	0.924	0.888	0.855	0.823	0.792	0.763	0.735	0.708	0.683
5	0.951	0.906	0.863	0.822	0.784	0.747	0.713	0.681	0.650	0.621
6	0.942	0.888	0.837	0.790	0.746	0.705	0.666	0.630	0.596	0.564
7	0.933	0.871	0.813	0.760	0.711	0.665	0.623	0.583	0.547	0.513
8	0.923	0.853	0.789	0.731	0.677	0.627	0.582	0.540	0.502	0.467
9	0.914	0.837	0.766	0.703	0.645	0.592	0.544	0.500	0.460	0.424
10	0.905	0.820	0.744	0.676	0.614	0.558	0.508	0.463	0.422	0.386
11	0.896	0.804	0.722	0.650	0.585	0.527	0.475	0.429	0.388	0.350
12	0.887	0.788	0.701	0.625	0.557	0.497	0.444	0.397	0.356	0.319
13	0.879	0.773	0.681	0.601	0.530	0.469	0.415	0.368	0.326	0.290
14	0.870	0.758	0.661	0.577	0.505	0.442	0.388	0.340	0.299	0.263
15	0.861	0.743	0.642	0.555	0.481	0.417	0.362	0.315	0.275	0.239

Periods (n)	Discount rate (r)									
	11%	12%	13%	14%	15%	16%	17%	18%	19%	20%
1	0.901	0.893	0.885	0.877	0.870	0.862	0.855	0.847	0.840	0.833
2	0.812	0.797	0.783	0.769	0.756	0.743	0.731	0.718	0.706	0.694
3	0.731	0.712	0.693	0.675	0.658	0.641	0.624	0.609	0.593	0.579
4	0.659	0.636	0.613	0.592	0.572	0.552	0.534	0.516	0.499	0.482
5	0.594	0.567	0.543	0.519	0.497	0.476	0.456	0.437	0.419	0.402
6	0.535	0.507	0.480	0.456	0.432	0.410	0.390	0.370	0.352	0.335
7	0.482	0.452	0.425	0.400	0.376	0.354	0.333	0.314	0.296	0.279
8	0.434	0.404	0.376	0.351	0.327	0.305	0.285	0.266	0.249	0.233
9	0.391	0.361	0.333	0.308	0.284	0.263	0.243	0.225	0.209	0.194
10	0.352	0.322	0.295	0.270	0.247	0.227	0.208	0.191	0.176	0.162
11	0.317	0.287	0.261	0.237	0.215	0.195	0.178	0.162	0.148	0.135
12	0.286	0.257	0.231	0.208	0.187	0.168	0.152	0.137	0.124	0.112
13	0.258	0.229	0.204	0.182	0.163	0.145	0.130	0.116	0.104	0.093
14	0.232	0.205	0.181	0.160	0.141	0.125	0.111	0.099	0.088	0.078
15	0.209	0.183	0.160	0.140	0.123	0.108	0.095	0.084	0.074	0.065

8.7 Model answers to activities

Learning activity 8.1.6

Your answer could include:

- The costs of development of a new product;
- A marketing campaign designed to increase long-term brand awareness;
- Investment in training and management development;
- Acquisitions of other businesses;
- Reorganisation and rationalisation costs;
- Research and development costs incurred in developing a strategic advantage.

Self-assessment question 8.1 answer

(a) Relevant
(b) Relevant
(c) Irrelevant
(d) Relevant
(e) Relevant
(f) Relevant
(g) Irrelevant
(h) Relevant

You should note that: (c) is irrelevant because it will not be affected by the decision; (g) is irrelevant because it is an accounting entry and not a cashflow; (h) is relevant because it is an opportunity cost, a contribution which the company will forego if it takes the decision to go ahead with the new product.

Self-assessment question 8.2 answer

Terminal values:

(a) After 1 year   £4,000 x (1.05)   = £4,200.00
(b) After 2 years   £4,000 x (1.05 x 1.05)   = £4,410.00
(c) After 3 years   £4,000 x (1.05 x 1.05 x 1.05)   = £4,630.50

Present values:

(d) After 1 year   £5,000 * $(1 / 1.05^1)$ = £4,761.90
     After 2 years   £5,000 * $(1 / 1.05^2)$ = £4,535.15
     After 3 years   £5,000 * $(1 / 1.05^3)$ = £4,319.19

Learning activity 8.3

Self-assessment question 8.3 answers

PAYBACK PERIOD

Project Y	1,616,000 / 500,000	=	**3.2 years**
Project Z	556,000 / 200000	=	**2.8 years**

ACCOUNTING RATE OF RETURN

Project Y

Depreciation	= (1,616,000 – 301,000) / 5	= £263,000
Profit	= 500,000 – 263,000	= £237,000
Average capital employed	= (1,616,000+301,000) / 2	= £958,500
ARR	= (237,000 / 958,500) * 100	= **24.7%**

Project Z

Depreciation	= (556,000 – 56,000) / 5	= £100,000
Profit	= 200,000 – 100,000	= £100,000
Average capital employed	= (556,000 + 56,000) / 2	= £306,000
ARR	= (100,000 / 306,000) * 100	= **32.7%**
Net Present Value at 15%		

Project Y

End of year	Cashflow	Discount Factor	Present Value
0	(1,616,000)	1.0000	(£1,616,000)
1–5	500,000	3.3520	£1,676,000
5	301,000	0.4970	*£149,597*
Net Present Value			**£209,597**

Project Z

End of year	Cashflow	Discount Factor	Present Value
0	(£556,000)	1.0000	(£556,000)
1–5	£200,000	3.3520	£670,400
5	£56,000	0.4970	*£27,832*
Net Present Value			**£142,232**

INTERNAL RATE OF RETURN

Project Y at 30% gives a negative NPV of £317,132
IRR = 15% + ((209,597 / 526,729) * 15) = 21.0%
Project Z at 30% gives a negative NPV of £53,798
IRR = 15% + ((142,232 / 186,021) * 15) = 25.90%

Viewed in isolation, both projects would be acceptable based on the discounting techniques; both give a positive NPV and an IRR above the cost of capital (15%). Whether the payback periods and ARR are acceptable is more a matter of managerial judgement.

However, these are mutually exclusive projects and if the company proceeds with one, it will have no use for the other. Here, there is a conflict between the two projects: Project Y has the higher NPV, whilst Project Z has the higher IRR.

The higher NPV will achieve the higher increase in shareholders' wealth, therefore Project Y should be selected.

9 Capital investment appraisal

Further considerations

9.1 Introduction

This chapter introduces greater reality and practicality to the investment appraisal process. Businesses are often faced with more than one alternative to solve a problem or achieve an objective. This can take place where the capital available is limited or rationed. It is also complicated to identify what costs and revenue information are relevant to the decision. Additionally, available projects do not all have the same duration. Taxation and inflation implications also need to be considered. The business also needs to review how the proposed investment and decision fits into the long-term strategic plan.

9.1.1 Capital rationing

Chapter 8 introduced the decision rule that if a project had a positive Net Present Value it should be accepted. In theory, all such projects should be accepted by the managers of limited companies, as the acceptance of each one will increase shareholders' wealth.

However, in practice, the resulting gearing of the company and whether the company can cope with a number of large projects may put a limit on the amount of finance made available for investment projects. These issues are known a 'soft rationing'.

In publically funded organisations such as the NHS and local government, central government often restricts the amount of capital available for investment. These types of businesses are therefore faced with a prioritisation decision: finance is limited ('hard rationing') so those projects with the highest profitability index are prioritised alongside other factors such as implications around value for money, targeted return on assets, private finance initiatives (PFI) and risk/uncertainty.

Self-assessment question 9.1

Identify a scenario within an organisation you work in or are familiar with. A choice has to be made between investment projects because of capital rationing. Were the rationing factors 'hard' or 'soft'?

9.2 Relevant costs and revenues

Costs and revenues which are relevant to an investment appraisal decision are those which will change as a result of the decision, the incremental future costs and revenues. Sunk costs (past costs) are ignored in the decision-making process and include expenses such as depreciation, which is an accountancy adjustment and not a flow of cash.

Examples of relevant costs for capital investment decisions include incremental costs which will be incurred or avoided as a result of making a decision and opportunity costs which is the benefit foregone by choosing one investment opportunity instead of the next best alternative.

Self-assessment question 9.2

A construction company is considering whether to open a bookshop at its staff training centre in the North West of England. The bookshop would sell books, stationery and newspapers. The company has an area of unused space available within the main entrance hall of the training centre. The entrance hall is well lit and adequately heated and no additional cost would be involved for these services.

No other useful purpose is foreseen for the area. However, a major redevelopment is planned for the whole training centre so the bookshop facility will only be available for five years.

To convert an area of the main entrance of the training building into a bookshop, with all the fitting, would cost £80,000.

The cost of capital at the construction company is 10% per annum. A survey of sales potential and estimated costs has produced the following annual figures:

Total sales	£400,000
Cost of sales	£320,000
Staff salaries	£52,000
Depreciation of bookshop	£16,000
Interest on capital	£8,000

Share of existing overheads

Administration	£4,000
Building repairs	£12,000
Heat and light	£4,000
Total costs	£416,000
Estimated annual loss	£16,000

During discussions with a local book wholesaler on the supply of books, the wholesaler offered to pay half of the cost of building the workshop and an annual rent of £16,000 inclusive of lighting and heating in return for a five-year tenancy.

(a) Critically review the appropriateness of the annual figures supplied for the bookshop as a basis for appraising the investment.
(b) Evaluate the two proposed methods of operating the bookshop by calculating the NPV.
(c) Recommend a course of action you think should be taken using the data established in part (b), stating any assumptions you make.

9.3 Projects with unequal lives

When businesses are deciding between mutually exclusive projects, it may well be that the alternatives have unequal lives. For example, one supplier may promote its equipment an expected life of five years, whereas another supplier may specify that its equipment will only last for four years.

One solution to this problem is to compare the projects over a common period, the replacement chain approach. But in this example the common period would be (5 * 4) = 20 years. The usual solution, therefore, is to calculate the equivalent annual annuity (EAA), also known as the annual equivalent cost (AEC).

Self-assessment question 9.3

Roberts Ltd runs a cement-processing business called Dallas Builders. Dallas Builders prepares all the cement for Roberts Ltd whose main clients are in the housebuilding industry.

Dallas Builders is in the process of purchasing an industrial mixing machine. A competitive tendering process is used to purchase the mixing machine. Dallas Builders has received quotes from two suppliers, A and B, and both quotes meet the necessary requirements.

Supplier A is offering a mixing machine that will cost £900,000 and has an expected life of 10 years, after which it is expected to have a £20,000 scrap value. A warranty is included for the first year; however, after that continuing maintenance is available at £540,000 for a single payment to cover the remaining nine-year period, or alternatively by an annual charge of £75,000, payable in advance of the cover.

Supplier B is offering a mixing machine at £755,000. This machine is only expected to last for five years, after which it will have a scrap value of £110,000. The supplier offers a maintenance service at an annual cost of £72,000, payable annually in advance. The first year's maintenance is covered by the manufacturer's warranty.

Dallas Builders' cost of capital is 6%.

The purchase of the new mixing machine is essential to meet the group's expansion plans.

You are required to appraise the proposals from the two suppliers and recommend the most appropriate proposal.

9.4 Inflation

Because investment appraisal decisions have long-term implications for a business, forecast revenues and costs will be affected by the rate of inflation over the life of the project, so this rate will need to be forecast as well. Either the cost of capital can be adjusted to take into account inflationary changes or the cashflows themselves can be adjusted. Whichever approach is taken, the answer is the same. But it is important to realise that this analysis will only be as good as the forecast of the inflation rate.

For example, a machine costs £36,000 and is projected to produce, in current prices, cashflows of £12,000, £20,000 and £14,000, respectively, over the next three years. The expected rate of inflation is 6% and the firm's cost of capital is 16.6%.

We can adopt either:

- Forecast cashflows in monetary terms and discount at the nominal or money cost of capital including inflation (16.6%); or
- Forecast cashflows in constant (current) monetary terms and discount at the real cost of capital.

The monetary terms approach

Year	Cashflow Current Prices (£)	Actual money Prices (£)	Discount factor at 16.6%	Present Value (£)
0	18,000 * 1.0	(18,000)	1.00	(18,000)
1	6,000 * 1.06	6,360	1 / 1.166	5,454
2	10,000 * (1.06)^2	11,236	1 / 1.166^2	8,264
3	7,000 * (1.06)^3	8,337	1 / 1.166^3	5,259
NPV				977

Monetary values here means the actual price levels that are forecast to obtain at the date of each cashflow; constant terms covers the price level prevailing today and the real cost of capital covers the cost net of inflation.

The real cost of capital approach

	Cashflow	Real discount	Present
Year	Current Prices (£)	Rate at 10%	Value (£)
0	(18,000)	1.00	(18,000)
1	6,000	1 / 1.10	5,454
2	10,000	$1 / 1.10^2$	8,264
3	7,000	$1 / 1.10^3$	5,259
NPV			977

The real cost of capital discounted rate is calculated by (1.166 / 1.06) * 100 = 10%.

9.5 Taxation

If a business is subject to taxation on its profits, this is a cashflow and requires to be taken into account when determining the cashflows relevant to an investment appraisal.

All commercial organisations' profits are normally subject to corporation tax, the rate varying over time as set by central government.

Current corporation tax payment rules are more complex than previous tax legislation. Previously, corporation tax was due nine months after the end of the accounting period. The required payment date now differs according to the size of the entity. In NPV calculations you should normally assume that the tax is paid at the end of the year following the period in which the profit is earned.

Taxation legislation treats depreciation as a 'non-allowable expense' – instead the government allows capital allowances. These are calculated using pre-set rates: for example, 25% reducing balance is a common rate to be used for plant and equipment. In tax calculations, businesses will have to add back depreciation costs onto profits and then deduct the appropriate rate for capital allowances for that year.

Often capital allowances are at a different write-off rate than depreciation. This creates timing differences known as 'crystallisation of liabilities'. This is purely an accountancy adjustment dealt with through deferred taxation and will have no impact on cashflows and subsequent NPV calculations.

When an asset is eventually disposed of, a business will need to calculate the gain/loss on disposal by comprising the net book value (based on the cost of the asset less any capital allowances to date) less any proceeds received on disposal. If a gain arises this will give rise to extra profits and tax, whilst a loss will result in reduced profits and tax.

9.6 Conclusions

One of the most difficult aspects of capital investment appraisal decisions is identifying and gathering the relevant information for analysis. This chapter has examined the incremental cashflow approach to project decision making. Specific attention has been given to the relevant income/cost approach and to the impact of inflation and taxation on investment decisions.

A business decision maker would need to use the investment appraisal techniques examined in chapters 4 and 5 against a strategic framework, paying particular attention to new technology and environmental/ethical considerations.

The resource allocation process is often the mechanism by which business strategy can be implemented. Investment decisions are not simply the result of project evaluations. Investment appraisal is essentially a search process: a search for ideas, for information and for decision criteria. Shareholder wealth depends more on an organisation's ability to create profitable investment opportunities than on its ability to merely appraise them.

9.7 Model answers to questions

Self-assessment question 9.1

You may have identified a particular choice between two or more projects where there are insufficient funds available to do all the projects; or you may have identified the annual capital budgeting process where a decision has had to be made on which projects to include in the current year budget to match funds available and which projects to defer until next year's budget. These would be examples of hard rationing.

You might also have identified projects in a department where there are staff shortages or there is limited staff time available to absorb the changes caused by new projects. This would be an example of soft rationing.

Managers would need to take an holistic view of all the 'hard' and 'soft' rationing factors and then rank each project in terms of their profitability index scores before any investment decisions were implemented.

Externally imposed constraints are referred to as 'hard capital rationing' and internally imposed constraints are known as 'soft rationing'.

Further examples of soft capital rationing include:

- Management sets maximum limits on borrowing.
- Additional equity finance is unavailable internally.
- Management pursues a policy of slow and stable growth.
- Management imposes divisional ceilings within capital budgets.
- Management is highly risk-averse and operates a rationing policy to select only highly profitable projects.

Self-assessment question 9.2

(a) The annual figures are not appropriate for the following reasons:

- Apportioned costs have been included, which will be incurred even if the space is unused. The annual figures should have considered the annual contribution of £4,000 (£400,000 – £396,000) and ignored the £20,000 share of existing overheads.

- The figures don't take into account the time value of money, a critical factor when looking at cashflow impacts over five years of this project.
- Depreciation has been included, but we should be focusing on the cashflow impact of the capital expenditure and not accounting adjustment entries.
- Staff salaries have been charged as an annual cost. The business would need to check whether they would have been incurred anyway or whether they are additional costs specific to this project. If so, the potential contribution could be £56,000 (£52,000 + £4,000).

(b) Bookshop operated by the construction company is:

Total Revenue	£400,000
Cost of Sales	£320,000
Staff Salaries (assuming incremental)	£52,000 (TC £372,000)
Potential Surplus	£28,000

The interest on capital is not included as the discounting process takes account of the interest rate, and it would be double counted if shown as a cashflow.

Year	Cashflow (£)	Discount factor	Present value (£)
0	(80,000)	1.00	(80,000)
1–5	28,000	3.79 (10% over five yrs)	106,120
Net Present Value			26,120

Bookshop rented to wholesaler

Year	Cashflow (£)	Discount factor	Present value (£)
0	(40,000)	1.00	(40,000)
1–5	16,000	3.79 (10% over five yrs)	60,640
Net Present Value			20,640

(c) On the basis of the NPV calculations it is recommended that the construction company operate the bookshop themselves, as it will generate the greater shareholder wealth.

This is on the basis of the following assumptions:

- There is no residual value of the capital expenditure items.
- There are no dismantling costs.
- The cost of financing inventories (stock) is negligible.

Self-assessment question 9.3

There are three alternatives:

1   Supplier A, single payment warranty

Year 0	£900,000 * 1.0000	£900,000
Year 1	£540,000 * 0.9434	£509,436
Year 10	(£20,000) * 0.5584	(£11,168)
Net Present Cost	£1,398,268	
Annual equivalent (10 yrs discount factors added together at 6%)	7.36	
Annual equivalent cost	(£1,398,268 / 7.36)	£189,982

2. Supplier A annual charge warranty

Year 0	£900,000 * 1.0000	£900,000
Year 1–9	£75,000 * 6.802	£510,150
Year 10	(£20,000) * 0.5584	(£11,168)
Net Present Cost	£1,398,982	
Annual equivalent (10 yrs discount factors added together at 6%)	7.36	
Annual equivalent cost (£1,398,982 / 7.36)		£190,079

2   Supplier B

Year 0	£735,000 * 1.0000	£735,000
Year 1–4	£72,000 * 3.465	£249,280
Year 4	(£110,000) * 0.7473	(£82,203)
Net Present Cost	£902,227	
Annual equivalent (5 yrs discount factors added together at 6%)	4.212	
Annual equivalent cost (£902,227 / 5)		£214,216

Option 1 – supplier A with a single payment warranty should be selected, as it has the lowest annual equivalent net present cost.

10 Corporate accounts

10.1 Interpretation of accounts

There are a number of accounting 'ratios' that can be used to help assess the performance of a company.

Different user groups will require different information. For example:

- Investors: for making decisions on buying and selling shares in particular companies;
- Providers of finance: whether or not to make a loan to a company;
- Management: for assessing the current performance of a company compared with both past performance and the performance of competitor.

The main areas for assessment are:

- Profitability: assessing the profit of a company in relation to, say, the capital employed to generate that profit, or in relation to revenue;
- Efficiency: assessing the efficiency of various aspects of a company, say, inventories (stock) turnover or trade receivables (debtors) collection period;
- Liquidity: assessing a company's ability to pay its way;
- Investment: measuring a company's performance, for example, its earnings per share.

Although the term 'ratio analysis' is often used for the assessment of the above, a better term would be the use of 'key performance indicators', as many of the techniques provide information in percentage or other terms and not in the form of ratios. When using information obtained from a company's annual report, any results must be treated with caution. Different companies will adopt different accounting policies and so inter-company comparison may be difficult. A company's Statement of Financial Position (Balance Sheet) represents a picture of the situation on one particular day, and the picture might be quite different in, say, a month's time.

10.1.1 Profitability key performance indicators

Return on capital employed

This indicator measures the profit for the year as a percentage of the funds that have been provided by the shareholders and other providers of long-term finance.

It is important to identify clearly what is to be included in the 'capital employed' figure.

It is usually taken to be shareholders' funds, which includes any reserves, and any long-term borrowing such as loans and debentures.

It could be argued that any 'permanent' bank overdraft should also be included. If long-term borrowing is included as part of the capital employed, the profit before interest and tax figure should be used.

$$\frac{Profit \ before \ Interest \ and \ Tax}{Shareholders' funds \ + \ Long\text{-}term \ borrowing} \text{ x } 10$$

Return to shareholders

This measures the profit available to the ordinary shareholders in relation to the funds they have provided.

$$\frac{Profit \ after \ tax \ (and \ Preference \ dividends)}{Ordinary \ share \ capital + reserves} \text{ x } 100$$

Net profit margin

$$\frac{Profit \ before \ Interest \ and \ Tax}{Sales \ (Revenue)} \text{ x } 100$$

Gross profit margin

$$\frac{Gross \ Profit}{Sales \ (Revenue)} \text{ x } 100$$

10.1.2 Efficiency key performance indicators

Trade receivables/debtor days

This provides information relating to a company's efficiency in collecting money owed by its customers. It may be expressed in ratio terms, but is more usually calculated as the number of days taken, on average, to receive money from customers.

$$\frac{\text{Trade Receivables}}{\text{Credit Revenue}} \times 365$$

Trade payables/creditor days

This provides information relating to a company's efficiency in paying money owed to its customers for purchases on credit. It may be expressed in ratio terms, but is more usually calculated as the number of days taken, on average, to pay money to customers. Credit purchases are sometimes shown as the cost of sales in the accounts.

$$\frac{\text{Trade Payables}}{\text{Credit Purchases}} \times 365$$

Fixed asset turnover period

This ratio compares the sales revenue of a company to its fixed assets. This ratio tells us how effectively and efficiently a company is using its fixed assets to generate revenues. This ratio indicates the productivity of fixed assets in generating revenues. If a company has a high fixed asset turnover ratio, it shows that the company is efficient at managing its fixed assets. Fixed assets are important because they usually represent the largest component of total assets.

$$\frac{\text{Sales Revenue}}{\text{Total Fixed Assets}}$$

Inventory turnover in days

This is a measure of the number of times inventory is sold or used in a given time period such as one year. It is a good indicator of inventory quality (whether the inventory is obsolete or not), efficient buying practices and inventory management. This ratio is important because gross profit is earned each time inventory is turned over.

Inventory turnover is calculated by dividing the cost of goods sold by the average inventory level ((beginning inventory + ending inventory) / 2):

$$\frac{\text{Cost of Goods Sold}}{\text{Average Inventory}}$$

10.1.3 Liquidity key performance ratios

Current ratio

The current ratio measures the ability of an organisation to pay its expenses in the short term (usually the next 12 months). The ratio is often used by analysts and banks to determine whether they should invest in or lend money to a business. An ideal benchmark is 2:1 – but this can vary dependent on the sector the business is trading in.

$$\frac{\text{Current assets}}{\text{Current liabilities}}(\text{expressed x: 1})$$

Acid test ratio or quick ratio

Liquidity is often measured as the relationship between current assets and current liabilities. As current assets are theoretically either already in the form of cash or can be converted into cash quite quickly, they provide the funds available for a company to pay its way on a daily basis (i.e. to cover its current liabilities). However, a more critical measure excludes stock from the current assets, on the basis that it may take some time for stock to be sold and converted into cash. An ideal benchmark is 1:1 – but this can vary dependent on the sector the business is trading in.

$$\frac{\text{Current assets less stock}}{\text{Current liabilities}}(\text{expressed x: 1})$$

10.1.4 Investment key performance indicators

Earnings per share

Companies are required to show the basic earnings per share on the face of the profit and loss account (income statements). It shows the amount earned by the company for its ordinary shareholders and is expressed in 'pence per share'.

The formula is:

$$\frac{\text{Profit after tax (and any preference dividends)}}{\text{Number of ordinary shares in issue}}$$

Price to earnings ratio (P/E ratio)

Given in the financial pages along with the share price and the dividend yield, this ratio indicates the number of years of current earnings required to recover the price paid for a share.

The ratio is usually taken as an indication of the stock market's confidence in a share/business.

$$\frac{\text{Market price of share}}{\text{Earnings per share}}(\text{expressed as a number of times})$$

Dividend yield

Dividend yield expresses, as a percentage, the return actually received (or receivable) in the form of a dividend payment in relation to the amount invested (the share price).

$$\frac{\text{Dividend per share}}{\text{Share price}} \times 100$$

This can be used for comparison with yields from other investments with a portfolio of shares for an investor.

Gearing

Gearing measures the relationship between shareholders' funds (equity) and borrowed funds (debt). It is usually expressed in percentage terms by taking:

$$\frac{\text{Debt (Long-term borrowing + Preference shares)}}{\text{Debt + Equity (Ordinary share capital + reserves)}} \times 100$$

The above method gives gearing on a scale of 0–100%.

The level of gearing can be of great importance to investors. A highly geared company will have a large interest charge to cover, which will reduce the profit figure.

However, if profits more than adequately cover the interest charge, the return to shareholders can be quite high.

10.1.5 Ratio analysis exercise

Bridge Ltd income statement for the year ended 31 October 2015

	£	£
Sales revenue		1,062,000
Cost of sales:		
Opening inventory	56,000	
Purchases	*680,000*	
	736,000	
Closing inventory	54,000	*682,000*
Gross profit		380,000
Less expenses:		
Heating and lighting	18,375	
Insurances	15,000	
Directors' fees	30,000	
Audit fee	10,000	
Depreciation	*30,800*	

Bridge Ltd income statement for the year ended 31 October 2015

	£	£
		104,175
Profit from operations		*275,825*
Finance costs: interest payable		*11,000*
Profit before tax		264,825
Tax expense		*63,000*
Net profit for period		*201,825*

Statement of changes in equity for the year ended 31 October 2015

	Issue Share Capital	Share Capital	Retained Earnings	Total
	£	£	£	£
Bal. at 31.10.14	500,000	100,000	31,570	631,570
Net profit for period			201,825	201,825
Dividends paid			(21,000)	(21,000)
Bal. at 31.10.14	500,000	100,000	212,395	812,395

Bridge Ltd Statement of Financial Position as at 31 October 2015

	£	£	£
Assets	Costs	Dep	NBV
Non-current Assets			
Freehold land	350,000		350,000
Buildings	400,000	30,000	370,000
Equipment	*228,000*	*33,800*	*194,200*
	978,000	*63,800*	914,200
Current assets			
Inventories		54,000	
Receivables		140,500	
Prepayments		*5,000*	
			199,500
Total assets			1,113,700
Equity and liability			
Equity			
Ordinary share capital			400,000
6% preference share capital			100,000
Share premium			100,000
Retained earnings			212,395
			812,395

Bridge Ltd Statement of Financial Position as at 31 October 2015

	£	£	£
Assets	*Costs*	*Dep*	*NBV*
Non-current Assets			
Non-current liabilities			
10% debentures 2017			110,000
Current liabilities			
Bank overdraft		26,935	
Payables		89,870	
Accruals		11,500	
Corporation tax		*63,000*	
			191,305
Total equity and liabilities			**1,113,700**

Bridge Ltd has an authorised share capital of 1,000,000 ordinary shares of 50p each and 100,000 6% preference shares of £1. At the balance sheet date the ordinary shares had a market valuation of 25p. Calculate appropriate ratios to assess the financial performance of the business.

10.1.6 Solution to ratio analysis exercise

Liquidity ratios

CURRENT RATIO

$$= \frac{\text{Current assets}}{\text{Current liabilities}}$$

$$= \frac{£199,500}{£191,305}$$

$$= 1.04 : 1$$

ACID TEST RATIO

$$= \frac{\text{Current assets less inventories}}{\text{Current liabilities}}$$

$$= \frac{£199,500 - 54,000}{£191,305}$$

$$= 0.76 : 1$$

Profitability ratios

GROSS PROFIT MARGIN

$$= \frac{\text{Gross profit}}{\text{Sales revenue}} \times 100$$

$$= \frac{£380,000}{£1,062,000} \times 100$$

$$= 35.78\%$$

NET PROFIT MARGIN

$$= \frac{\text{Net profit before tax}}{\text{Sales revenue}} \times 100$$

$$= \frac{£264,825}{£1,062,000} \times 100$$

$$= 24.94\%$$

MARK-UP

$$= \frac{\text{Gross profit}}{\text{Cost of sales}} \times 100$$

$$= \frac{£380,000}{£682,000} \times 100$$

$$= 55.72\%$$

EXPENSES TO SALES RATIO

$$= \frac{\text{Expenses}}{\text{Sales revenue}} \times 100$$

$$= \frac{£104,175}{£1,062,000} \times 100$$

$$= 9.81\%$$

RETURN ON CAPITAL EMPLOYED

$$= \frac{\text{Net profit before tax}}{\text{Shareholders' funds}} \times 100$$

$$= \frac{£264,825}{£812,395} \times 100$$

$$= 32.60\%$$

RETURN ON NET ASSETS

$$= \frac{\text{Net profit before tax}}{\text{Total assets less liabilities}} \times 100$$

$$= \frac{£264,825}{£1,113,700 - £110,000 - £191,305} \times 100$$

$$= 32.60\%$$

RETURN ON NON-CURRENT ASSETS

$$= \frac{\text{Net profit before tax}}{\text{Non-current assets at NBV}} \times 100$$

$$= \frac{£264,825}{£914,200} \times 100$$

$$= 28.97\%$$

Efficiency ratios

INVENTORIES TURNOVER RATIO

$$= \frac{\text{Cost of sales}}{\text{Average inventory}}$$

$$= \textit{Average inventory} = \frac{£56,000 + £54,000}{2} = £55,000 = \frac{£682,000}{£55,000}$$

$$= 12.4 \textit{ times}$$

NON-CURRENT ASSETS TURNOVER RATIO

$$= \frac{\text{Sales revenue}}{\text{Non-current assets (at NBV)}}$$

= £1,062,000
£914,200
= 1.16 times

TRADE RECEIVABLES COLLECTION PERIOD

= Trade receivables   x 365

SALES REVENUE

$$= \frac{£1,062,000}{£1,062,000} \times 365 \,\text{days}$$

$$= 48.29 \text{ days}$$

TRADE PAYABLES COLLECTION PERIOD

$$= \frac{\text{Trade payables}}{\text{Purchases}} \times 365$$

$$= \frac{£89,870}{£680,000} \times 365 \text{ days}$$

$$= 48.24 \text{ days}$$

Shareholder ratios

DIVIDEND PER ORDINARY SHARE

= Ordinary share dividend
  Number of ordinary shares
= £15,000
£800,000
= 1.875p per share

DIVIDEND YIELD

Dividend per share x 100
Price per share

$$= \frac{1.875\text{p} \times 100}{25\text{p}}$$

$$= 7.5\%$$

DIVIDEND COVER

Net profit after tax less preference dividend
Paid and proposed ordinary dividends

$$= £201,825 - £6,000$$

£15,000

$$= 13.06 \text{ times}$$

BASIC EARNINGS PER SHARE

$$= \frac{\text{Net profit after tax less preference dividend}}{\text{Number of ordinary shares}}$$

$$= \frac{£201,825 - £6,000}{800,000}$$

$$= 24.48\text{p per ordinary share}$$

PRICE/EARNINGS (P/E) RATIO

$$= \frac{\text{Market price per share}}{\text{Basic earnings per share}}$$

$$= \frac{25\text{p}}{24.48\text{p}}$$

$$= 1.02 \text{ times or } 1.02 \text{ years}$$

GEARING RATIO (METHOD 1)

$$= \frac{\text{Loans}}{\text{Shareholders' funds}} \times 100$$

$$= \frac{£110,000 + £26,935}{£812,395} \times 100$$

$$= 16.86\%$$

GEARING RATIO (METHOD 2)

$$= \frac{\text{Preference shares and long-term loans}}{\text{Shareholders' funds and long-term loans}} \times 100$$

$$= \frac{£100,000 + £110,000}{£812,395 + £110,000} \times 100$$

$$= 22.77\%$$

Key analysis points

When interpreting ratios it is important to make calculations over a number of years to establish a trend in the data. Managers within organisations will also want to compare an organisation's performance across sectorial best practice. This is usually done by identifying sector benchmarks across key performance indicators in areas such as profitability, efficiency, liquidity and shareholder value.

10.2 Income statement (pro forma)

Income statement/profit and loss account

Basically shows:	
Revenue (Sales)	x
Cost of sales	(x)
Gross profit	x
Administration expenses	(x)
Distribution costs	(x)
Operating profit	x
Net interest	(x)
Profit before taxation	x
Taxation on profit	(x)
Profit after tax	x
Other comprehensive profit	x
Total comprehensive profit	x
Basic earnings per share	'x' pence

The operating profit for the year is analysed between:
Continuing operations, discontinued operations, and acquisitions
This enables relevant comparisons to be made between:
Previous year   →   Current year   →   Forecast for following year

Certain items, such as profit or loss on sale of fixed assets and 'Exceptional' items, are required to be shown separately in order to provide users with more information on the current year's activities.

10.2.1 The income statement

This shows the calculation of the profit or loss of the company for the period in question. It starts with the INCOME (usually referred to as 'Revenue') for the period, generated from the company's main operations.

Deducted from this is the cost of those operations (usually referred to as 'cost of sales'); also deducted are the expenses incurred during the period for administration, selling and distribution. Any other sources of income (such as interest received) are added, and any finance expenses (such as interest paid) are deducted.

Any profit that the company makes is subject to 'Corporation Tax'.

The profit, after tax has been deducted, belongs to the owners – the equity shareholders.

Some of this profit may be passed on to the shareholders in the form of 'dividends'.

Any remaining profit may be set aside for some specific purpose in a 'reserve', but will be added to the shareholders' funds in the balance sheet.

10.3 The statement of financial position (balance sheet)

The statement of financial position is simply a list of the company's assets and liabilities on a specific date – usually the year end of the company.

It is based on the accounting equation which states that at any point in time:

ASSETS = LIABILITIES

The liabilities can be sub-divided into Equity and other LIABILITIES. Therefore we have:

ASSETS = EQUITY + LIABILITIES

This can be re-stated as:

ASSETS – LIABILITIES = EQUITY

ASSETS are the resources owned by the business and may be split into two categories:

NON-CURRENT ASSETS are assets which are owned by the business and which are intended for long-term use within it, e.g. machinery, buildings, etc.

CURRENT ASSETS are assets which are owned by the business that are in the form of cash or are intended to be turned into cash within the next 12 months, e.g. trade receivables, inventories, etc.

10.4 Equity and liabilities

CURRENT LIABILITIES are amounts that have to be paid within the next 12 months.

NON-CURRENT LIABILITIES are amounts which fall due for payment over the longer term (usually over 12 months).

EQUITY represents the amount which the business owes to its owners.

SHARE CAPITAL is the amount of money invested into the business by shareholders.

The SHARE PREMIUM ACCOUNT balances the difference between the par value of a company's shares and the amount that the company actually received for newly issued shares.

Suppose a company issues 500 shares of £1 each, but is paid £3 per share. It then has £1,500 of equity capital. Only £500 of this is share capital. The balance (£1,000) is shown as share premium.

RETAINED EARNINGS are the accumulated profits earned from company inception.

REVALUATION RESERVES are created when assets are revalued at a different amount than their historical (original cost).

10.3.1 Statement of financial position as at: (pro forma)

Assets
Non-current assets:
Property, plant and equipment

Current assets:
Inventories
Trade and other receivables
Cash and cash equivalents

Total assets
Equity and liabilities
Share capital
Share premium
Retained earnings
Revaluation reserve
Total equity

Non-current liabilities
Bank loans

Current liabilities
Trade and other payables
Tax liability
Total liabilities

Total equity and liabilities

---

11 Raising capital and managing liquidity

11.1 Capital for small- and medium-sized enterprises

SMEs can be defined as having three main characteristics:

- Companies are not quoted on a stock exchange – they are 'unquoted'.
- Ownership of the business is typically restricted to a few individuals. Often this is a family connection between the shareholders.
- Many SMEs are the means by which individuals (or small groups) effectively achieve self-employment.

The SME sector is a vital one in the UK economy. In 2009, the Department for Trade and Industry (DTI 2009) estimated that there were 3.7 million SME businesses in the UK. Sole traders account for the majority of the businesses in the UK (63%) but a smaller proportion of the number of employees (23%) and an even smaller proportion of turnover (9%). As a proportion of all businesses in the UK, SMEs account for some 55% of employment and 45% of turnover.

11.1.1 Why do SMEs find financing a problem?

The main problem faced by SMEs when trying to obtain funding is that of uncertainty:

- SMEs rarely have a long history or successful track record that potential investors can rely on in making an investment.
- Larger companies (particularly those quoted on a stock exchange) are required to prepare and publish much more detailed financial information – which can actually assist the finance-raising process.
- Banks are particularly nervous of smaller businesses due to a perception that they represent a greater credit risk.

Because the information is not available in other ways, SMEs will have to provide it when they seek finance. They will need to give a business plan, list of the company assets, details of the experience of directors and managers and demonstrate how they can give providers of finance some security for amounts provided.

Prospective lenders – usually banks – will then make a decision based on the information provided. The terms of the loan (interest rate, term, security and repayment details) will depend on the risk involved and the lender will also want to monitor their investment.

A common problem is often that the banks will be unwilling to increase loan funding without an increase in the security given (which the SME owners may be unable or unwilling to provide).

A particular problem of uncertainty relates to businesses with a low asset base. These are companies without substantial tangible assets which can be used to provide security for lenders.

When an SME is not growing significantly, financing may not be a major problem. However, the financing problem becomes very important when a company is growing rapidly, for example when contemplating investment in capital equipment or an acquisition.

Few growing companies are able to finance their expansion plans from cashflow alone. They will therefore need to consider raising finance from other external sources. In addition, managers who are looking to buy in to a business ('management buy-in' or 'MBI') or buy out ('management buy-out' or 'MBO') a business from its owners may not have the resources to acquire the company. They will need to raise finance to achieve their objectives.

11.1.2 Sources of finance for SMEs

There are a number of potential sources of finance to meet the needs of small and growing businesses:

- Existing shareholders and directors funds ('owner financing');
- Overdraft financing;
- Trade credit;
- Equity finance;
- Business angel financing;
- Venture capital;
- Factoring and invoice discounting;
- Hire purchase and leasing;
- Merchant banks (medium- to longer-term loans).

A key consideration in choosing the source of new business finance is to strike a balance between equity and debt to ensure the funding structure suits the business.

The main differences between borrowed money (debt) and equity are that bankers request interest payments and capital repayments, and the borrowed money is usually secured on business assets or the personal assets of shareholders and/or directors. A bank also has the power to place a business into administration or bankruptcy if it defaults on debt interest or repayments or its prospects decline.

In contrast, equity investors take the risk of failure like other shareholders, whilst they will benefit through participation in increasing levels of profits and on the eventual sale of their stake. However, in most circumstances venture capitalists will also require more complex investments (such as preference shares or loan stock) in addition to their equity stake.

The overall objective in raising finance for a company is to avoid exposing the business to excessive high borrowings, but without unnecessarily diluting the share capital. This will ensure that the financial risk of the company is kept at a minimal level.

11.2 Equity finance

What is equity?

'Equity' is the term commonly used to describe the ordinary share capital of a business.

Ordinary shares in the equity capital of a business entitle the holders to all distributed profits after the holders of debentures and preference shares have been paid.

11.2.1 Ordinary (equity) shares

Equity shares are issued to the owners of a company. The equity shares of UK companies typically have a nominal or 'face' value (usually something like £1 or 5Op, but shares with a nominal value of 1p, 2p or 25p are not uncommon).

However, it is important to understand that the market value of a company's shares has little (if any) relationship to their nominal or face value. The market value of a company's shares is determined by the price another investor is prepared to pay for them. In the case of publicly quoted companies, this is reflected in the market value of the ordinary shares traded on the stock exchange (the 'share price').

In the case of privately owned companies, where there is unlikely to be much trading in shares, market value is often determined when the business is sold or when a minority shareholding is valued for taxation purposes.

Why might a company issue ordinary shares?

A new issue of shares might be made for several reasons:

1   The company might want to raise more cash.

    For example cash might be needed for the expansion of a company's operations. If, for example, a company with 500,000 ordinary shares in issue decides to issue 125,000 new shares to raise cash, should it offer the new shares to existing shareholders, or should it sell them to new shareholders instead?

Where a company sells the new shares to existing shareholders in proportion to their existing shareholding in the company, this is known as a **'rights issue'**.

2   The company might want to issue new shares partly to raise cash but more important to 'float' its shares on a stock market.

When a UK company is floated, it must make available a minimum proportion of its shares to the general investing public.

3   The company might issue new shares to the shareholders of another company, in order to take it over.

There are many examples of businesses that use their high share price as a way of making an offer for other businesses. The shareholders of the target business being acquired receive shares in the buying business and perhaps also some cash.

Sources of equity finance

There are three main methods of raising equity:

1   **Retained profits:** i.e. retaining profits, rather than paying them out as dividends. This is the most important source of equity.
2   **Rights issues:** i.e. an issue of new shares. After retained profits, rights issues are the next most important source.
3   **New issues of shares to the public:** i.e. an issue of new shares to new shareholders. In total in the UK, this is the least important source of equity finance.

11.3 Raising capital for larger organisations (PLCs)

Equity finance can be provided by the sale of ordinary shares to investors. This may be a sale of shares to new owners, perhaps through the stock market as part of a company seeking a quotation, or it may be a sale of shares to existing shareholders, for example by means of a rights issue.

Ordinary shares are bought and sold on a regular basis on stock exchanges all over the world. By law, the ordinary shares of a company must have a nominal or par value and cannot be issued for less than this amount. The nominal value of a share bears no relation to its market value.

New shares, whether issued at the foundation of a company or subsequently, are almost always issued at a premium above their nominal value.

11.3.1 New aspects in equity markets

There have been many changes and innovations over the past decades in the financial markets, especially in developed economies, regarding which shares are traded and how they are traded.

(a) The Official List (OL)

Companies which wish to be listed have to sign a Listing Agreement which commits directors to certain high standards of behaviour and levels of reporting to shareholders. This is a market for medium and large established firms with a reasonably long trading history. The cost of listing is so high that small companies are unable to afford or to justify a full market listing.

(b) Unlisted Securities Market (USM)

This is a second-tier market which was introduced to assist small and medium-sized firms to raise capital and to provide a liquid secondary market. The requirements for admission to the USM are less onerous than for the OL.

(c) The Alternative Investment Market (AIM)

The driving philosophy behind AIM is to offer young and developing companies access to new sources of finance, whilst providing investors with the opportunity to buy and sell shares in a trading environment. Costs are down and the rules as simple as possible. AIM companies are expected to comply with strict rules regarding the publication of price-sensitive information and the quality of annual and interim reports. Upon floatation a detailed prospectus is required.

(d) Trade points

Trade points allow those who want to buy or sell shares to bypass the usual system of going through a market maker. On trade points they can advertise their orders directly and anonymously via computer screens. Share dealing is carried out automatically by a simple mouse-click in a Windows-based computer system. Trade points' central computer is able to match, buy and sell orders automatically and at relatively low cost.

(e) Internet

The Internet has led to two main changes. First, real-time prices of shares and financial software to analyse equities markets, which were once exclusively available to the large, well-funded financial institutions, are today accessible through a modestly priced personal computer, modem and software. This has put millions of small investors in up-to-the-second contact with the markets. Second, trading in shares over the Internet is now possible.

11.3.2 Trading systems

There are basically two different trading systems: the quote-driven and the order-driven systems.

Quote-driven systems

This is a share trading system in which one market maker quotes a bid and another market maker quotes an offer price for shares. This remains the dominant trading system in the world. The 'bid' price is the price at which the market maker is willing to buy and the 'offer' price is the price at which the market maker will sell.

The difference between the two prices, called 'the spread', represents a hoped-for return to the market maker. The market makers are obliged to deal (up a certain number of shares) at the price quoted, but they have the freedom to adjust prices after deals are completed. The investor or broker (on behalf of an investor) is able to see the best price available and is able to make a purchase or sale.

Order-driven systems

This is a share trading system in which investors' buy and sell orders are matched without the intermediation of market makers.

This is the most used system in the world that does not require market makers to act as middlemen. They allow buy and sell orders to be entered on a central system, and investors are automatically matched (they are sometimes called 'matched-bargain systems').

The system works as follows: a subscriber (say, a broker acting for an investor client) advertises on their computer screen the price at which they will buy or sell a block of shares. If another subscriber wishes to accept the deal, all they have to do is tap their keyboard or mouse-click and the bargain will be struck. All trades are reported to all users on a 'tape' which runs continuously along Trade point screens.

11.3.3 Raising equity capital

There are a number of alternative ways of raising finance by selling shares.

Authorised, issue and par values

When a firm is created the original shareholders will decide the number of shares to be authorised (the authorised capital). This is the maximum amount of share capital that the company can issue (unless shareholders vote to change the limit). In many cases firms do not issue up to the amount specified.

Floating on the official list

To 'go public' and become a listed company is a major step for a firm. The substantial sums of money involved can lead to a new, accelerated phase of business growth. Obtaining a quotation is not a step to be taken lightly; it is a major legal undertaking.

The Stock Exchange rigorously enforces a set of demanding rules and the company will be put under the strain of new and greater responsibilities both at the time of floatation and in subsequent years.

Prospectus

A company wishing to go to the public is required to prepare a detailed prospectus to inform potential shareholders about the company. This may contain far more information about the firm than it has previously dared to put into the public domain.

A successful flotation can depend on the prospectus acting as a marketing tool as the firm attempts to persuade investors to apply for shares.

11.3.4 The cost of new issues

1 Administrative/transaction costs

The amount of this cost may very much depend on the size of issue and the method used.

2 The equity cost of capital

Shareholders suffer an opportunity cost. By holding shares in one company they are giving up the use of that money elsewhere. The firm, therefore, needs to produce a rate of return for those shareholders which is at least equal to the return they could obtain by investing in other shares of a similar risk class. If the firm does not produce this return then shares will be sold and the firm will find raising capital difficult.

3 Market pricing costs

This has links with the possibility of under-pricing new shares. It is a problem which particularly affects offers for sale. The firm is usually keen to have the offer fully taken up by public investors. To have shares left with the underwriters gives the firm a bad image because it is perceived to have had an issue which 'flopped'.

Rights issues

A rights issue is an invitation to existing shareholders to purchase additional shares in the company. It is easy and relatively cheap (compared with new issues). Rights issues are generally very successful as shareholders are usually given strong incentives to act. The shares are usually offered at a significantly discounted price from the market value.

Shareholders can either buy these shares themselves or sell the 'right' to buy to another investor. For further reassurance that the firm will raise the anticipated finance, rights issues are usually underwritten by institutions.

EXAMPLE

Suppose Bolt Systems Ltd, which has 10 million shares in issue, wants to raise £25 million for expansion but does not want to borrow it. Assume that its existing shares are quoted on the stock market at £12. Assume further that XYZ Ltd has decided that the £25 million will be obtained by issuing 2.5 million shares at £10 each. Thus the ratio of new shares to old is 2.5:10. In other words, this issue is 'one-for-four' rights issue. Each shareholder will be offered one new share for every four already held.

11.3.5 Other equity issues

Open offer

In an open offer, new shares are sold to a wide range of external investors on the condition that existing shareholders have the right to buy them at the same price instead. There are no nil paid rights to sell.

Acquisition for shares

Shares are often issued to purchase business or assets. This is usually subject to shareholder approval.

Vendor placing

If a company wishes to pay for an asset such as a subsidiary of another firm or an entire company with newly issued shares, but the vendor does not want to hold the shares, the purchaser could arrange for the new shares to be bought by institutional investors for cash. In this way the buyer gets the asset, the vendor (for example shareholders in the target company in a merger or take-over) receives cash and the institutional investor makes an investment.

Script issues

Script issues do not raise new money: a company simply gives shareholders more shares in proportion to their existing holdings. The value of each shareholding does not change, because the share price drops in proportion to the additional shares. They are also known as capitalisation issues or bonus issues. The purpose is to make shares more attractive by bringing down the price. Note that with a script issue there will be some adjustment necessary to the balance sheet.

Warrants

Warrants give the holder the right to subscribe for a specified number of shares at a fixed price at some time in the future.

11.3.6 Equity finance for unquoted firms

Not every company (especially the relatively small and medium-sized firms – SMEs) has the opportunity to have access to the Stock Exchange. Thus, it is in line to consider some of the ways that unquoted firms can raise equity capital.

The financing gap

Unquoted firms usually rely on retained earnings, capital injections from the founder and bank borrowings but in most cases these are not enough to finance growth aspirations. Thus, between large firms with access to the stock market and small firms financed by internally generated funds and personal and bank loans, there is a financing gap. This aspect confronts intermediate businesses which find themselves too large or too fast-growing to ask the individual shareholders for more funds or to obtain sufficient bank finance, and they are not ready to launch on the stock market.

How the financing gap can be filled?

BUSINESS ANGELS

Business angels are wealthy individuals, generally with substantial business and entrepreneurial experience, who usually invest substantial amounts primarily in start-up, early stage or expanding firms.

The majority of investments are in the form of equity finance but they do purchase debt instruments and preference shares. They usually do not purchase a controlling interest, and they are willing to invest at an earlier stage than most formal venture capitalists (sometimes they call themselves 'informal venture capitalists' instead of business angels). They are generally looking for entrepreneurial companies which have high aspirations and potential for growth. Business angels are generally patient investors willing to hold their investment for at least a five-year period. The main way in which firms and angels find each other is through friends and business associates, although sometimes informal networks may be of help.

VENTURE CAPITAL

Venture capital funds provide finance for high-growth-potential unquoted firms. Venture capital is a medium- to long-term investment and can consist of a package of debt and equity finance. Venture capitalists take high risks by investing in the equity of young companies often with a limited (or no) track record. Many of their investments are into little more than a team of management with a good idea – which may not have started selling a product or even developed a prototype.

There are a number of different types of venture capital (VC):

1   *Seed corn* – this is financing to allow the development of a business concept. Development may also involve expenditure on the production of prototypes and additional research.
2   *Start-up* – a product or idea is further developed and/or initial marketing is carried out. Companies are very young and have not yet sold their product commercially.
3   *Other early-stage* – funds for initial commercial manufacturing and sales. Many companies at this stage will remain unprofitable.

4   *Expansion* – companies at this stage are on to a fast-growth track and need capital to fund increased production capacity, working capital and for further development of the product or market.

5   *Management buy-outs (MBO)* – here a team of managers make an offer to their employees to buy a whole business, a subsidiary or a section so that they own and run for it for themselves. Large companies are often willing to sell to these teams, particularly if the business is under-performing and does not fit with the strategic core business. Usually the management team have limited funds of their own and so call on venture capitalists to provide the bulk of finance.

6   *Management buy-ins (MBI)* – a new team of managers from outside an existing business buy a stake, usually backed by a venture capital fund. A combination of an MBO and MBI is called a BIMBO – buy-in management buy-out – where a new group of managers joins forces with an existing team to acquire a business.

There are a number of different types of VC providers:

1   *The independents* – these can be firms, funds or investment trusts which have raised their capital from more than one source. The main sources are pension and insurance funds, but banks, corporate investors and private individuals also put money into these funds.

2   *Captives* – these are funds managed on behalf of a parent institution (banks, pension funds, etc.).

3   *Semi-captives* – these invest funds on behalf of a parent and also manage independently raised funds.

11.3.7 Venture capital trusts (VCTs)

These are special tax-efficient vehicles for investing in small unquoted firms through a pooled investment. They offer investors a way of investing in a broad spread of small firms with high potential, but with greater uncertainty, in a tax-efficient manner.

Corporate venturing

Larger companies sometimes foster the development of smaller enterprises. This can take numerous forms, from joint product development work to an injection of equity finance.

Government sources

The government may set up VC-type funds in order to attract and encourage industry. Equity, debt and grant finance may be available from these sources.

11.4 Working capital – liquidity management

One year after the Late Payment of Commercial Debts (Interest) Act 1998 was passed, market information specialists Experian reported that British companies are now taking two days longer to settle their bills with suppliers than before the legislation was introduced. The average time taken to pay for credit purchases by British companies is now 74 days. Although the 1998 legislation enables companies employing fewer than 50 staff to levy an 8% interest charge above the base rate on late-paying larger clients, few have done so in fear of alienating the enterprises on whom they frequently so heavily rely. The study also found that most large businesses now insist on a 60-day payment period. Reliant upon cash from trade debtors to pay suppliers, wages and other costs, the failure to receive the amounts owing from credit customers on the due dates creates enormous problems for businesses in paying their own way.

11.4.1 Assessing the credit worthiness of customers

Before extending credit to a customer, a supplier should analyse the five Cs of credit worthiness, which will provoke a series of questions. These are:

- **Capacity:** will the customer be able to pay the amount agreed within the allowable credit period? What is their past payment record? What is the financial health of the customer? Is it a liquid and profitable concern, able to make payments on time?
- **Character:** do the customers/management appear to be committed to prompt payment? Are they of high integrity? What are their personalities like?
- **Collateral:** what is the scope for including appropriate security in return for extending credit to the customer?
- **Conditions:** what are the prevailing economic conditions? How are these likely to impact on the customer's ability to pay promptly?

Whilst the materiality of the amount will dictate the degree of analysis involved, the major sources of information available to companies in assessing customers' credit worthiness are:

- **Bank references.** These may be provided by the customer's bank to indicate their financial standing. However, the law and practice of banking secrecy determines the way in which banks respond to credit enquiries, which can render such references uninformative, particularly when the customer is encountering financial difficulties.
- **Trade references.** Companies already trading with the customer may be willing to provide a reference for the customer. This can be extremely useful, providing that the companies approached are a representative sample of all the clients' suppliers. Such references can be misleading, as they are usually

based on direct credit experience and contain no knowledge of the underlying financial strength of the customer.

- **Financial accounts.** The most recent accounts of the customer can be obtained either direct from the business, or for limited companies, from Companies House. Whilst subject to certain limitations *past accounts* can be useful in vetting customers. Where the credit risk appears high or where substantial levels of credit are required, the supplier may ask to see evidence of the ability to pay on time. This demands access to internal *future* budget data.
- **Personal contact.** Through visiting the premises and interviewing senior management, staff should gain an impression of the efficiency and financial resources of customers and the integrity of its management.
- **Credit agencies.** Obtaining information from a range of sources such as financial accounts, bank and newspaper reports, court judgements and payment records with other suppliers, in return for a fee, credit agencies can prove a mine of information. They will provide a credit rating for different companies. The use of such agencies has grown dramatically in recent years.
- **Past experience.** For existing customers, the supplier will have access to their past payment record. However, credit managers should be aware that many failing companies preserve solid payment records with key suppliers in order to maintain supplies, but they only do so at the expense of other creditors. Indeed, many companies go into liquidation with flawless payment records with key suppliers.
- **General sources of information.** Credit managers should scout trade journals, business magazines and the columns of the business press to keep abreast of the key factors influencing customers' businesses and their sector generally. Sales staff who have their ears to the ground can also prove an invaluable source of information.

11.4.2　Credit terms granted to customers

Although sales representatives work under the premise that all sales are good (particularly, one may add, where commission is involved!), the credit manager must take a more dispassionate view. The credit manager must balance the sales representative's desire to extend generous credit terms, please customers and boost sales with a cost/benefit analysis of the impact of such sales, incorporating the likelihood of payment on time and the possibility of bad debts. Where a customer does survive the credit checking process, the specific credit terms offered to them will depend upon a range of factors. These include:

- **Order size and frequency:** companies placing large and/or frequent orders will be in a better position to negotiate terms than firms ordering on a one-off basis.
- **Market position:** the relative market strengths of the customer and supplier can be influential. For example, a supplier with a strong market share may be able to impose strict credit terms on a weak, fragmented customer base.

- **Profitability:** the size of the profit margin on the goods sold will influence the generosity of credit facilities offered by the supplier. If margins are tight, credit advanced will be on a much stricter basis than where margins are wider.
- **Financial resources of the respective businesses:** from the supplier's perspective, it must have sufficient resources to be able to offer credit and ensure that the level of credit granted represents an efficient use of funds. For the customer, trade credit may represent an important source of finance, particularly where finance is constrained. If credit is not made available, the customer may switch to an alternative, more understanding supplier.
- **Industry norms:** unless a company can differentiate itself in some manner (e.g. unrivalled after sales service), its credit policy will generally be guided by the terms offered by its competitors. Suppliers will have to get a feel for the sensitivity of demand to changes in the credit terms offered to customers.
- **Business objectives:** where growth in market share is an objective, trade credit may be used as a marketing device (i.e. liberalised to boost sales volumes).

The main elements of a trade policy are:

- **Terms of trade:** the supplier must address the following questions: which customers should receive credit? How much credit should be advanced to particular customers and what length of credit period should be allowed?
- **Cash discounts:** suppliers must consider whether to provide incentives to encourage customers to pay promptly. A number of companies have abandoned the expensive practice of offering discounts as customers frequently accepted discounts without paying in the stipulated period.
- **Collection policy:** an efficient system of debt collection is essential. A good accounting system should invoice customers promptly, follow up disputed invoices speedily, issue statements and reminders at appropriate intervals and generate management reports such as an aged analysis of debtors. A clear policy must be devised for overdue accounts, and followed up consistently, with appropriate procedures (such as withdrawing future credit and charging interest on overdue amounts). Materiality is important. Whilst it may appear nonsensical to spend time chasing a small debt, by doing so, a company may send a powerful signal to its customers that it is serious about the application of its credit and collection policies. Ultimately, a balance must be struck between the cost of implementing a strict collection policy (i.e. the risk of alienating otherwise good customers) and the tangible benefits resulting from good credit management.

11.4.3 Problems in collecting debts

Despite the best efforts of companies to research the companies to whom they extend credit, problems can, and frequently do, arise. These include disputes over invoices, late payment, deduction of discounts where payment is late and the troublesome issue of bad debts. Space precludes a detailed examination of debtor finance, so this next section concentrates solely on the frequently examined method of factoring.

11.4.4 Factoring: an evaluation

Key features

Factoring involves raising funds against the security of a company's trade debts, so that cash is received earlier than if the company waited for its credit customers to pay. Three basic services are offered, frequently through subsidiaries of major clearing banks:

- Sales ledger accounting, involving invoicing and the collecting of debts;
- Credit insurance, which guarantees against bad debts;
- Provision of finance, whereby the factor immediately advances about 80% of the value of debts being collected.

There are two types of factoring service. *Non-recourse factoring* is where the factoring company purchases the debts without recourse to the client. This means that if the client's debtors do not pay what they owe, the factor will not ask for his money back from the client.

   Recourse factoring, on the other hand, is where the business takes the bad debt risk. With 80% of the value of debtors paid up front (usually electronically into the client's bank account, by the next working day), the remaining 20% is paid over when either the debtors pay the factor (in the case of recourse factoring), or when the debt becomes due (non-recourse factoring). Factors usually charge for their services in two ways: administration fees and finance charges. Service fees typically range from 0.5 to 3% of annual turnover. For the finance made available, factors levy a separate charge, similar to that of a bank overdraft.

Advantages

- It provides faster and more predictable cashflows.
- Finance provided is linked to sales, in contrast to overdraft limits, which tend to be determined by historical balance sheets.
- Growth can be financed through sales, rather than having to resort to external funds.
- The business can pay its suppliers promptly (perhaps benefiting from discounts) and because they have sufficient cash to pay for stocks, the firm can maintain optimal stock levels.
- Management can concentrate on managing rather than chasing debts.
- The cost of running a sales ledger department is saved and the company benefits from the expertise (and economies of scale) of the factor in credit control.

Disadvantages

- The interest charge usually costs more than other forms of short-term debt.
- The administration fee can be quite high depending on the number of debtors, the volume of business and the complexity of the accounts.
- By paying the factor directly, customers will lose some contact with the supplier. Moreover, where disputes over an invoice arise, having the factor in the middle can lead to a confused three-way communication system, which hinders the debt collection process.
- Traditionally the involvement of a factor was perceived in a negative light (indicating that a company was in financial difficulties), though attitudes are rapidly changing.

11.4.5 Conclusion

Working capital management is of critical importance to all companies. Ensuring that sufficient liquid resources are available to the company is a pre-requisite for corporate survival. Companies must strike a balance between minimising the risk of insolvency (by having sufficient working capital) with the need to maximise the return on assets, which demands a far less conservative outlook.

11.5 Acquisitions and mergers

An acquisition usually takes place when a larger company (predator) takes over a smaller, targeted company by either purchasing the business's assets or acquiring the business, goodwill, etc.

A merger usually involves the pooling of interests by two or more business entities which leads to a common ownership.

There are three types of merger:

- A **horizontal integration** refers to two or more companies in the same industry, whose operations are very closely related. Examples of these include the banking merger of Lloyds and TSB alongside AECOM's acquisition of the Hunt Construction Group. The main motives for these activities taking place usually involve economies of scale, increased market power and improved product diversification. The main disadvantage involves possible referral to competition authorities.
- A **vertical integration** involves two or more companies from the same industry, but from different stages of the production chain. Large oil companies tend to be highly vertically integrated. The main motives for these activities taking place include increased certainty of supply or demand and consolidation of just-in-time systems leading to savings in inventory holding costs.
- A **conglomerate integration** occurs through a combination of unrelated businesses. The main synergies usually fall within managerial skills (GEC and Tomkins) and brand name (Virgin).The main motives for these activities are risk management through diversification of product portfolios, cost

reduction through the pooling of resources or improved revenue streams through re-branding.

11.5.1 Advantages of business organic growth

Organic growth allows a business to plan its strategic growth in line with short- and long-term business objectives. It is less risky than growth by acquisitions, provides less uncertainty for existing staff and is cheaper in both the short term and long term.

11.5.2 Advantages of growth by acquisition

This is often the quickest way to enter into a new product or geographical market. It reduces the risk of over-supply and excessive competition creating fewer competitors. Increased market power usually follows in that the combined business can exercise some control over the price of the product, e.g. monopoly or by collusion with other producers.

11.5.3 The regulation of takeovers

The regulation of takeovers varies from country to country but focuses primarily on controlling the directors. Typical factors include the following:

- Directors should always act in the best interests of their shareholders and should disregard their personal interests, treating all shareholders equally.
- The board should not take any action without the prior approval of shareholders, which could result in the offer being defeated.
- Any assumptions on which profit forecasts are based and accounting policies used should be examined and reported on by accountants with an independent valuer supporting asset valuations.
- The acquisition of quoted companies in the UK is regulated by the 'City Code' which is underpinned by the Stock Exchange Yellow Book.

11.5.4 A defence against hostile takeovers

Every company is potentially subject to a takeover when the price is attractive enough to induce shareholders to sell their shares. When a bid is received, the directors will consider it from the shareholders' perspective to assess if it will increase shareholder wealth.

A problem could arise if a company receives an unwelcomed and predatory bid, with the clear objective of buying the company at a price below the value that management put on it. Defence mechanisms can be put in place to ward off the bid to remain independent. Predator companies may be looking to buy at less than fair/full value because the share price is depressed because of ineffective management at the company.

The group could also occupy a strong position in one or more markets, seeing the purchase as an opportunity to increase market share fairly quickly or the group's future prospects are much better than the share price would currently indicate.

The board of directors could attack the logic of the bid, find a friendlier bidder (white knight strategy) or improve the current image of the company through revaluations, profit projections, dividend promises and the employment of public relations and marketing consultants.

11.5.5 Finance available for cash-based acquisitions

* **Retained earnings** – often the cheapest option and generally restricted to smaller acquisitions.
* **Debt financing** – larger companies may use the bond market which is government-backed, but this will have an impact on the share price – a bridging loan to reduce the impact on share price followed by a bond issue could be instigated. Bank loans are an option for smaller companies with terms fluctuating depending on the risks involved. Mezzanine finance could also be used where borrowing lies somewhere between equity (shareholder funds) and debt (borrowing) finance.
* **Sale of existing assets** – an attractive proposition when businesses are trying to establish a good fit within the new group.
* **Issuing of shares** – to existing shareholders through a rights issue or to potentially new shareholders. The latter may result in a dilution of control for existing shareholders.

Index

Page numbers in *italics* indicate figures and tables.

Printed in the United States
by Baker & Taylor Publisher Services